Atmospheric Change
An Earth System Perspective

T. E. GRAEDEL

AT&T Bell Laboratories

PAUL J. CRUTZEN

Max-Planck-Institut für Chemie

W. H. Freeman and Company

New York

Cover images courtesy of Arlin J. Krueger, Laboratory for Atmospheres, Goddard Space Flight Center, NASA, Greenbelt, Maryland

Library of Congress Cataloging-in-Publication Data

Graedel, T. E.
Atmospheric change: an Earth system perspective / T. E. Graedel and Paul J. Crutzen.
p. cm.
Includes bibliographical reference and index.
ISBN 0–7167–2334–4. — ISBN 0–7167–2332–8 (pbk.)
1. Climactic changes. I. Crutzen, Paul J., 1933– . II. Title.
QC981.8.C5G69 1992
551.6—dc20   92–38511
                                                                      CIP

Printed in the United States of America

1 2 3 4 5 6 7 8 9 0 VB 9 9 8 7 6 5 4 3

To Susannah and Terttu—
our best friends

CONTENTS

Preface

The aim of the book is to explore the physical and chemical workings of Earth's planetary system, recognizing that human development of the planet has a variety of impacts at every scale—locally, regionally, and even globally. Development should and will occur, but our central premise is that it needs to proceed in such a way as to ensure that the basic properties of climate and system are retained. The book got its start with manuscripts that we prepared for three efforts in interdisciplinary studies of Earth and its changes. The first was a meeting entitled "Sustainable Development of the Biosphere," organized under the auspices of the International Institute for Applied Systems Analysis (IIASA) in Laxenburg, Austria. The second was a two-year project entitled "Retrospective and Prospective Global Development and Its Impact on the Atmosphere," done with colleagues at IIASA, Carnegie-Mellon University, Duke University, and Resources for the Future, Inc. The third was a conference at Clark University in Worcester, Massachusetts entitled "The Earth as Transformed by Human Action." Dr. William Clark, currently at Harvard University, was a principal organizer and participant in all three efforts, and this book is due in great measure to his encouragement.

There are five distinct parts of *Atmospheric Change*. An introductory chapter provides a perspective on the topics to be discussed and the following six chapters introduce the reader to the basic operation of Earth as a system. Then, because of their intrinsic interest and because it is only by understanding the present situation that the past and future can be put into perspective, the chemistry of today's atmosphere, freshwaters, seawater, and underground and undersea systems are discussed. The four chapters that follow deal with the climate and chemistry of the past, measured and inferred, on three very different time scales. The important concept of chemical trace species budgets is introduced in Chapter 14, and concepts and techniques for modeling in Chapter 15. In Chapters 16 through 18, we present possible future states of the planet, as derived by ourselves and others. In the final chapter we discuss and summarize these topics, with particular attention to questions of stability and change.

This book is written at a technical level suitable for undergraduate students in the sciences and is designed to be suitable for a one-semester course in Earth system science or environmental chemistry. Alternatively, it may be used as a supplementary text in courses dealing in more detail with one aspect of Earth system science, such as atmospheric chemistry, climate, or ecology. The book presumes introductory college-level mathematics, chemistry, and physics as prerequisites, but high school level training is probably adequate for all but part of the treatment of the radiation budget in Chapter 3, where the concept of the integral is vital, and portions of the discussions of atmospheric motions in Chapter 4 and computer models in Chapter 15, where differential equations come into play. To make the book accessible to people with different levels of training, these more advanced topics appear as boxes in smaller type, and can be included or omitted at the discretion of the instructor. Chemical reactions and kinetics are, in our view, so central to many of the issues that are discussed that they are thoroughly integrated into the text. We have, however, attempted to present these topics in sufficient pedagogical detail to make them approachable for those without extensive chemical backgrounds.

Exercises, both problems and projects, are included for most of the chapters, because working through such examples is inevitably the most efficient way to learn a subject. In the early chapters, relatively straightforward problems arise naturally from the introduction of material concerning the solid, liquid, and gaseous parts of Earth and of its chemical nature. In later chapters dealing with atmospheric trends, projects are presented in addition to problems. These projects are representative of many of the activities Earth system scientists: they require assumptions and may not lend themselves to rigorous calculations. Various approaches are suitable, and there may be a range of possible answers because not all the information needed to solve the problem is available. These circumstances are often unsettling to students, but demonstrate vividly that important scientific problems are often not presented in the straightforward way that would best facilitate analysis. As with an English theme or a senior thesis, the project results will benefit from broad outside reading, diligence, and initiative. We have provided a few readily available references for most of the projects and chapters; citations in those references will lead to additional information. As such, the exercises encourage the familiarity of students with the diverse literature of atmospheric chemistry and global change.

It is our strong conviction that permitting the stresses now being placed on the Earth system to be continued and intensified is injudicious and that the development of the planet in a demonstrably sustainable fashion can occur only if global, regional, and local development decisions are approached logically and comprehensively and with restraint. Such actions will also require improved knowledge of the interactions among biological, physical, chemical, and social processes on Earth. Heightened research efforts and new young scientists will be needed. We hope that readers of this book will share our goals and carry them further by their research and personal actions.

We have referred often in this book to meritorious work done by our colleagues in the relevant Earth sciences, indicating in each case the institution at which the work that is cited was done. Many of these scientists and many others have provided assistance in the preparation of this book; we thank the following for the use of unpublished material, for useful comments, and for reviews of the various chapters: Orson Anderson, Greg Ayres, Andre Berger, Paul Boyer, Peter Brimblecombe, Richard Brost, Christoph Brühl, Kevin Burke, Jack Calvert, Julius Chang, Robert Charlson, William Clark, Robert Delmas, Russell Dickerson, Jack Fishman, Geoff Harris, Albrecht Hofmann, Leen Hordijk, Carlton Howard, Rupert Jaenicke, Michael Keller, Carl Kisslinger, Johannes Lilieveld, Donald Lenschow, Craig Lindberg, Claude Lorius, Sasha Madronich, Robbin Martin, Loretta Mickley, Geert Moortgat, Jozef Pacyna, V. Ramanathan, Henning Rodhe, Eugenio Sanhueza, Manfred Schidlowsky, Robert Sievers, Karen Valentin, Peter Warneck, and Thompson Webb. Most of the illustrations were prepared at AT&T Bell Laboratories by Danuta Sowinska-Kahn and Mona Hennes. For providing special figures we thank Robert Chervin, Robert Delmas, Jack Fishman, Susannah K. Graedel, Kyung-Ryul Kim, Arlin Krueger, Lawrence Mauch, Michael Prather, Henry Reichle, David Rind, and Christopher Scotese. We are assisted in text processing and in many administrative matters by Robin Ward. We also express our gratitude to Jerry Lyons, Diane Maass, Renata Gomes, and Christine Rickoff, of W. H. Freeman and Company, for their enthusiastic and efficient efforts on behalf of this book.

Introduction

THE SCIENCE OF GLOBAL CHANGE

The Ozone Hole

On May 16, 1985, J. C. Farman, B. G. Gardiner, and J. D. Shanklin from the British Antarctic Survey published in the British journal *Nature* an article that instantly became famous. The article reported that the total amount of ozone in the atmosphere over the observational site in Antarctica showed rapid losses during the austral spring. Those data, updated with more recent measurements, are shown in Figure 1.1. Future trends in atmospheric ozone had already been a concern of scientists, who had predicted that the injection of chlorofluorocarbons (CFCs or "Freons") into the atmosphere would

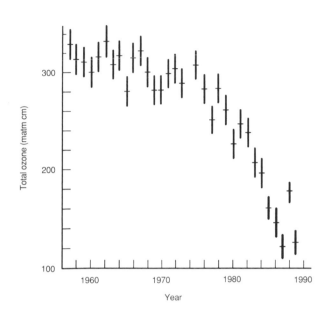

Figure 1.1 Monthly means (horizontal lines) and ranges (vertical bars) of total ozone over Halley Bay, Antarctica, for October of the years 1957 through 1989. A layer of gaseous ozone 1 mm thick at 1 atmosphere pressure and 0 °C corresponds to 100 matm cm. (Adapted by permission from J. C. Farman, B. G. Gardiner, and J. D. Shanklin, Large losses of total ozone in Antarctica reveal seasonal interaction, *Nature, 315*, 207–210. Copyright 1985 Macmillan Magazines Limited.)

gradually cause losses in ozone over a period of decades. Farman's observations, however, raised some major new questions—principally, why no one else had seen the decrease and why it was happening so rapidly.

The first question was answered promptly: The effect had been detected elsewhere but had not been noted. The data processing system for the US *Nimbus* 7 satellite, which had been measuring ozone over Antarctica several times per day, had been programmed to ignore "aberrant" data points having what were thought to be unrealistically low values. Once this limitation was removed from the computer program and the data reanalyzed, the British result was confirmed. In fact, the data showed that the ozone depletion covered the entire Antarctic region.

Understanding the cause of the ozone hole was not so quickly accomplished. Several theories were proposed, and experiments to test them were designed. Atmospheric scientists then mounted a series of experiments in September and October of 1986 and 1987 by aircraft (Figure 1.2) and from the Antarctic continent to study the meteorology and chemistry of the Antarctic atmosphere in detail. A year and a half later, these and subsequent field studies, together with supporting laboratory experiments, had established without question

Figure 1.2 The ER-2 research plane was part of the aircraft fleet for the 1986 National Ozone Expedition to Antarctica. A modified version of the U-2 spy plane, it carried its single crewman and computerized instruments into the lower stratosphere to make measurements within the ozone hole. Some of the sampling nozzles are visible in the wing pod at the left. (Courtesy of Michael Prather, NASA.)

that the decrease was due to the chemical activity of the chlorine atoms contained in the CFCs, but with a few special twists produced by the air motions, low temperatures, and ice clouds peculiar to the Antarctic.

The ozone hole provides a classic case of the workings of science at its best: the unexpected discovery of an important effect, proposals of theories to explain it, quick mounting of a logistically difficult experimental field program to test the theories, and blending of laboratory data, field observations, and computer models to achieve understanding, in this case within only two years.

Drilling into Climates of the Past

While the air over Antarctica is currently the venue of rapid change, the ice of Antarctica and elsewhere is the repository of a record of change over a much longer period of Earth's history. The source of this ice is snow that falls each year on the glacier or ice cap, thaws in the sun, recrystallizes, and subsequently consolidates (Figure 1.3). As it freezes, bubbles of gas are trapped in the ice. If those bubbles are recovered, extracted, and analyzed, they provide samples of ancient air. With the realization that this rich record exists, teams of scientists have drilled into glaciers, braving the elements to bring to refrigerated laboratories samples that reveal the chemical compositions and climatic conditions of the past.

The longest continuous record of the ancient atmosphere is from an ice core drilled by a Russian-French team on the Antarctic glacier; its data span 160,000 years. When the results of the studies began to

Figure 1.3 An ice core from the Antarctic glacier. The regions of annual surface thawing and recrystallization are clearly visible. (Courtesy of Robert Delmas, CNRS, France.)

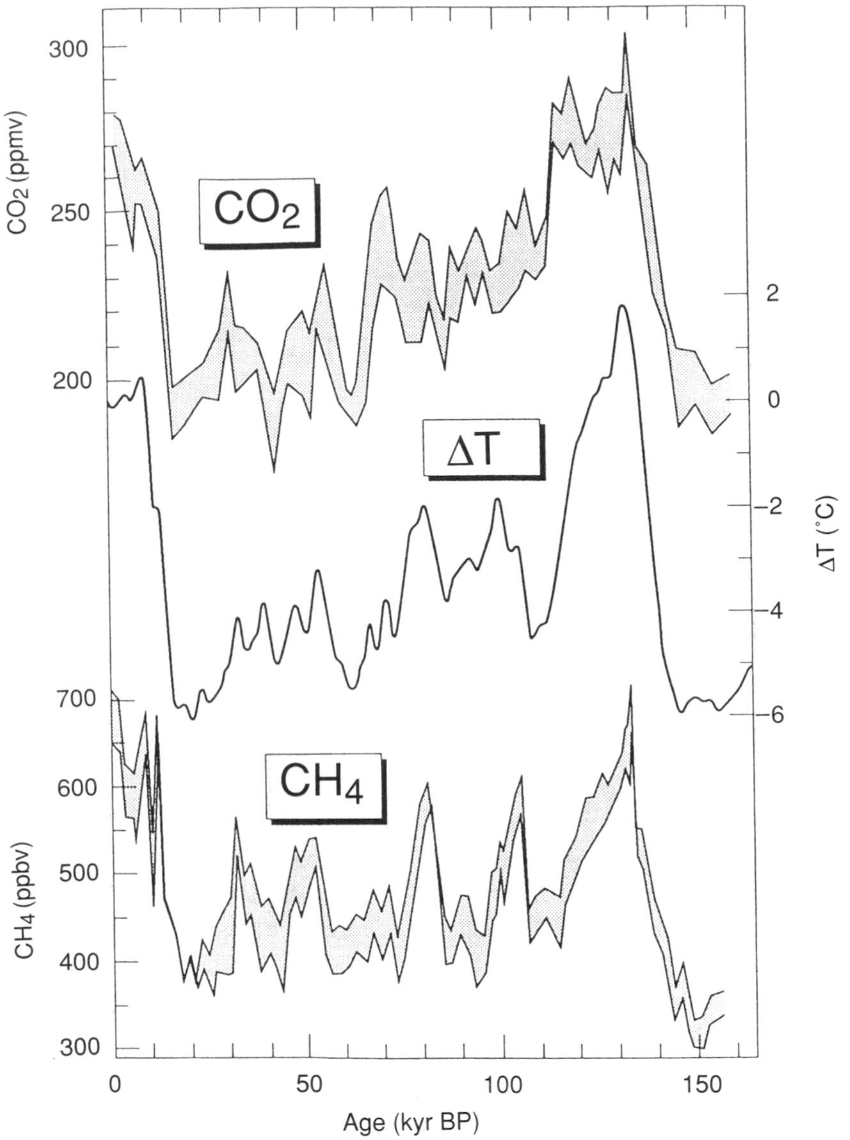

Figure 1.4 Antarctic ice core records of local atmospheric temperature and atmospheric carbon dioxide and methane volume mixing ratios for the past 160,000 years. (Mixing ratios and other units are discussed in the appendix on units of measurement.) Note that for most of the period temperatures were lower than the present world average (denoted as $\Delta T = 0$ at 0 Kyr BP). The present warm period (interglacial) started some 20,000 years ago. The previous period with comparable temperatures occurred between 120 and 140 kyr BP. (C. Lorius, private communication, 1990.)

be published in 1987, Earth scientists received a number of surprises. Rather than demonstrating a stable or smoothly evolving Earth system, the record resembled erratic behavior more like "spit on a hot griddle," as one scientist put it. Furthermore, the carbon dioxide and methane records tracked those of planetary temperature changes to a remarkable degree (Figure 1.4). Climatologists and Earth scientists of all kinds have been forced to ask themselves how they can predict the characteristics of the planet in the future if they are unable to understand its past.

The Origin and Evolution of Life

Other problems in atmospheric science have much longer time spans, an extreme example being the study of the origin of life. As we understand it, Earth formed from lifeless masses of interstellar gas and dust, gradually cooling, sorting its initial material through volcanic eruption, crystallization, and heat loss, and evolving into a planet with the general characteristics so familiar to us today. Somewhere along the line, quite early in the planet's history, in fact, life was formed at or near Earth's surface. The manner in which this could have occurred remains one of the great scientific questions of all time, but it is known that the atmosphere and its chemistry must have played major roles in defining how life could have been formed and what paths it might have been able to take. This same life affected the chemical composition of the atmosphere, thereby influencing the conditions for its own evolution. The study of the atmosphere thus encompasses portions of the study of the origin and evolution of life.

Climate Changes of the Present and the Future

Climate is most succinctly defined as the statistics of meteorological conditions (temperature, precipitation, winds, etc.) over long time periods (30-year averages are often used as a modern measure of climate). As we will see, climate has changed considerably over centuries or millenia throughout Earth's history, but concerns are now arising that climate changes may be proceeding extremely rapidly under the influence of humanity's activities. An example of the evidence leading to these concerns is the temperature record for the past century, shown in Figure 1.5. This record shows that about 0.5 °C of warming has occurred in the last 130 years and that the decade of the 1980s was the warmest during that period. The pattern roughly parallels that of the use of fossil fuels and the injection into the atmosphere of gases that can absorb radiation and thus produce warming. On the basis of this information, calculations suggest that

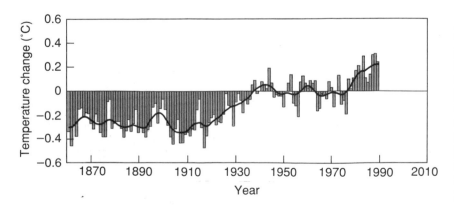

Figure 1.5 Annual deviation of global mean (land and sea) temperature over the past century relative to the average for the period 1951–1980 (here defined as 0). The curve shows the results of a smoothing filter applied to the annual values. (Policymakers' Summary, *Climate Change: The IPCC Scientific Assessment*, Cambridge University Press, 1990.)

much more warming is still to come, warming that will change in major ways the existing patterns of habitation, agriculture, and vegetation on the planet.

Science and Society

The examples of global change discussed above were chosen from a much longer list including photochemical smog, lake acidification, and radioactive fallout and demonstrate that the Earth sciences are connected with some of the most intellectually interesting and societally critical issues anywhere. As is now becoming obvious, the stresses on many aspects of the Earth system are strongly influenced by the needs of a population of human beings that must be provided for and by the standard of living those people desire. Population and population change are thus of vital importance to discussions of the sustainability of the planet and its ecology; Figure 1.6 shows that Earth's population is very rapidly increasing. The population explosion, coupled with the universal ambition for a high standard of living,

Figure 1.6 The population of Earth over the past three centuries. Industrialized countries are the United States, Canada, western Europe, the Soviet republics, eastern Europe, Japan, Australia, and New Zealand.

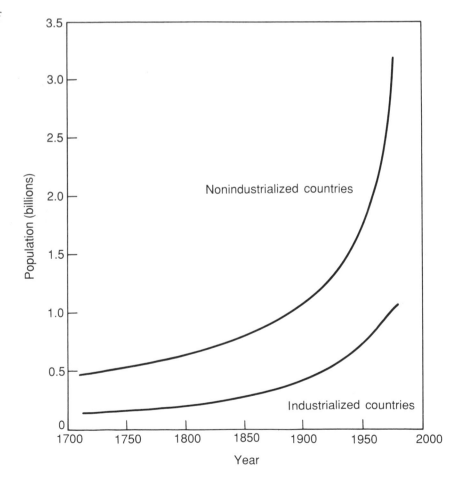

calls into question the ability of the planet to provide the resources necessary for the population's sustenance.

Because change has such important social, political, and scientific consequences, some knowledge of the Earth system's characteristics and trends should be part of the arsenal of every educated citizen. This book is therefore designed to involve each of its readers in a most exciting and important scientific endeavor—the study of the Earth system, past, present, and future.

OPENING COMMENTS

As a result of diligent effort since the eighteenth century and especially over the past several decades, Earth scientists understand much of the workings of the system with which they deal; yet there is much they do not understand. Substantial recent efforts have been directed to the atmosphere. It is known that its makeup is very complex and that it is not static. The atmosphere's present composition can be summarized succinctly. Disregarding highly variable amounts of water vapor, more than 99.9% of the other molecules constituting Earth's atmosphere are nitrogen (N_2), oxygen (O_2), and the chemically inert noble gases (largely argon). All of these gases may have been present at nearly constant levels during much of the past billion (10^9) years.* The remaining atmospheric constituents, representing less than 0.1% of the atmospheric molecules, are diverse but important because they influence or control a number of crucial atmospheric processes. For example, carbon dioxide, the chemical feedstock for the photosynthesis of organic matter, is an important factor in Earth's radiation balance. It is chemically unreactive in the troposphere (the lowest portion of the atmosphere; see Chapter 3) and is currently (1992) present at an average concentration of about 0.035% by volume. The most abundant of the reactive gases is methane, which represents less than 0.0002% of the tropospheric gas. Other reactive species are still less prevalent; the combined concentration of all of the reactive trace constituents in the lower atmosphere seldom totals 0.001%, even in the most polluted environments. Table 1.1 gives the concentrations of representative atmospheric constituents. How has this composition changed over the eons of existence and evolution of atmospheric gases, particles, and droplets? As we will see, it is possible to answer this

---

* Common European and American definitions of the terms *billion* and *trillion* are not the same. For the former, Europeans usually mean 10^{12}, whereas Americans usually mean 10^9. For the latter, Europeans usually mean 10^{18}, whereas Americans usually mean 10^{12}. Earth system scientists generally adhere to the American convention regardless of their nationality, and we will do so throughout this book. The prefixes n (nano- = 10^{-9}), μ (micro- = 10^{-6}), m (milli- = 10^{-3}), c (centi- = 10^{-2}), k (kilo- = 10^3), M (mega- = 10^6), G (giga- = 10^9), T (tera- = 10^{12}), and P (peta- = 10^{15}) will also be used.

TABLE 1.1. Composition of Dry Air at Ground Level in Remote Continental Areas		
CONSTITUENT	FORMULA	CONCENTRATIONS
Nitrogen	N_2	78.1%
Oxygen	O_2	20.9%
Argon	Ar	0.93%
Carbon dioxide	CO_2	0.035%
Neon	Ne	0.0018%
Helium	He	0.0005%
Methane	CH_4	0.00017%
Krypton	Kr	0.00011%
Hydrogen	H_2	0.00005%
Ozone	O_3	0.000001– 0.000004%

question with rather greater assurance for the current century and with progressively less certainty for earlier times.

To present information on these topics in a logical sequence, we begin this book with five chapters (2–6) discussing several basic topics in Earth system science—the solid Earth and its atmospheric interactions, atmospheric energy cycles and motions, particles in air and water, and the water cycle and climate. Chapter 7 presents the important chemical principles used in Earth system science, and Chapters 8 and 9 present the chemistry of different Earth system regimes. This presentation encompasses most of the traditional chemical fields, including analytical, inorganic, and organic chemistry, photochemistry, and biochemistry. The diversity of fields is complemented by the diversity of the atmosphere's constituents—several thousand different chemical compounds are known to be present at one time or another in the gas, liquid, and solid phases of Earth's various regimes.

In different ways, studying the chemistry of the past is equally daunting, because aged and altered samples and inferential analyses play such a large role. There are two reasons why one might choose to attempt it. The first is purely intellectual, that of finding out how Earth came to be, how it works, and how the parts of the Earth system—the oceans, the rocks, the rivers, living matter, the Sun's radiation, and outer space—interact. The other reason is more practical: by studying the past, we may better understand the workings of the present and future. Four chapters, 10 through 13, present these histories of climate and chemistry.

Just as one might not expect that knowledge would be available on the past chemical constitution of Earth, so one might not anticipate that Earth system scientists could describe the Earth of the future. Such efforts can be made with computer models incorporating assumptions about the changes anticipated for input fluxes of particles and gases as a consequence of human activities. The process is often aided by the input-output analysis technique of budgeting, which is discussed in Chapter 14. In Chapter 15, we show how different types of models are formulated and give some idea of the results each can be expected to produce. Model studies are inevitably contentious, because it is often difficult to agree on the choices that must be made to describe a global system that is imperfectly understood and dependent in large part on human societal and technical developments. In addition, because the workings and interactions of environmental systems are so complex, one can never be certain that all significant processes are included. Comparisons with detailed observations can reveal discrepancies. In that sense, models are important tools for research as well as for predictions, but their inherent limitations suggest that the results should be treated as "warning signals" and not as certainties. Nonetheless, models often provide us with our only means of looking into the future, and we need to pay careful attention to the warning signals they are sending us. Indeed, in many cases model results have been sufficiently successful in simulating current physical and chemical properties of Earth, and especially its atmosphere, to justify their application for policy decisions. A sampling of computer model predictions and other thoughts concerning the regional and global characteristics of the future is given in Chapters 16 through 18. Finally, some of the implications of these results and related studies are discussed in the concluding chapter.

THE INTERDISCIPLINARY NATURE OF EARTH SYSTEM SCIENCE

Earth system science has proved to be highly interdisciplinary because the atmosphere's composition is closely tied to its interactions with solar radiation, winds and clouds, ice sheets, oceans, flora, fauna, and the solid Earth. A perspective on these intricate relationships is shown in Figure 1.7. A few moments of reflection on this figure suggests that the innovative Earth scientist is likely to be more of a generalist than a specialist. Many disciplines and activities make vital contributions to the field of Earth system science, including those listed below.

Chemistry and Photochemistry Many chemical reactions taking place in Earth's air, water, and soil are studied in detail in laboratories, whenever possible under pressure and temperature conditions that are typical of the actual situation. Such studies may also lay the foundation for the development of specific instrumentation for measurements of important constituents in the environment itself.

Meteorology and Climatology Gases and particulate matter emitted into or created in the atmosphere are not stagnant but are transported by atmospheric winds. Their radiative absorption properties strongly influence the atmosphere's temperature structure and, through it, the climate. Eventually they, or their chemical transformation products, are brought back to Earth's surface by direct contact with it or by precipitation processes. Chemical reactions in clouds also play important roles in atmospheric chemistry. Atmospheric motions, precipitation processes (rain, snow, hail), cloud cycles, and climatology are all components of classical meteorology.

Biology and Ecology   Many compounds are emitted into and absorbed from the environment by plants or animals or as a consequence of microbial action in the soil, the oceans, or other aquatic environments. In turn, many industrial emissions or their reaction products can interact with the natural environment to produce impacts over a variety of temporal and spatial scales.

Geology and Geochemistry   Many aspects of the Earth system's evolution and operation are revealed by studies of rock formations, mineral properties, and sediment chemistry. Such studies provide the only direct information we have on the environment of Earth until relatively recent times.

Solar and Space Physics   The Sun is the source of energy for all meteorological and biological processes, and the composition of the atmosphere is strongly influenced by solar radiation, both in the short term and over very long time scales. Solar ultraviolet radiation causes biological mutations and thus influences the evolution of new life forms. Also of importance is knowledge of the characteristics of energetic particle emissions, primarily proton and helium nuclei from the Sun or the cosmos. Such emissions can have a strong influence on upper atmospheric chemical processes.

Figure 1.7 Processes affecting trace gases in Earth's atmosphere. The arrows indicate net fluxes. (Adapted by permission from R. Prinn and B. Fegley, Jr., The atmospheres of Venus, Earth, and Mars: A critical comparison, *Annu. Rev. Earth Planetary Sci., 15*, 171–212. Copyright 1987 by Annual Reviews, Inc.)

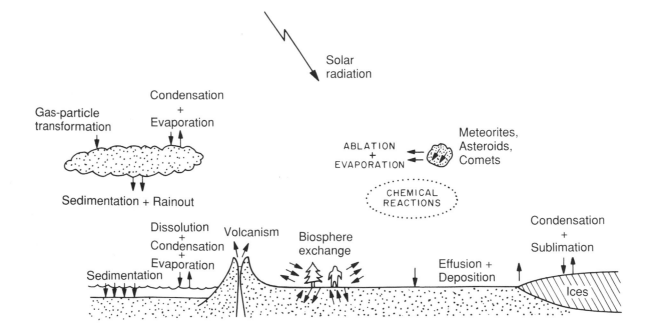

Social and Behavioral Sciences   Because human conditions are strongly influenced by environmental factors and because human activity now has important impacts on the planet, studies of human practices of the past, present, and future are relevant to the Earth system and its changes. Recent investigations have included topics such as the history of agriculture, deforestation, land use, and trends in urban development.

Instrument Development and Measurement Networks The development of appropriate instrumentation for measurements of critical environmental constituents, often present in extremely low concentrations, is of critical importance and requires the most advanced experimental techniques. Prior to their deployment in major measurement campaigns, instruments built by various research groups are brought together and operated simultaneously on the ground, on ships, in aircraft, or on balloons to intercompare results and confirm reliability. Thereafter, the instruments can be used in comprehensive measurement campaigns for the study of interacting chemical processes. In addition, observations from space can provide global coverage of atmospheric chemical composition and of factors related to it, such as vegetation type and sea-surface temperature.

Monitoring of Trends   For studies of long-term trends of several chemically and climatically important gases, or of particulate matter, the establishment of stations around the world is very important. Long-term observations have led to the discovery of upward trends for carbon dioxide, methane, nitrous oxide, and other species. With the addition of more monitoring stations and more sensitive equipment, prospects exist for the careful study of trends in many more trace chemical species. Also, optical equipment on spacecraft has recently proved to be highly successful for monitoring certain constituents. Satellite studies will increasingly be used to complement ground-based measurements of the constituents of the atmosphere and surface.

Computer Modeling   It has become increasingly common over the past two decades to combine apposite scientific knowledge from various disciplines into models consisting of mathematical descriptions of known or postulated physical and chemical processes. The results of such models must be tested against observations. Disagreements between theoretically expected results from the models and measurements often lead to new scientific postulates and discoveries.

SPATIAL SCALES IN EARTH SYSTEM SCIENCE

Scientists who deal with Earth are confronted every day with an enormous range of spatial scales (Figure 1.8). The distance scale in the figure is logarithmic, so the top to bottom difference is 10^{17}! The smallest scales are those of atoms and molecules; these are the scales on

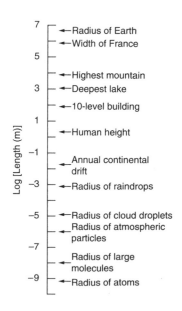

Figure 1.8 Spatial scales in Earth system science. The vertical scale is the base 10 logarithm of length.

which chemical reactions occur and on which laboratory experiments are performed. Larger than atoms and molecules, but still at small fractions of a millimeter, are the atmosphere's solid and liquid phases—particles, cloud droplets, and raindrops. The typical dimension of continental drift is a centimeter or two per year; amazingly, this procession of the continents has about the same dimension as that of a large snowflake. Lengths near a meter are those typical of humans and their structures. Much larger lengths describe major physiographic features—deep lakes, tall mountains, and ocean canyons. Finally, the dimensions of countries, continents, and the planet itself appear at the largest scale.

It is worth thinking about Figure 1.8 in the context of familiar objects. Perhaps the biggest message that it contains is that the scale of the physiographic features of Earth is very large, whereas the scale of the processes that drive the Earth system is very small. The creation of drops of acid from nonacidic reactants, the weathering of rocks, the eruption of volcanoes, and the formation and dispersal of clouds are all products of vast numbers of individual physical and chemical acts on a tiny scale. We cannot hope to understand the large, obvious features and processes of Earth unless we understand the miniscule but vital processes that bring them about.

TIME SCALES IN EARTH SYSTEM SCIENCE

One intuitively thinks of the chemical composition of the atmosphere and oceans as being among those few constant properties of a changing world. It is now clear, however, that they must be regarded, like Earth itself, as having evolved through a number of different stages over a long period of time. This evolution has occurred as a result of variations in chemical cycling among the atmosphere, the solid surface, and the oceans, by tectonic activity and modification due to volcanic eruptions, by changes in the intensity of solar radiation, and by the interplay of the atmosphere with flora and fauna. Thus, it is necessary to understand how these factors themselves have evolved since the beginning of the planet. Furthermore, the impact of meteorites from outer space is thought to have intermittently led to major perturbations in the chemical composition of the atmosphere and in Earth's climate, with serious repercussions for the evolution of life on Earth.

A variety of evidence tells us that the Sun, Earth, and the remainder of the solar system were created by the gravitational collapse of a huge nebula of dust and gas between 4.5 and 4.7 thousand million years ago (referred to as gigayears before the present, or Gyr BP). The oldest Earth rocks have been radioactively dated at about 4.2 Gyr BP. The dates of different rock sections and the sediments in which they were identified serve to define for geologists the epochs in Earth history, as pictured in Figure 1.9. The time scale on the figure is not linear because the more plentiful evidence on Earth's recent history allows scientists to make much finer divisions for the recent past than can be made for the more distant past with its sparser data base.

A few major geological and paleontological events are listed in Figure 1.9. These events are crucial for the study of the relationships among Earth's solid surface, its oceans, its atmosphere, and its plant and animal life. Just as students of Anglo-Saxon or European history should know the dates of William the Conqueror's invasion of England, the start of Columbus's voyage to the New World, and the French Revolution, so students of Earth system science should know the information in Figure 1.9. Perhaps the most significant dates and events are those of the formation of Earth, the appearance of the earliest known life, the Proterozoic-Cambrian transition, the Permian-Trias-

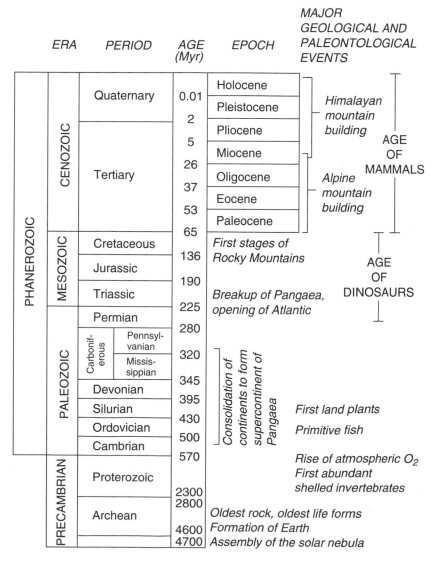

Figure 1.9 The geological time scale, expressed in millions of years (Myr) before the present; the vertical scale is not linear. The boundaries between the eras and periods are defined by changes in rock or fossil records. A few of the major events that have shaped Earth and its living things and that define much of Earth system science are indicated. (Adapted from T. J. Crowley, The geologic record of climatic change, *Rev. of Geophys. Space Phys.*, 21, 828–877, 1983.)

sic transition, the Cretaceous-Tertiary transition, and the Holocene era, together with the geological and paleontological events that define those boundaries. We will refer again and again throughout the book to events and time scales contained in Figure 1.9.

A perspective on the interplay between space and time on Earth is shown in Figure 1.10, where the time scales of interest extend from minutes to billions of years and the spatial scale from less than a kilometer to the dimensions of the planet. The figure emphasizes the grand sweep of Earth system science, from the local movement of air past a building to the long, slow preparation of conditions suitable for the origin of life.

Figure 1.10 Characteristic spatial and temporal scales for Earth system processes. The ordinate is logarithmic in distance; the abscissa is arbitrarily spaced for clarity.

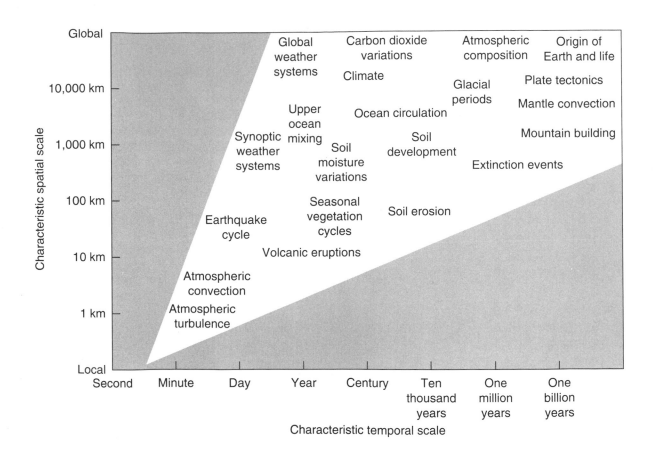

STABILITY AND INSTABILITY

Concepts of stability and instability are crucial to the study of Earth system science. Although assessing the stability of complicated systems can be a demanding task, understanding what is meant is straightforward if we discuss the concept by analogy with a mechanical system, such as an object on a hilly surface (see Figure 1.11). Begin by considering a system at equilibrium. In such a system, the

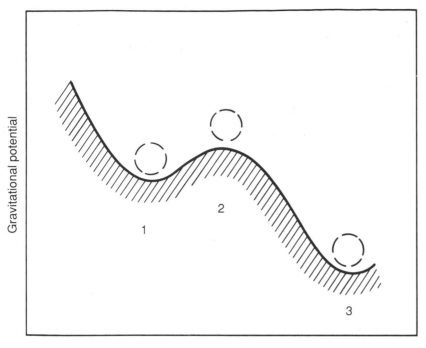

Gravitational potential

Position

1

2

3

Figure 1.11 Stability diagram for a mechanical system. The solid line indicates the boundary of the locations ("states") allowed by the system. The ordinate parameter is the internal potential energy.

vector sum of forces is zero and is characterized by the gravitational potential of the object. Now consider what happens to the system if a small perturbing force is applied. A ball placed in a deep depression and given a push will not keep moving; it is *stable*. The same ball, if started rolling down a hillside of steady slope, will not stop rolling; it is *unstable*. If a ball is placed in position ("state") 1 and to it a small force is applied, the ball will oscillate for a while and then settle back into its original position; this is termed a *metastable state*, which can only be escaped if a larger force is imposed. If the ball happens to be temporarily at position 2, however, a small force applied either to the right or left is sufficient to move the ball to another state; position 2 is said to be an *unstable state*. A ball in position 3 is still metastable, but in a higher stability state: it will remain where it is in the absence of the application of a very large perturbing force.

The concepts of stability and instability can be looked at in a more general way with the help of Figure 1.12. Here we illustrate systems with four types of stability characteristics, the instability increasing from bottom to top, the gravitational potential thus being a paradigm for stability. The concept of a geometric figure on hilly terrain is again useful in picturing the behavior of the different options. Also shown are three degrees of sensitivity for each type of stability. Sensitivity is a measure of the intrinsic resistance of a system to change; we indicate this property schematically here: a hexagon is more stable on a hillside than is a ball, and a cube is more stable still.

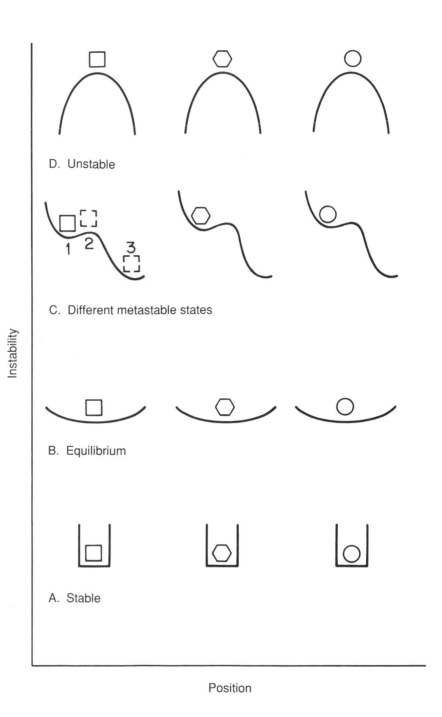

Figure 1.12 Conceptual diagrams of stabilities for systems of different types and sensitivities. A. *Stable*. Nothing except system destruction can cause a change of state; B. *Equilibrium*. A perturbation can temporarily modify a state, but the system can never move far from its position on the vertical axis; C. *Different metastable states*. If a large enough perturbation is applied, then a new metastable state is achieved; D. *Unstable*. A small perturbation causes a radical change of state. Instability increases from bottom to top; sensitivity to perturbation (suggested by the shapes of the figures) increases from left to right (at least for B and C). The meanings of "small" and "large" are functions of the sensitivity of the system. (Adapted from a diagram communicated by A. Henderson-Sellers, 1988.)

Whether or not a system is stable to a modest perturbation is seen to be characteristic not only of the allowed energy paths of the system, but also of the sensitivity of all parts of the system to perturbation.

How do stability and instability enter into discussions of Earth system science? The answer to that question depends on the particular system under consideration, on the strength and type of forcing (which might be, for example, increases in solar radiation intensity or changes in atmospheric composition), and on the sensitivity of the system (for example, an equivalent addition of heat causes a greater temperature change in the atmosphere than in the ocean). The time scale is also important. The bed of a river is quite stable over a few decades or perhaps a few centuries, but it assumes a new state if one allows it to evolve for a few millennia. During the Pleistocene and Holocene, for example, global climate, as measured by the area covered by the polar ice packs, swung from glacial to interglacial states a number of times in each million years. Climate, therefore, appears to be of stability type C, a system with several different metastable states. Some natural systems, such as the flow patterns of upper atmospheric winds, are so easily perturbed that they must be regarded as very sensitive.

This book is, in a sense, a story of positions within various stability diagrams. Earth systems are much more complex than mechanical systems, so we are often uncertain of properties equivalent to the types, sensitivities, and topologies of Figure 1.12. The perturbing forces may either be internal to Earth's system (e.g., volcanic eruptions, tectonic activity, human industrial and agricultural activity) or external (e.g., solar radiation variations, impacts by meteorites, supernova explosions). As the Earth system evolves, its stability states may be changing. For example, continental drift and mountain building may influence the stability of Earth's climate. The Earth system clearly has exhibited internal instabilities. As a dramatic example, we may point to the appearance of oxygen as a major constituent in the atmosphere some 1.5 Gyr ago, as a consequence of the introduction of photosynthesis by biological evolutionary processes. This event led to the demise of much of the biosphere (the composite of vegetative and animal life on the planet), which was previously based on anoxic (oxygen-poor) processes.

The most important message, and one that will take the remainder of the book to develop fully, is that if the Earth system has never before passed from a stability position of type 1 or 2 to one of type 3, we may have no way of knowing what the characteristics of the metastable state represented by position 3 may be. This could be the case, for example, for future climatic and chemical disturbances caused by human activities that are now known to affect global-scale processes. Hence, even if information is incomplete, decisions may have to be made promptly about the prudent use of the planet's resources; we can no longer afford to defer such important decisions to future times and future generations.

EXERCISES

(Note: The appendix entitled "Units of Measurement in Environmental Chemistry" will be helpful in completing the exercises in this and following chapters.)

1.1 Position the following on Figure 1.8: the distance from Tokyo to Santiago, the height of a pine tree, the length of your foot, the size of a grain of sand.

1.2 Ammonia gas (NH_3) is present over the open oceans only in very low mixing ratios, typically 0.4 pptv. Express this value in concentration units of molecules per cubic centimeter if the atmospheric density at sea level and 25 °C is 2.48×10^{19} molecules per cubic centimeter.

1.3 Given the carbon dioxide concentration in volume units in Table 1.1, find the carbon dioxide concentration in the mass units $\mu g/cm^3$ at standard temperature and pressure (STP).

1.4 The hydrocarbons that make up plant waxes are only moderately volatile. As a consequence, many of them exist in the atmosphere partly as gases and partly as constituents of aerosol particles. If tetradecane ($C_{14}H_{30}$, molecular weight 198) has a gas phase mixing ratio over the North Atlantic Ocean of 250 ppt and an aerosol phase concentration of 180 ng m^{-3}, in which phase is it more abundant?

FURTHER READING

J. G. Anderson, D. W. Toohey, and W. H. Brune, Free radicals within the Antarctic vortex: The role of CFCs in Antarctic ozone loss, *Science, 251,* 39–52, 1991.

T. J. Crowley, The geologic record of climatic change, *Reviews of Geophysics and Space Physics, 21,* 828–877, 1983.

C. Lorius, J. Jouzel, D. Raynaud, J. Hansen, and H. LeTreut, The ice-core record: Climate sensitivity and future greenhouse warming, *Nature, 347,* 139–145, 1990.

W. A. Shear, The early development of terrestrial ecosystems, *Nature, 351,* 283–289, 1991.

CHAPTER 2

Earth and Its Driving Forces

We begin our substantive discussions with the solid planet and its surface, because the planet's system cannot adequately be discussed in the context of isolated regimes. Rather, each regime needs to be understood well enough to take into account its interactions with the others.

With our goal of studying the changes taking place at the planet's surface firmly in mind, we will not discuss here the details of Earth's interior, its mechanisms of rock formation, earthquakes, or the many other fascinating aspects of geology and geophysics that are extraneous to our purpose. We will look instead at those aspects of geology and geophysics that in some way relate to observable surface and atmospheric effects, both currently and historically, for only in so doing can we hope to understand how the atmosphere, the oceans, and other features came to be, how surface features and climates have evolved throughout Earth's history, and how factors such as volcanic eruptions, changing land use, ocean depth, and evolving continental margins and positions have influenced the past and may influence the future.

THE SURFACE OF THE PLANET

Figure 2.1 illustrates schematically the present surface structure of the planet Earth. This picture is worth close examination, because it provides important clues to the nature of the planet. Data on the areas of the continents and oceans and the types of surface features are given in Table 2.1.

About 70% of the planet's surface is covered by the oceans, a striking property in a solar system without another water-covered planet and perhaps without another planet covered by liquid of any kind (although liquid methane may cover the surface of Titan, Saturn's largest moon). The continents, which make up about 30% of the surface area, are distributed quite unequally over Earth, more than

TABLE 2.1. Areas of Continents, Oceans, and Surface Regimes

FEATURE	AREA (10^6 km^2)	PERCENTAGE OF LAND OR OCEAN	PERCENTAGE OF EARTH'S SURFACE
Earth's surface	510	—	100
All continents	149	100	29.2
All oceans	361	100	70.8
Asia	45	30	8.8
Africa	30	20	5.9
North America	22	15	4.3
South America	18	12	3.5
Antarctica	16	11	3.1
Europe	10	7	2.0
Australia	8	5	1.5
Pacific Ocean	181	50	35.5
Atlantic Ocean	94	26	18.4
Indian Ocean	74	21	14.5
Arctic Ocean	12	3	2.4
Forests	50	34	9.8
Grass, cropland	33	22	6.5
Mountains	23	16	4.5
Deserts	21	14	4.1
Urban areas	17	11	3.3
Wetlands	5	3	1.0

65% of the land lying in the northern hemisphere. The land in the midlatitudes is generally the most suitable for agriculture and has become the most intensively developed. Land near the equator is now coming under similarly heavy developmental pressure.

Forested regions constitute about a third of the land area, with most of the more heavily forested regions lying in either tropical or temperate latitudes. Grassland accounts for another quarter of the land area. Mountains and deserts cover similar areas, and together total about a third of the continents. It is interesting that in this time of rapid population expansion, urban and suburban geography constitutes about a tenth of the total land area.

Several natural processes result in emissions of chemical species to and uptake from the atmosphere. Among the most significant of these is the assimilation-respiration cycle of vegetation, in which atmospheric carbon dioxide is utilized to form plant material during the growth part of the annual cycle and is released to the atmosphere during the period of decay. The amount of carbon dioxide cycled through this process is a function of the density of biomass, so it is highest in the lush tropics and also large in the forested areas of the midlatitudes. Additional geographical regimes of interest from an atmospheric viewpoint are the wetlands and agricultural areas, which emit substantial fluxes of a variety of important gases. Finally, we note

Figure 2.1 The positions of the continents and the oceans on the surface of Earth (Mercator projection).

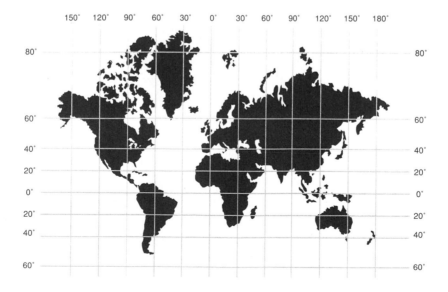

that many gases and particles are produced by the burning of biomass, sometimes through natural forest fires, but more often as a human land-clearing measure.

Other geographical regimes are of lesser consequence as far as short time scale emissions are concerned: these areas include steppes and mountains, deserts, and glaciated areas. Sea salt aerosol injection from the oceans into the lower troposphere is important to the atmospheric chlorine cycle, and gaseous sulfur emissions are of importance to cloud formation (of which more later). In addition, the ability of the oceans to absorb carbon dioxide plays a crucial role in the global carbon cycle.

ISOTOPE GEOCHRONOLOGY

Basic Concepts of Radioactivity

In studying any aspect of the history of the planet's system, accurate dates for the ages of rock features, fossils, and other samples are of crucial importance. The dating is generally accomplished by measurements of the products of radioactive decay. Recalling a few of the properties of atoms and their nuclear physics is useful in this discussion. An atom of element E has Z protons (the atomic number) and a mass number (the sum of protons and neutrons) denoted by A. Most elements have several forms, each with a different value of A; these are known as the *isotopes* of the element. Particular isotopes are designated $_Z^A E$, although the atomic number is often omitted.

Radioactivity or radioactive decay is the spontaneous emission of particles or electromagnetic radiation by metastable or unstable isotopes as a consequence of the disintegration of active atomic nuclei. All of the elements with $Z > 83$ are radioactive, as are some of the isotopes of the lighter elements. When isotopes undergo radioactive decay, one of the common processes is the emission of an alpha [α] particle (a nucleus with two protons and two neutrons). The alpha particle emission thereby reduces the Z value of the emitting nucleus by 2 and the A value by 4 and produces a new chemical element. An alternative radioactive process is the emission of an electron (termed a beta [β] particle), a process that results in the conversion of a neutron to a proton. Because an electron has negligible mass, Z increases by one in this process while A remains the same; again the result is the formation of a new element from the old.

Radioactive Lifetimes

The likelihood of α or β emission from a radioactive isotope is determined by a probability that is characteristic of the isotope. The probability is an internal property of the atom, is unaffected by physical conditions such as temperature or pressure, and is independent of time. For n atoms of a specific isotope at time t, the number of disintegrations per unit time can be expressed as

$$\frac{dn}{dt} = -\lambda n, \qquad (2.1)$$

where λ is a constant termed the *decay constant*. By integration,

$$n_t = n_0 e^{-\lambda t}, \qquad (2.2)$$

where n_0 is the number of atoms present at time $t = 0$. Another useful number referring to radioactive isotopes is the *half-life*, the time required for the disintegration of half the original number of atoms of a particular isotope. Rewriting Equation (2.2) gives

$$\frac{n_0}{2} = n_0 e^{-\lambda t_{1/2}} \qquad (2.3)$$

or, alternatively,

$$t_{1/2} = \frac{\ln 2}{\lambda} \ . \qquad (2.4)$$

When alpha or beta particles are emitted from a radioactive atom into the surrounding rock, they come to rest within a distance of a few millimeters or so, their kinetic energy being converted to heat. As will be seen, this contribution to the internal heating of Earth has important geophysical consequences.

The technique of radioactive dating consists of measurement of the relative concentrations of parent nuclei (the species that undergo disintegration) and daughter nuclei (the species produced by the disintegration). Knowledge of the rate of decay for the parent isotopes permits the calculation of the time when only parent isotopes were present; this is the age of the rock from which the sample was taken.

Radioactive isotopes also can be used for dating features on Earth that have lifetimes much shorter than those of rocks: plants, organic deposits, even air parcels. Different isotopes are useful for dating different time periods, because the decay constants vary widely. The most frequently used pairs and their relevant properties are given in Table 2.2.

TABLE 2.2. Geochronologically Useful Isotope Decay Systems			
PARENT/ DAUGHTER	EMITTANTS	HALF-LIFE (YR)	USEFUL AGE RANGE[a]
$^{147}Sm/^{143}Nd$	α	1.1×10^{11}	10^8 to $A_\odot$ yr
$^{238}U/^{206}Pb$	$8\alpha, 6\beta$	4.5×10^9	10^7 to $A_\odot$ yr
$^{235}U/^{207}Pb$	$7\alpha, 4\beta$	7.1×10^8	10^7 to $A_\odot$ yr
$^{232}Th/^{208}Pb$	$6\alpha, 4\beta$	1.4×10^{10}	10^7 to $A_\odot$ yr
$^{87}Rb/^{87}Sr$	β	4.9×10^{10}	10^7 to $A_\odot$ yr
$^{40}K/^{40}Ar$	EC^b	1.3×10^9	5000 to $A_\odot$ yr
$^{14}C/^{14}N$	β	5370	0 to 70,000 yr
$^{222}Rn/^{210}Pb$	$3\alpha, 2\beta$	3.8 da	0 to 40 da
$^7Be/^7Li$	EC^b	53.4 da	0–1 yr

a. $A_\odot$ indicates the age at which Earth was formed.
b. EC indicates that an electron is captured during this process, rather than α or β particles being emitted.

The Uranium-Thorium-Lead System

Most of the epochs pictured in Figure 1.9 were dated through the use of the uranium-thorium-lead system. This system involves three separate decay schemes, beginning with ^{238}U, ^{235}U, and ^{232}Th, respectively. The decay

Figure 2.2 The decay scheme for the radioactive isotope ^{238}U. The parent nucleus is indicated by the double box, the final daughter nucleus by the double circle. The numbers within the symbols are the atomic masses of the isotopes, and the times indicated are the decay half-lives (M, minute; D, day; Y, year; 4.5E9 means 4.5×10^9). $^{214}_{83}Bi$ has two radioactive decay paths; both are shown on the diagram, together with the percentages of occurrence for each branch. This parent-daughter decay scheme is used extensively in the dating of ancient rock samples.

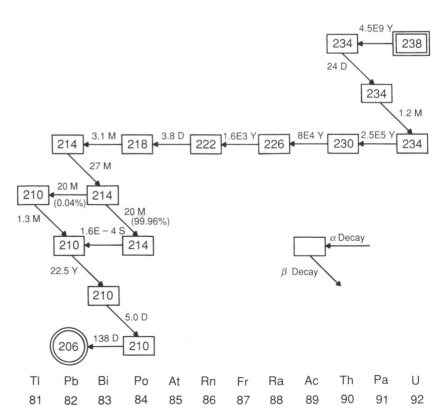

scheme for ^{238}U is shown in Figure 2.2. It consists of a series of 16 radioactive decay processes, resulting in the emission of 8 α and 6 β particles. The decay schemes for ^{235}U and ^{232}U are similarly complex. The presence of more than one radioactive isotope in a mineral, each with its own decay time, allows dating with much more certainty than when data for only one radiometric decay series are available.

The Carbon-14 System

The most extensive use of radioactive isotopes for direct determinations of atmosphere-surface interactions has involved the isotope ^{14}C. This isotope is being continually created in the atmosphere by the reaction of nitrogen with cosmic ray-produced neutrons:

$$^{14}_{7}N + {}^{1}_{0}n \rightarrow {}^{14}_{6}C + {}^{1}_{1}H. \qquad (2.5)$$

The carbon atoms form carbon monoxide by combining with molecular oxygen,

$$^{14}_{6}C + O_2 \rightarrow {}^{14}_{6}CO + O, \qquad (2.6)$$

are oxidized to carbon dioxide by reaction with the hydroxyl radical,

$$^{14}_{6}CO + HO\cdot \rightarrow {}^{14}_{6}CO_2 + H\cdot, \qquad (2.7)$$

and eventually decay by emission:

$$^{14}_{6}CO_2 \rightarrow {}^{14}_{7}NO_2 + \beta. \qquad (2.8)$$

Before the carbon isotopes decay, however, the radioactive carbon dioxide may be utilized by growing vegetation in the same way as is the more common carbon dioxide containing ^{12}C. As a consequence, the $^{14}C/^{12}C$ isotope ratio can be used to date any living thing or any artifact containing or made from any living thing. Alternatively, if independent dating techniques are available, for example, with tree rings, then the ratio measurement permits analyses to be made of the amount of ^{14}C present in the atmosphere at the time of uptake into the trees, thus providing a marker of solar activity. This information provides an important calibration, because the cosmic ray flux that creates ^{14}C is known to vary. As a consequence, radiocarbon dates older than a few thousand years require correction for solar activity variations.

As with most other subdisciplines, carbon dating has a language of its own. The only part of that language needed here is that involving the way in which isotopic abundances are expressed. The relative abundance of the radioactive isotope of carbon is defined as the number of parts of the isotope per thousand atoms, compared with that same ratio in a standard:

$$\delta^{14}C = \frac{(^{14}C/^{12}C)_{sample} - (^{14}C/^{12}C)_{std}}{(^{14}C/^{12}C)_{std}} \times 1000, \qquad (2.9)$$

and the results are expressed as a number *per mil*, abbreviated ⁰/₀₀. Sometimes it is convenient to compare the isotope ratios of two materials; in that case the notation is $\Delta_{12} = \delta_1 - \delta_2$. Examples of the use of isotope abundance notation and of the usefulness of carbon isotope analyses are given in subsequent chapters.

Radioactive Isotopes as Tracers of Air Parcels

Just as long-lived isotopes are important for dating old rocks, short-lived isotopes can be used to date and trace atmospheric air parcels. One of the systems useful in such efforts is the radon-lead system, a by-product of the uranium-thorium-lead system. The latter is useful for dating only if the rocks within which the isotopes occur are sufficiently impenetrable that parent and daughter isotopes are contained. In the radon-lead system, the technique requires that the radon, a daughter of the uranium-thorium chain, escape from the continental crust containment structure, a process aided by the fact that radon is a gas. Because ^{222}Rn and ^{210}Pb, its longest lived radioactive daughter, are readily detected in the air, they constitute a radioactive system useful to atmospheric scientists.

The rate of emission of ^{222}Rn from the continental crust is constant with time for any particular crustal structure, or nearly so. As shown in Figure 2.2, the radon decays within a few days to ^{210}Pb, which rapidly associates itself with submicrometer-sized aerosol particles. The short half-life of ^{222}Rn renders it suitable for use in monitoring the patterns of rapid vertical air motions from the lower to the upper atmosphere and the exchange of air from continents to oceans. In contrast, because the aerosol particles have lifetimes of months or years, ^{210}Pb measurements are used to monitor longer range atmospheric circulation, such as transport between the northern and southern hemispheres, as well as the rate of scavenging of aerosol particles by precipitation.

Beryllium-lead is a second isotope system applicable to atmospheric studies. ^{7}Be is a radioactive isotope formed in the atmosphere as a result of the interaction of high-energy cosmic rays from outer space with atmospheric nitrogen and oxygen. ^{7}Be decays by electron capture, an electron from the K shell of the beryllium atom being transferred into the atomic nucleus. The product is the principal natural isotope of lithium.

Because most cosmic rays that interact with atmospheric molecules do so at relatively high altitudes, most of the ^{7}Be is formed high in the atmosphere. Some, however, clearly forms at lower altitudes and results from deeply penetrating cosmic rays. Like ^{210}Pb, ^{7}Be in the atmosphere is primarily associated with small aerosol particles. Lower atmospheric ^{7}Be is therefore useful as a tracer of air motions and aerosol scavenging, and its lifetime of about two months complements the longer life of ^{210}Pb.

The high-atmospheric component of ^{7}Be is particularly valuable for tracing exchanges of air between the upper and lower atmosphere and has been used to demonstrate that certain instances of high concentrations of ozone in the lower atmosphere have been caused by air exchange and not primarily by photochemical smog processes involving ground-level emissions.

PLATE TECTONICS AND MODERN PLATE MOTIONS

The Interior of the Planet

On the scale of human lifetimes, Earth's geography is comfortably constant. From the perspective of the long-term evolution and modification of chemistry and climate, however, Earth's geography is not static but is involved in gradual change. The driving force for this evolution turns out to be heat generated in Earth's interior.

The guiding synthesis for understanding many of the characteristics of the solid Earth and its evolution is that of plate tectonics and seafloor spreading. Figure 2.3 pictures the major divisions of Earth's interior, as revealed by analyses of the delay times of waves from earthquakes. At the center is a solid, very hot *inner core*; it is virtually pure iron. Surrounding it is the *outer core* of liquid, impure iron. The existence of these phases is a function of both the high temperatures and the high pressures that occur there. Surrounding the core is the *lower mantle*, which is significantly cooler and is rich in silicates. Next is the *upper mantle*, also rich in silicates and semiplastic in structure. Near the upper mantle's outer edge is the *asthenosphere*, a structurally weak region, perhaps partly molten, that readily allows shearing motion. The top 100 km of solid Earth is termed the *lithosphere*. This rigid, strong shell of rock contains the *crust*, the continents, and the plates.

As heat is produced within the inner parts of Earth, it is distributed outward by convection in the mantle. The mantle thickness is nearly half the planet's radius, and its ability to support sluggish convective flow is crucial to the processes of solid Earth geophysics. The figure illustrates a process termed *whole-mantle convection*, but that process may in fact be a *layered convection*, in which the upper and lower mantle function relatively independently; the details of mantle convection remain to be resolved. In any case, heat rising through the asthenosphere moves matter upward, expanding the rock and increasing its buoyancy. The continuing reduction in pressure from the overlying load eventually results in partial melting and magma (melted rock) formation. Where rifts occur in the lithosphere, the magma rises to the surface. As it emerges, the lithospheric plates are pushed aside and the continents atop them experience an applied force known as *ridge-push*. A second force contributing to the motions of the plates is provided by *plate-pull* at the opposite end of the lithospheric

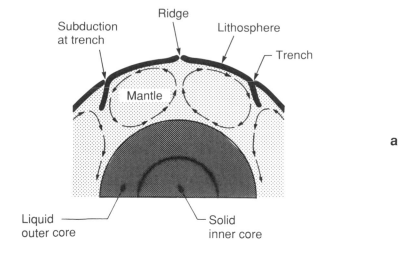

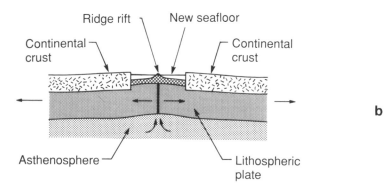

Figure 2.3 Modern plate
tectonics processes. (a) Convection
currents in the mantle produce
molten material (magma), which
ascends under a structurally weak
ocean ridge, rises through rifts, and
flows away, carrying the plates
along. As the plates cool and
become denser, they may sink at
subduction zones. (b) The seafloor
spreading process. (c) When plates
collide, one plate usually buckles
downward. The overriding plate is
crumpled and uplifted. Deep-sea
trenches (sites of the greatest ocean
depths), high mountain chains,
volcanoes, and the greatest
earthquakes are associated with
such regions of plate convergence.

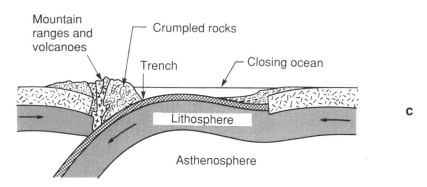

plates. In these locations, cooling and increasing density cause the
slowly moving and spreading seafloor to sink upon encountering
another plate. As this oceanic crust sinks into the asthenosphere, it is
reheated, and magma emerges through cracks and fissures in the
covering lithosphere to form mountains and volcanoes.

Heat Flow Within Earth

Substantial sources of heat are needed to drive mantle convection. One of these sources is the decay and energy release of the radioactive isotopes that were incorporated into the planet when it was formed. The heat productivity of a specific isotope is a combination of the probability of disintegration and the kinetic energy of the emitted particles. In Earth, the isotopes of primary interest are two of uranium (^{238}U and ^{235}U), one of thorium (^{232}Th), and one of potassium (^{40}K). As they have decayed over the eons, the internal heat production of Earth has decreased to perhaps 20% of its value at the time of Earth's formation.

A second source of heat energy arising from processes in Earth's interior is the heat released as the outer core of Earth gradually freezes. This crystallization occurs at temperatures of about 6300 K and pressures of about 330 gigapascals, so innovative laboratory experiments are required to determine the precise heat of crystallization. Estimates must then be made for the volume of rock involved and the time scale for the crystallization.

A third heat source is gravitational in nature and is manifested in the conversion of the kinetic energy of infalling objects—the molecules and particles that were drawn together to constitute the planet—into heat. This initial *heat of aggregation* is no longer a significant factor, gravitational heating now being produced as crystallization at the inner core boundary ejects lighter elements into the outer core. This process is termed *zone refining* and has been used in industry to purify silicon wafers for microelectronics. As a consequence of zone refining in the core, the planet's center has become increasingly dense, and the corresponding increase in gravitational energy produces heat.

The present fluxes of heat from the different sources are summarized in Table 2.3, which indicates that the major source of heat within Earth is radioactive decay. Although the heat flux is far from uniform over Earth's surface, it is useful to note that its globally averaged value is about 0.08 J m^{-2} s^{-1}. This heat engine, which moves the continents of Earth at their stately pace and produces volcanoes and earthquakes of awesome intensity and power, is driven by a flux less than a thousandth of the magnitude of the heat energy received at the surface of the planet from the Sun!

The Surface in Motion

The manner in which the heat engine manifests itself on the surface of today's Earth is shown in Figure 2.4a, where the major tectonic features of the world are indicated. Several active rift zones (or *divergent boundaries*) are known, the best studied being one down the center of the Atlantic Ocean and one in the Pacific Ocean off South America. Where plates moving together overlap each other, *convergent boundaries*

TABLE 2.3. Sources of Internal Heat for Earth

HEAT SOURCE	HEAT FLUX ESTIMATE (J s^{-1})
Radioactivity	34×10^{12}
Crystallization	5×10^{12}
Gravitation	3×10^{12}
Total	42×10^{12}

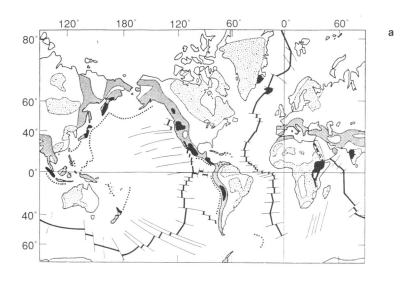

Figure 2.4a Major tectonic features of Earth. Key: heavy lines, active rift systems of oceanic ridges; light lines, oceanic faults; dotted lines, oceanic trenches; ornamented, continental shields (exposed very old crystalline rocks); unshaded, continental platforms (shield rocks overlain by sediment); dark gray, folded mountain chains; black, volcanic regions. (Reprinted by permission from P. J. Wyllie, *The Dynamic Earth: Textbook in Geosciences.* Copyright 1971 by John Wiley & Sons, Inc.)

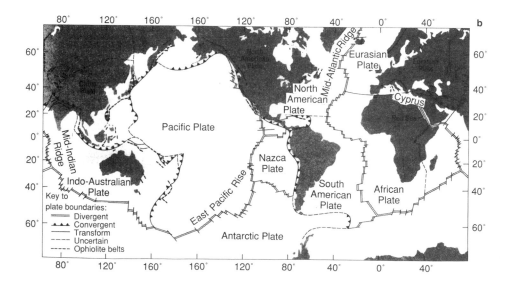

Figure 2.4b The major lithospheric plates and plate boundaries currently present on Earth. (P. A. Rona, *Transactions of the American Geophysical Union*, 58, 629–639, 1977.)

occur. If the converging plates have the same density, then a *collisional boundary* results, at which features such as folded mountain ranges are built. If the plates have different densities, then one will sink beneath the other, forming a *subduction zone*.

Mountain ranges and subduction zones suggest the boundaries of the tectonic plates. Depending on one's definition, there are seven or eight major plates and a number of minor ones. Figure 2.4b shows the principal plates and plate boundaries.

Above the moving plates and the spreading oceans, the chemical composition of the atmosphere is affected in several ways by the evolving planet beneath it. Perhaps the most obvious is the intermittent injection into the atmosphere of vast quantities of gases and particles from volcanic eruptions. As seen in Figure 2.5, the earthquakes and volcanoes outline the tectonically active areas of the Earth, especially the subducting plate boundaries and the spreading centers, or undersea ridges. Submarine volcanism has a major impact on ocean water chemistry. Volcanoes erupting directly into the air have more immediate and obvious influences. If the eruption is large enough, then, as sometimes happens, volcanoes can affect the atmosphere of the entire Earth and Earth's climate.

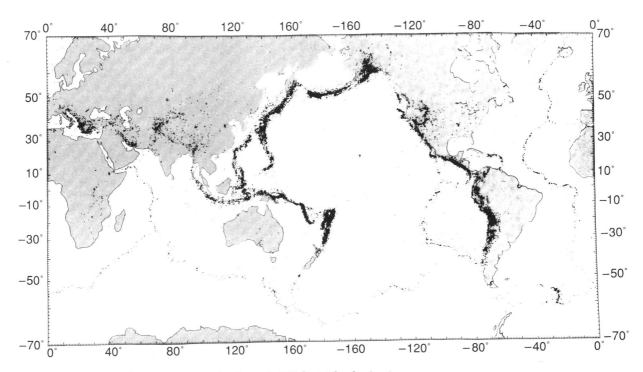

Figure 2.5 Earthquake epicenter map for the period 1961–1967, for depths to 700 km. Epicenters are the surface locations above the rupture point. The most common locations are at convergent zones where subduction of oceanic crust is occurring. (M. Barazangi and J. Dorman, *Bulletin of the Seismological Society of America*, 59, 369–380, 1969.)

TECTONIC HISTORIES

Plate tectonics is not a new phenomenon, and plates have been in motion on the surface of the planet for a very long time. How much can we say about the locations of the continents in the past and in the future? To answer this question, one must search for clues in and on Earth to past tectonic processes. It turns out that useful, if perhaps not ideal, clues are available in the rock sequences known as *ophiolites*, which are thought to be slices of oceanic crust formed during deep seafloor spreading and which occur in concert with overlying layers of deep-sea sediment. The ophiolite sequence is a reliable signature of a region in which subducting ocean crust has occurred. When such structures are found on continents, generally as features of collisional mountain belts, they are termed *suture belts*, and define where zones of collision between moving continents occurred in the past. The Alpine-Himalayan mountain zone is one such collisional belt, formed (and currently forming) as a result of the collision of the Eurasian landmass with that of India and Africa. At some time in the future, it will reflect the same suture belt characteristics as those recognized today at locations far removed from subduction zones.

Several ancient suture belts have been identified in North America from ophiolites and other geological features; they are shown in Figure 2.6. In order of estimated age, they are the Precambrian terranes (sutures older than 570 Myr), the Appalachians (sutures aged from perhaps 500 to 250 Myr BP), the Cordillera (sutures aged up to 400 Myr BP), and the Quachitas-Marathons (sutures aged from perhaps 250 to 300 Myr BP). Studies of the types of rocks, the fossils, and the remnant magnetic fields in these suture belts have established with fair certainty that part of Florida was once part of the African continent, coastal New England and Newfoundland part of Europe, and parts of Nevada, British Columbia, and Alaska part of Asia.

By combining suture-belt and related information with studies of plate motions, we can determine many features of historical plate tectonics. For example, rocks on other continents that are similar to modern rock structures in composition and sequence suggest that processes bearing some similarity to modern plate tectonics have been present on Earth for 2.5 Gyr. Rock structures older than that age are rather different, a finding suggesting that early tectonic processes were significantly different from those found today. Although the evidence is sketchy, it is currently believed that tectonics has proceeded in three stages. The first was in operation from about 3.8 to 2.6 Gyr BP, during which time the heat flow from radioactive decay was several times greater than its present value. This higher heat flow produced a less dense, more buoyant lithosphere, with vigorous convection, little subduction, and many relatively small plates that collided frequently. In the second tectonic stage, occurring between about 2.6 and 1.3 Gyr BP, the gradual decrease in heat flow resulted in the development of

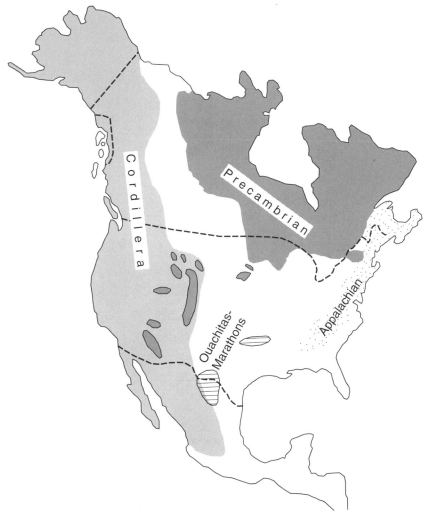

Figure 2.6 The suture belts of North America. The principal belts, indicated by the shaded areas, are discussed in the text. Along the Pacific Coast many small remnants of other continents, thought to have accumulated over the eons, have been identified. (Adapted with permission from E. M. Moores, Ancient suture zones within continents, *Science, 213*, 41– 46. Copyright 1981 by the AAAS.)

a few larger plates, but with too little difference between the densities and buoyancies of oceanic and continental crust to provide substantial tectonic activity. Finally, during the most recent 1.3 Gyr, the decrease in buoyancy of the oceanic crust relative to continental crust resulted in a gradual transition to the processes of modern plate tectonics, with crust subduction and regeneration.

The consequences of the tectonic process for climate and chemistry are related to the ability of continents in midlatitudes and the tropics to support vegetative activity and to the propensity of polar landmasses to serve as receptors for highly reflective snow and ice. To see how these continental locations have changed, one can plot continental locations at several epochs in the past and compare the plots with that of the present. Typical reconstructions are shown in Figure 2.7. We will return to these diagrams in our discussion of ancient climates, but it is sufficient here to note that the bulk of the

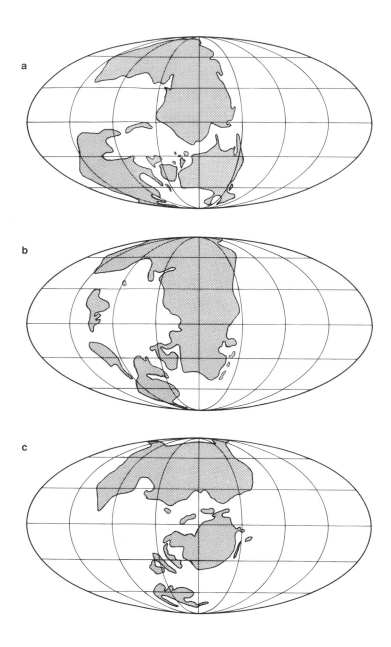

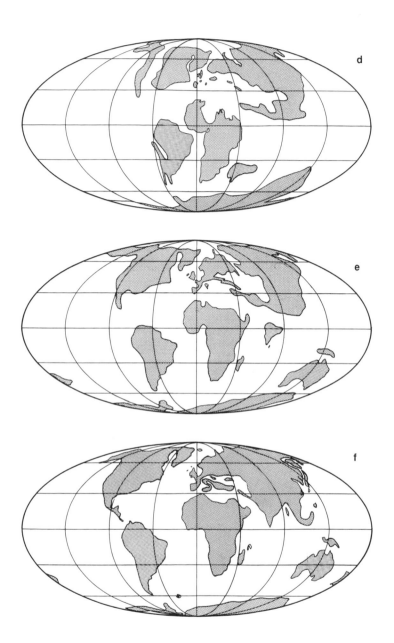

Figure 2.7 Reconstructed locations for the continents: (a) 320 Myr BP: Most of the land was contained in two supercontinents, Laurussia (combining Greenland, North America, Scandinavia, and most of Russia) and Gondwana (containing most of the rest of the landmass). (b) 250 Myr BP: The collision of the two supercontinents formed Pangaea, a new supercontinent, with the mountain ranges created in the suture belt including the Appalachians. (c) 135 Myr BP: Pangaea begins to break up, and the Atlantic Ocean starts to form between North America and Eurasia. (d) 100 Myr BP: Africa, South America, and India separate. (e) 45 Myr BP: Africa, South America, and India move northward toward Europe, North America, and Asia, respectively. Australia breaks off from the southern landmass. (f) Today: The American continents are joined, as are Africa and Eurasia. India has collided with Eurasia to form the Himalayan mountains. (K. Winn and C. R. Scotese, *Phanerozoic Paleographic Maps*, University of Texas Institute of Geophysics, Tech. Rpt. 84, 31 pp., 1987.)

land 320 Myr ago was in the southern hemisphere. Today the bulk of the land is in the north. In the intervening period, as the modern continents formed and moved toward their present positions, their climates underwent significant changes. Signatures of ancient climate, chemistry, and biology, such as the chemical nature of old rocks and the abundances of specific fossils within them, must be interpreted in the light of the knowledge of these changes in the locations of the continents.

SUMMARY

The chemical nature of Earth's surface and atmosphere is inextricably entwined with the evolving nature of Earth's surface and interior. On time scales of tens of millions of years, that surface undergoes noticeable change on a global scale. On very short time scales, volcanic eruptions produced as a by-product of plate motions affect local, regional, and sometimes global air quality and climate. On time scales much longer than the human experience, the evolving surface of the planet provides the base on which chemistry and climate are built.

EXERCISES

2.1 (a) What new isotope is formed by the radioactive decay of ^{226}Ra, if the decay results in α particle emission? (b) What new isotope is formed by the radioactive decay of ^{214}Pb, if the decay results in β particle emission?

2.2 The radioactive decay series for ^{232}Th is as follows 33.7% of the time:

Radioactive Decay Series for ^{232}Th		
ISOTOPE	EMITTED PARTICLE	HALF-LIFE
$^{232}_{90}$Th	α	1.4×10^{10} yr
$^{228}_{88}$Ra	β	6.7 yr
$^{228}_{89}$Ac	β	6.13 hr
$^{238}_{90}$Th	α	1.90 yr
$^{224}_{88}$Ra	α	3.64 da
$^{220}_{86}$Rn	α	54.53 s

Radioactive Decay Series for ^{232}Th (con'd)		
ISOTOPE	EMITTED PARTICLE	HALF-LIFE
$^{216}_{84}$Po	α	0.158 s
$^{212}_{82}$Pb	β	10.67 hr
$^{212}_{83}$Bi	α	60.48 min
$^{208}_{81}$Tl	β	3.1 min
$^{208}_{82}$Pb		Stable

The other 66.3% of the time, the ninth step produces a β particle and ^{84}Po. This step is followed by particle emission to produce the stable final product.

Diagram the decay scheme as was done for ^{238}U in Figure 2.2.

2.3 The heat productivity of uranium in its present isotopic distribution is 0.73 cal g^{-1} yr^{-1}. The corresponding numbers for thorium and potassium are respectively 0.20 cal g^{-1} yr^{-1} and 2.7 $\times$ 10^{-5} cal g^{-1} yr^{-1}. If a sample of oceanic crust (density = 3.2 g cm^{-3}) includes 0.42 ppm (by weight) uranium, 1.68 ppm thorium, and 0.69% potassium, calculate the heat production of the crust sample in units of J cm^{-3} s^{-1}.

2.4 The equator receives the most solar radiation, and continents absorb solar radiation much better than do oceans. How would you expect the positions of the continents over time to have influenced the global heat balance?

2.5 The half-life for the isotope ^{238}U is 4.5 $\times$ 10^9 yr. Compute the decay constant. If a certain rock was formed at 4.0 Gyr BP with 1000 atoms of ^{238}U, how many atoms are present now?

FURTHER READING

D. Gubbins, *Seismology and Plate Tectonics*, Cambridge University Press, Cambridge, UK, 1990.

F. Press and R. Siever, *Earth*, 4th Ed., W. H. Freeman, San Francisco, 1986.

B. J. Skinner and S. C. Porter, *The Dynamic Earth*, John Wiley, New York, 1989.

S. M. Stanley, *Earth and Life Through Time*, 2nd Ed., W. H. Freeman, New York, 1989.

CHAPTER *3*

The Atmospheric Radiation Budget

SOLAR RADIATION AND ITS ABSORPTION

Except for the relatively small amounts of energy provided to Earth's surface as a consequence of its internal composition and internal heating, all of the energy that drives biological, geophysical, and geochemical processes comes from the Sun in the form of radiation. The radiation can be sensed in various ways, some as visible radiation, some as heat radiation, some as radio waves, and so forth, but the basic mechanism involved in the absorption or emission of radiation is the same; it reflects the transition of an atom or molecule between a higher and a lower energy state. As a molecule gains energy, its electrons jump to higher (excited) energy states. As the molecule loses energy, that energy is emitted as radiation.

The hotter a radiating body is, the shorter is the wavelength of radiation that it emits. The Sun, with an effective surface temperature of 5780 K, has a maximum in its radiated energy at a wavelength of about 500 nm. The radiation spectrum received from the Sun at Earth's surface differs from that emitted by the Sun because some is absorbed in the atmosphere (Figure 3.1).

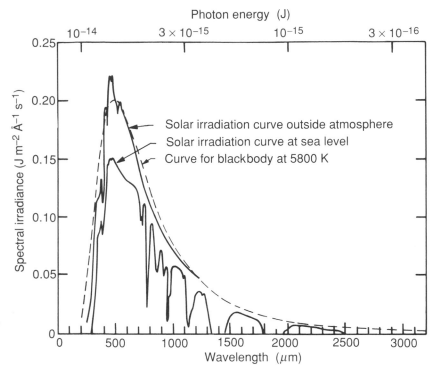

Figure 3.1 Solar irradiance curves for the top of the atmosphere and at Earth's surface, including comparison with blackbody radiation at a temperature of 5800 K. The absorption features at sea level are largely due to O_2, O_3, H_2O, and CO_2. Wavelength is shown at the bottom (in μm, 1 μm = 1000 nm), photon energy at the top.

Blackbody Radiation

Radiation is best understood as being both a particle and a wave, and a basic relationship exists between the wavelength of the radiation (λ) and the energy E of the particles, or *photons*, that make up the radiation:

$$E = hc/\lambda = h\nu, \qquad (3.1)$$

where h is a fundamental quantity of nature called *Planck's constant*, with a value of 6.626×10^{-34} J s, ν is the frequency of the radiation, and c is the velocity of light (2.998×10^8 m s^{-1}). A consequence of Equation (3.1) is that photons with short wavelengths have higher energies than photons with long wavelengths. The *radiant flux* (i.e., quantity of radiation emitted per second) per wavelength interval is described by *Planck's law*:

$$\Psi_\lambda = \frac{a}{\lambda^2[\exp(b/\lambda T) - 1]}, \qquad (3.2)$$

where Ψ_λ is the flux at a given wavelength λ, T is temperature in kelvins, and a and b are constants. For the wavelengths and temperatures of interest here, the exponential term in Equation (3.2) is much larger than unity, and the equation can therefore be approximated by

$$\Psi_\lambda = a\lambda^{-5} \exp(-b/\lambda T). \qquad (3.3)$$

Clearly, for an object of any given temperature, some radiation is emitted at all wavelengths. The quantity emitted, however, falls off steeply both above and below the peak emission wavelength, which is found by setting the derivative of Equation (3.3) to zero. The result, termed *Wien's displacement law*, is (for wavelengths in micrometers and temperature in kelvins)

$$\lambda_{peak} = 2897/T. \qquad (3.4)$$

In the case of the Sun, with its peak emission near 500 nm, about 10% of the solar radiation energy is in the ultraviolet region ($\lambda < 400$ nm), 45% is in the visible region ($\lambda = 400$ to 750 nm), and 45% is in the infrared ($\lambda > 750$ nm).

Each atmospheric atom and molecule has different efficiencies and different wavelength regions for absorption of radiation. At the shortest wavelengths, the most important absorbers of solar radiation in Earth's atmosphere are molecular oxygen and ozone. Figure 3.2a shows that molecular oxygen absorbs photons with wavelengths shorter than about 240 nm. The most important result of the absorption process is to break the bonds holding the oxygen atoms together:

$$O_2 + h\nu \rightarrow O + O. \qquad (3.5)$$

The free oxygen atoms that are produced react with other oxygen molecules to form ozone,

$$O + O_2 + M \rightarrow O_3 + M. \qquad (3.6)$$

[M denotes an air molecule, usually molecular (diatomic) nitrogen or oxygen, that acquires the excess energy liberated by the exothermic association reaction and disperses it to surrounding molecules by collisional processes, thus preventing the newly formed O_3 molecule from reverting to its precursors because of an excess of internal energy.] The absorption cross section (see box) of ozone, shown in Figure 3.2b, has a peak at wavelengths nearly 100 nm longer than those maximally absorbed by molecular oxygen. A photon with wavelengths of 200–300 nm or, with lesser efficiency, at wavelengths as long as 1140 nm, causes fragmentation when absorbed by a molecule of ozone:

$$O_3 + h\nu \rightarrow O_2 + O. \qquad (3.7)$$

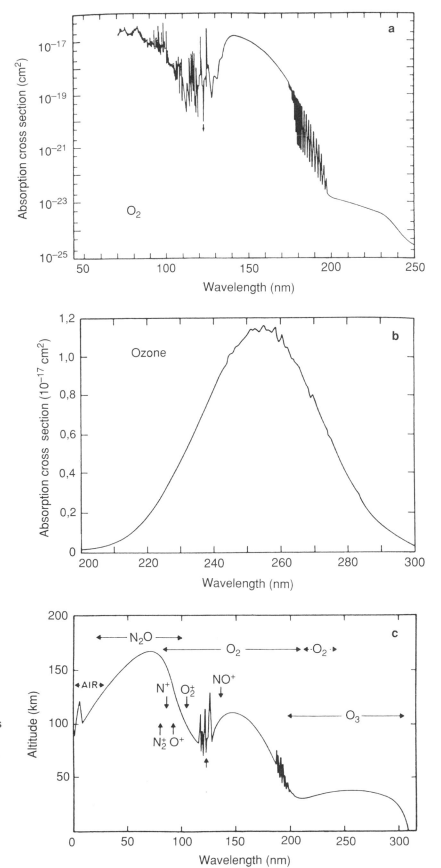

Figure 3.2 (a) The spectral distribution of the absorption cross section of molecular oxygen. (b) The spectral distribution of the Hartley band of ozone, the absorption feature responsible for ozone's filtering of solar ultraviolet radiation. (c) The depth of penetration of solar radiation as a function of wavelength. The principal absorbers and a selection of dissociation and ionization limits are indicated. (Reprinted by permission from G. Brasseur and S. Solomon, *Aeronomy of the Middle Atmosphere*, 2nd Ed. Copyright 1986 by Kluwer Academic Publishers.)

The penetration of solar radiation through the atmosphere as a function of wavelength from shortward of 310 nm is shown in Figure 3.2c. Most of the radiation with wavelengths shorter than about 100 nm is absorbed above 100 km in the atmosphere by N_2, O_2, N, O, and the ionic compounds of those species. At wavelengths longer than 100 nm, N_2, N, and O cease to absorb, and the radiation can penetrate more deeply. Strong absorption by molecular oxygen limits photons of wavelengths shorter than about 210 nm to a 50-km altitude and higher. Photons with wavelengths longer than about 210 nm are so weakly absorbed by O_2 that O_3 assumes the role of major absorber, which it retains over the band of 210–310 nm. This ozone absorption provides the energy that heats the stratosphere and much of the mesosphere (see Figure 3.5). It also provides the screening for the photons just shortward of 310 nm which are responsible for biological mutations, sunburn, and other physiological effects. Hence, a decrease in stratospheric ozone concentrations may lead to an increase in the intensity of the most energetic radiation reaching Earth's surface. At wavelengths longer than about 310 nm, only modest attenuation of radiation by air molecules, clouds, and particles interferes with the transport of solar photons to the surface, and it is this radiation that provides the visual illumination of the planet.

Solar Radiation Absorption in the Atmosphere

It is a straightforward process to calculate the solar energy absorbed by a specific region of the atmosphere. Each molecule of species i has its own characteristic absorption spectrum, which can be expressed as an *absorption cross section* as a function of wavelength: $\sigma_i(\lambda)$ (units: cm^2 per molecule); this quantity reflects the likelihood of effecting electronic state transitions in the molecule, and has a maximum numeric value of about 1×10^{-17}.

Consider an incident beam of solar radiation of intensity $I(\lambda)$ passing through a unit cross section of the atmosphere (units: photons per cm^2 per second). At a specific wavelength λ, the photon flux Ψ absorbed by a species i (number of absorbed photons per cm^3) is given by

$$\Psi_i(\lambda) = \sigma_i(\lambda) \, I(\lambda) \, [C_i], \qquad (3.8)$$

where $[C_i]$ is the concentration of the absorbing species (molecules cm^{-3}). If the full range of wavelengths is considered, Equation (3.8) becomes

$$\Psi_i = [C_i] \int_0^\infty \sigma_i(\lambda) \, I(\lambda) \, d\lambda. \qquad (3.9)$$

To calculate the absorbed energy, integrand (3.9) must be further multiplied by the photon energy. When all absorbing molecules are considered, the energy absorption in a unit volume is derived from Equations (3.1) and (3.9) as

$$\Psi = hc \sum_i [C_i] \int_0^\infty \sigma_i(\lambda) \, I(\lambda) \, d\lambda/\lambda. \qquad (3.10)$$

The solar radiation intensity decreases as one descends in the atmosphere, because of the absorption and scattering of photons by gas molecules and by particulate matter. This attenuation is governed by Beer's law, such that the intensity $I(\lambda)$ at height z is given by

$$I(\lambda, z) = I_{(\lambda, \infty)} \exp(-\tau_\lambda \cos \theta), \quad (3.11)$$

where $I_{\lambda, \infty}$ is the intensity at the top of the atmosphere, τ_λ is the *optical depth*, and θ is the solar zenith angle. The optical depth in Earth's atmosphere for incoming solar radiation can be expressed as the sum of several terms:

$$\tau_\lambda = T(O_2)\sigma_{O_2}(\lambda) + T(O_3)\sigma_{O_3}(\lambda) +$$
$$T(M)\sigma_s(\lambda) + T(p)\sigma_p(\lambda) \quad (3.12)$$

where $T(O_2)$, $T(O_3)$, and $T(M)$ are the numbers of O_2, O_3, and all molecules per unit area above height z, σ_{O_2} and σ_{O_3} are the absorption cross sections for molecular oxygen and ozone, σ_s is the cross section for light scattering by the ensemble of atmospheric molecules (mostly N_2 and O_2), and T_p and σ_p denote light scattering by particles in a similar fashion.

RADIATION FROM EARTH'S SURFACE

The portion of solar radiation that is not absorbed or scattered during its passage through the atmosphere reaches the ground and causes heating of soil, water, vegetation, and the adjacent air. Because any body not at absolute zero Kelvin radiates, emission of radiation from Earth to the atmosphere follows. Earth is much cooler than the Sun, so most of its radiation occurs at long wavelengths, in the infrared portion of the spectrum. Like the incoming solar radiation, the outgoing radiation is modified by absorption by atmospheric molecules. The energy of these infrared photons, however, is insufficient to cause chemical changes. Instead, the absorption merely increases the internal vibrational and rotational energy of the absorbing molecules. That excess energy is subsequently transferred to the atmosphere as kinetic energy (heat) by molecular collisions. Conversely, cooling of the atmosphere occurs by emission of radiation, a process involving molecular state transitions between vibrational and rotational energy levels. Because these radiative transitions require a change in the dipole moment of the molecule (that is, of the electrical charge separation) and because diatomic molecules containing only one kind of atoms have no dipole moment, the dominant atmospheric gases N_2 and O_2 are not involved in infrared energy transfer. That role is reserved for H_2O, CO_2, O_3, and a number of other minor atmospheric constituents.

In contrast to the relatively inefficient absorption of visible radiation by the atmospheric gases and the supremely efficient absorption of far ultraviolet radiation by molecular oxygen and ozone, the atmosphere is a good absorber of infrared radiation ($\lambda > 750$ nm). Figure 3.3 pictures the principal spectral absorption regions for some of the molecules responsible for the infrared absorption. By far the most important molecules are water vapor and carbon dioxide. They are sufficiently abundant that they limit the transmission of radiation to space in many wavelength regions, particularly in much of the 8000-20,000 nm (8–20 μm) region, where the principal radiation from Earth occurs. The lowest part of the diagram shows that terrestrial radiation has the best chance to escape to space in the so-called *infrared window* from 8–12 μm. Thus, any gas that absorbs strongly in that region will have a comparatively large warming effect. Among those species of interest are methane (CH_4), nitrous oxide (N_2O), CFC-11 ($CFCl_3$), and CFC-12 (CF_2Cl_2). Collectively, the absorbing molecules are termed *greenhouse gases*. Because of their absorptions, only a small fraction (about 5%) of the outward-directed radiation can directly escape from Earth's surface into space. The remainder is absorbed by gases and by clouds, and more than 90% of that absorbed radiation is radiated back to Earth's surface.

The radiation budget for Earth's atmospheric system is illustrated in Figure 3.4. An average of slightly less than 30% of the incoming solar radiation is returned into space: by reflection (back-scattering) from clouds (about 19%), backscattering by air molecules and particles in the air (together about 6%), and reflection at Earth's surface (about 3%). Almost 25% of the solar radiation is absorbed within the atmosphere, mostly by ozone in the stratosphere (about 3%), and by clouds (5%) and water vapor (17%) in the troposphere. The remaining 47% of the incoming solar radiation is absorbed at Earth's surface.

Of the solar energy absorbed at Earth's surface, a little more than half goes into *latent heat*, that is, heat absorbed by water as a consequence of its transformation from liquid form to vapor form at Earth's surface and released again into the atmosphere when water droplets condense in clouds. Other significant amounts of surface heat energy are transferred to the atmosphere by convection and turbulence (about 10%) and by the absorption of infrared radiation by the greenhouse gases. Of the 47% of the initial solar energy absorbed at Earth's surface, only 18% is lost by radiation, because the remainder is captured in the atmosphere. This atmosphere-surface cycling is the greenhouse effect; it causes Earth's surface to be about 33 °C warmer than would otherwise be the case.

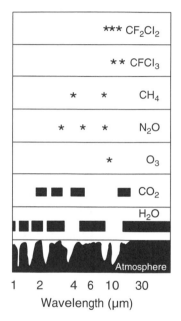

Figure 3.3 Regions of strong absorption as a function of wavelength for the atmosphere's principal greenhouse gases, H_2O and CO_2. Also shown are the absorption band centers for the gases now becoming increasingly important for the planetary radiation budget.

TEMPERATURE STRUCTURE OF THE ATMOSPHERE

Many properties of the atmosphere are determined by its pressure, which is highest at Earth's surface and decreases rapidly with increasing altitude, by about a factor of 2 in the first 5 km and by a factor of 10 in the first 16 km, as shown in Figure 3.5. The other crucial atmospheric property shown in the figure is the temperature; unlike the pressure variation, the temperature variation with height is quite complex. Atmospheric scientists use the inflection points on the temperature profile to distinguish the different regions for study and reference. Beginning at Earth's surface, these regions are called the *troposphere*, the *stratosphere*, the *mesosphere*, and the *thermosphere*, and their boundaries the *tropopause*, the *stratopause*, and the *mesopause*, respectively. In this book, we will restrict our discussions to the troposphere and stratosphere, which are the most important regions for climate and life on Earth. Both regions are strongly affected by anthropogenic and natural emissions at the surface. In addition, the stratosphere is influenced by volcanic explosions, nuclear explosions, aircraft emis-

Figure 3.4 The annual mean global energy balance for the Earth-atmosphere system. *Sensible heat* is that transferred to the atmosphere from the heated surface by turbulent eddies; *latent heat* is that supplied to the atmosphere upon condensation of water vapor. The figures are percentages of the energy of the incoming solar radiation. (Reprinted by permission from S. H. Schneider and R. Londer, *The Coevolution of Climate and Life.* Copyright 1989 by Schneider and Londer, Sierra Club Books, San Francisco.)

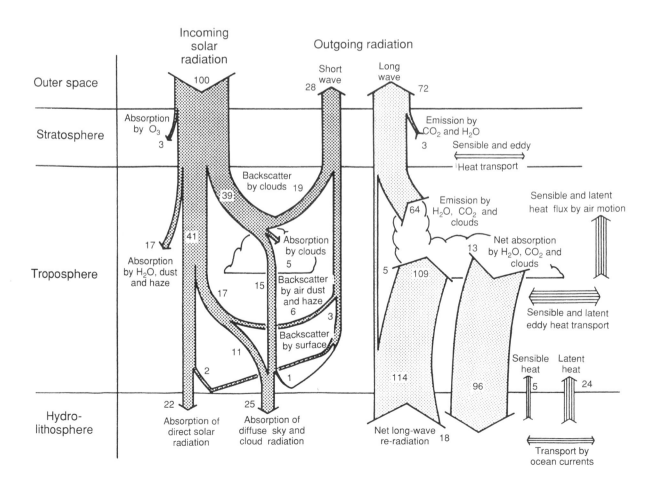

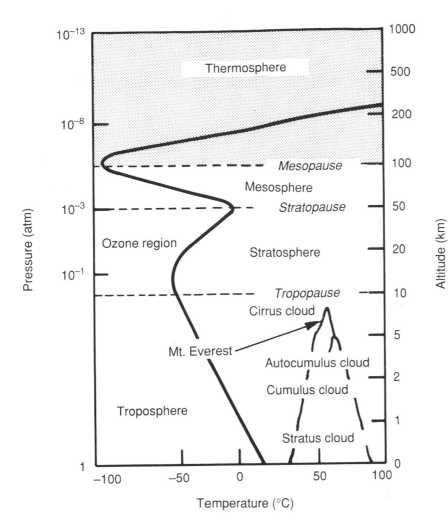

Figure 3.5 The variation of atmospheric pressure and temperature with altitude above Earth's surface. Typical heights of a few cloud types are indicated. (Adapted with permission from W. L. Chameides and D. D. Davis, *Chemistry in the troposphere, Chemical and Engineering News,* 60 (40), 38–52. Copyright 1982 by the American Chemical Society.)

sions, and solar eruptions. The higher levels of the atmosphere are much less affected by influences from below and are not addressed here.

The atmospheric temperature structure can ultimately be traced to the emission and absorption of radiation. The troposphere is heated from below. Convective processes, involving expansion and radiative cooling of air, explain the temperature decrease with altitude in the troposphere, to about −50 °C at altitudes around 10 km at middle to high latitudes and to about −80 °C at 17 km in the tropics. The increase in temperature in the stratosphere begins at the tropopause and is explained by the increasing importance of the absorption of downflowing solar ultraviolet radiation and upflowing infrared radiation from Earth's surface by stratospheric ozone. Because the product of the intensity of the solar ultraviolet radiation and the volume mixing ratio of ozone reaches a maximum at the stratopause height of about 50 km, maximum temperatures develop at this altitude. The volume mixing ratio of ozone decreases sharply above that height, and temperatures fall to nearly −100 °C in the mesosphere near 90 km

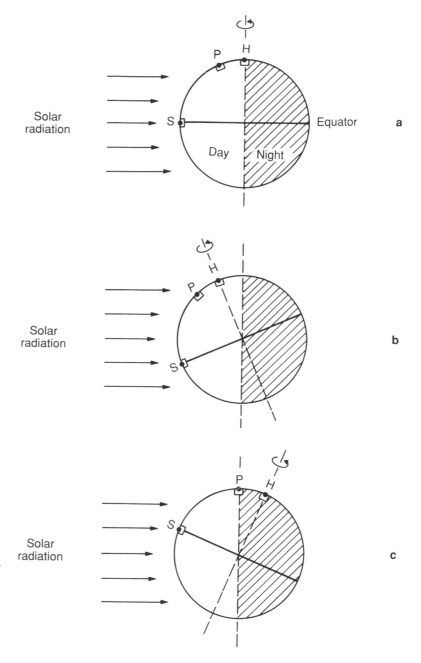

Figure 3.6 Schematic diagram of
sunlight received on the surface of
Earth as a function of latitude and
season: (a) at equinox, (b) during
the northern hemisphere summer,
(c) during the southern hemisphere
summer.

altitude. Higher up, temperatures again increase rapidly with altitude as
a result of the absorption by molecular oxygen of solar ultraviolet radiation
of wavelengths shorter than about 200 nm and, at still higher altitudes, also
by molecular nitrogen, atomic oxygen, and atomic nitrogen.

Up to this point, we have discussed atmospheric temperature without regard for its variation with latitude. However, this variation, resulting from the differing amounts of radiation falling on different portions of the atmosphere and surface, is crucial to an understanding of atmospheric motions, weather, and climate. In Figure 3.6a, compare the radiation received by a unit area at point S, the subsolar point, with that received at point P, a unit area at 66.5 °N on the polar circle and at point H, a unit area at the pole itself. The diagram shows the situation at one of the equinoxes when Earth's axis of rotation is perpendicular to the incoming solar radiation. Per unit area at Earth's surface, area S receives significantly more radiation than the unit area at point P, both because of the latitudinal variation and because at point P the scattering and absorption of a much longer passage through the atmosphere reduces the flux markedly. At this time of year, all points on Earth except exactly at the poles have 12 hours each of daylight and darkness. In the northern hemisphere summer (Figure 3.6b), the tilt of Earth increases the solar flux per unit surface area on the northern latitudes; in the southern hemisphere summer, the opposite is true (Figure 3.6c). The regions poleward of the polar circles have either 24 hours of day (during the six months surrounding the summer solstice) or six months of night (during the six months surrounding the winter solstice). As a result of these patterns, the amount of solar radiation absorbed at Earth's surface varies markedly with latitude and season, as shown in Figure 3.7.

Figure 3.7 Solar insolation at Earth's surface as a function of latitude and day of the year. The units are watts m^{-2}. (A. Berger, Milankovitch theory and climate, *Reviews of Geophysics*, 26, 624–657, 1988.)

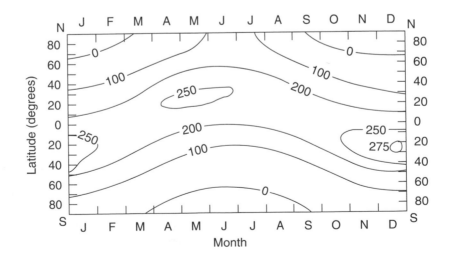

Earth as a Blackbody Radiator

When the properties of solar and terrestrial radiation and the atmospheric absorption are all taken into account, one can construct an energy budget for the atmosphere. The central feature of the budget is the balance struck by Earth and its atmosphere between the incoming solar radiation and the infrared radiation that is emitted back into space, expressed as

$$\frac{E_{SUN}}{4}(1 - \alpha) = \sigma T_e^4. \qquad (3.13)$$

This equation assumes that Earth radiates as a blackbody at an effective radiation temperature T_e. Such an approach, valid as an average for the entire Earth over several years, contains the following terms:

- E_{SUN} is the solar irradiance intercepted by Earth at the average Sun-Earth distance. (This is the so-called *solar constant*, having a value of $S = 1380$ W m^{-2}.)
- α is the planetary *albedo*, the fraction of the intercepted solar irradiance that is reflected into space, equal to about 0.28.
- σ is the *Stefan-Boltzman constant* = 5.67×10^{-8} W m^{-2} K^{-4}.

The factor 4 in Equation (3.13) is the ratio of the total surface to disk area of Earth, the latter being the area intercepting the solar radiation. The insertion of appropriate values into Equation (3.13) yields an effective radiation temperature for the Earth-atmosphere system of 255 K. This value is the average radiation temperature of Earth as seen from outer space.

The temperature at Earth's surface, T_s, is strongly influenced by the absorption of outgoing infrared radiation and downward backradiation. This influence may be denoted by inserting an effective atmospheric infrared transmission factor f into Equation (3.13) such that

$$\frac{E_{SUN}}{4}(1 - \alpha) = f\sigma T_s^4. \qquad (3.14)$$

For Earth's atmosphere, $f = 0.61$, giving $T_s = 288$ K or 15 °C, the observed global average surface temperature. If we take the logarithm of Equation (3.14) and differentiate, we get

$$\frac{\delta f}{f} + 4 \cdot \frac{\delta T_s}{T_s} = \frac{\delta E_{SUN}}{E_{SUN}} - \frac{\delta \alpha}{1 - \alpha}. \qquad (3.15)$$

If all other factors remain constant, a relative decrease in the atmospheric $\delta f/f$ of 0.01 as a consequence of increasing concentrations of greenhouse gases leads to a ground-level temperature increase of $\sigma T_s = 0.7$ K. Likewise, an increase of the same magnitude in the solar constant would lead to a 0.7 K increase. If the albedo were to increase by 1%, due to increased cloudiness, for example, the ground-level temperature would cool by about 1 K. These numbers are useful for perspective, but we note that f and α are not in fact independent of each other, especially under situations of changes in atmospheric water vapor or cloudiness.

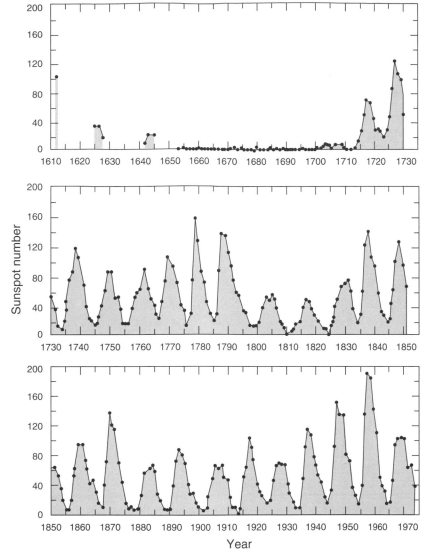

Figure 3.8 The time series of annual average sunspot number over the past several hundred years. The lack of sunspot activity during the seventeenth century is believed to be real. (Courtesy of J. A. Eddy, University Corporation for Atmospheric Research.)

VARIABILITY OF SOLAR RADIATION

The amount of radiation produced by the Sun is not constant, especially at short wavelengths, where variations of several percent occur as a function of solar activity. This activity, which is related to the changing magnetic structure of the gaseous Sun, is manifested in an 11-year cycle in the easily observed sunspots and other solar features. An historical chronology of sunspot frequency is reproduced in Figure 3.8. It shows clearly the 11-year solar cycle for most of the period studied and shows as well that sunspot numbers at solar maximum can vary by factors of 3 or more. The periods of increased sunspots are also times of increased ultraviolet radiation emissions, so the rates of photon absorption in Earth's atmosphere by reactions such as Equations (3.5) and (3.7) increase.

Many attempts have been made to link solar activity and its slightly enhanced energy deposition in the stratosphere to climate effects near Earth's surface. For example, many have noted that the dearth of sunspots during the seventeenth century coincide with the Little Ice Age. As yet, however, a firm theoretical coupling of solar activity with tropospheric meteorology and climate has not been made, and attempts at correlating different measured parameters remain problematical.

COMPARISON OF RADIATION BUDGETS OF THE TERRESTRIAL PLANETS

In Table 3.1, we list selected physical and chemical characteristics of Earth and of our neighboring planets Venus and Mars. As can be seen, the total mass and geometrical dimensions of Earth and Venus are very similar, whereas Mars has about 10% of Earth's mass (and a radius about half as large). The distance to the Sun is about 100 million km for Venus; Earth is about 1.5 times more distant, and Mars about 1.5 times further from the Sun than Earth. As a consequence, the solar constants at neighboring planets differ by about a factor of 2. On the basis of such a simple analysis, one would expect temperatures to be

TABLE 3.1. Physical Characteristics of the Terrestrial Planets			
CHARACTERISTIC	EARTH	VENUS	MARS
Total mass (10^{27} g)	6	5	0.6
Radius (km)	6371	6049	3390
Atmospheric mass (ratio)	1	100	0.06
Distance to Sun (10^6 km)	150	108	228
Solar constant (W m^{-2})	1367	2613	589
Albedo (%)	30	75	15
Cloud cover (%)	50	100	Variable
Eff. rad. temp. (°C)	−18	−39	−56
Surface temp. (°C)	15	427	−53
Greenhouse wrmg. (°C)	33	466	3
N_2 (%)	78	<2	<2.5
O_2 (%)	21	<1 ppmv	<0.25
CO_2 (%)	0.035	>98	>96
H_2O (range %)	$3 \times 10^{-4} - 4$	$1 \times 10^{-4} - 0.3$	<0.001
SO_2	<1 ppbv	150 ppmv	nil
Cloud composition	H_2O	H_2SO_4	Dust, H_2O, CO_2

warm at the surface of Venus, cooler on Earth, and cooler still on Mars, other factors being equal. The planetary albedos differ substantially, however, and the result is that the effective radiation temperatures are 234 K (−39 °C) for Venus, 255 K (−18 °C) for Earth, and 217 K (−56 °C) for Mars. In other words, as seen from space, Mars is the coldest of the three planets and Earth is the warmest.

The surface temperatures of the terrestrial planets bear little resemblance to their effective temperatures. Venus is the warmest by far, at 427 °C, with Earth at 15 °C and Mars at −53 °C. These data imply that Venus has an enormous greenhouse effect, Mars hardly any at all, and Earth a moderate one. The reason is that Venus has a very extensive atmosphere, about a hundred times greater in mass than that of Earth and consisting mainly of the greenhouse gases CO_2, H_2O, and SO_2. Although the atmosphere of Mars contains more CO_2 than does that of Earth, it has more than a hundred times less water vapor. This composition, in combination with the much lower solar irradiance, explains the low surface temperature.

It is clear from this comparative discussion that the chemical compositions of atmospheres have a major influence on the radiation budgets of the terrestrial planets and that changes in those compositions can produce dramatic climate effects.

EXERCISES

3.1 Using Planck's law, compute and plot the distribution of radiative emission from Earth without the absorbing infrared gases in the atmosphere (255 K) and with the gases (288 K). What is the relative difference between the radiated energy in the two cases?

3.2 Miranda, the innermost of the Uranian moons that were known prior to the *Voyager 2* flyby, has a surface temperature of about 86 K. At what wavelength does the peak blackbody emission from Miranda occur? Repeat the calculation for Neptune's moon Triton, the coldest known body in the solar system at 38 K.

3.3 If the content of the atmosphere were to double, how would this affect the absorptivity of the atmosphere as a whole at a wavelength of 35 μm? Why?

3.4 The table on the right lists solar flux intensities (unit: photon cm^{-2} sec^{-1} 10 nm^{-1}) at an altitude of 1 km and north latitude 40° at the summer solstice (June 21), together with absorption cross sections (unit: cm^2 molecule^{-1}) for the formaldehyde molecule. If the concentration of formaldehyde at 1 km on June 22 at that location is 5 ppbv, calculate the total photon flux deposited into a cubic centimeter of air at midday as a consequence of formaldehyde absorption of solar radiation.

Photoprocess Data for Formaldehyde			
WAVELENGTH (nm)	$I(\lambda)$	σ	$\phi(HCO\cdot + H\cdot)$
295–305	2.66×10^{13}	2.62×10^{-20}	0.79
305–315	4.20×10^{14}	2.45×10^{-20}	0.79
315–325	1.04×10^{15}	1.85×10^{-20}	0.64
325–335	1.77×10^{15}	1.76×10^{-20}	0.31
335–345	1.89×10^{15}	1.18×10^{-20}	0
345–355	2.09×10^{15}	0.42×10^{-20}	0
355–365	2.15×10^{15}	0.06×10^{-20}	0

3.5 The energy contained in a photon depends on its wavelength according to

$$W = hc/\lambda$$

where h is Planck's constant (6.63×10^{-34} J s) and c is the speed of light (3×10^{10} cm s^{-1}). Compute the approximate rate of energy deposition for the absorption of solar radiation by formaldehyde as specified in Exercise 3.4. Is this amount of energy significant? Why?

FURTHER READING

G. Brasseur and S. Solomon, *Aeronomy of the Middle Atmosphere*, 2nd Ed., D. Reidel, Dordrecht, 1986.

V. Ramanathan, B. R. Barkstrom, and E. F. Harrison, Climate and the Earth's radiation budget, *Physics Today, 42* (5), 22–32, 1989.

The Atmospheric Circulation: Transporter of Chemical Constituents

ATMOSPHERIC HEAT TRANSPORT AND DYNAMICS

Patterns in the Troposphere

In the troposphere, more energy is received from the Sun at tropical (0°–30°) and subtropical (30°–40°) latitudes than is given off by outgoing terrestrial radiation (Figure 4.1). This is the case not only for land areas, but also for the oceans. As a consequence, there is year-round energy transfer from low to high latitudes by both air and ocean currents, as shown in Figure 4.2. It is these processes that drive the general circulation of the atmosphere.

Because the heating of Earth is greatest over the equator, heated equatorial air expands and moves upward to a greater degree than does

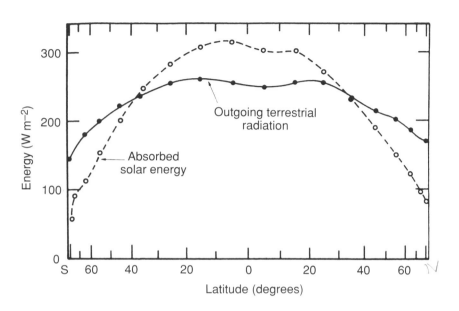

Figure 4.1 Absorbed solar radiation and outgoing terrestrial radiation as a function of latitude. (T. H. Von der Haar and V. E. Suomi, Measurement of the Earth's radiation budget from satellites during a five year period. Part I. Extended time and space means, *Journal of the Atmospheric Sciences, 28*, 305–314. Copyright 1971 by the American Meteorological Society.)

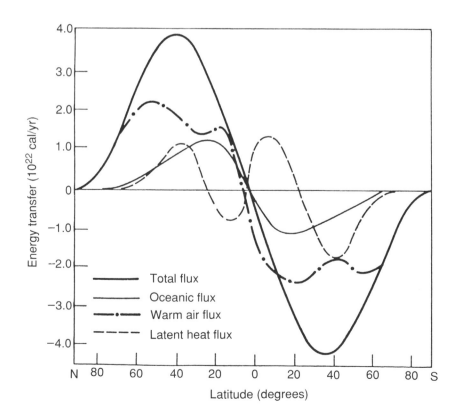

Figure 4.2 Rates of poleward energy transport by various mechanisms. Positive energy transfer values denote northward transport and negative values denote southward transport. (Adapted with permission from W. D. Sellers, *Physical Climatology*. Copyright 1965 by the University of Chicago Press.)

the air at other latitudes. As the equatorial air rises, air at low altitudes and from higher latitudes moves toward the equator to take its place. This flow is balanced by a return flow at higher altitudes, as seen in Figure 4.3. While flowing poleward, the air cools by radiation to space. At about 30° latitude in both hemispheres (the so-called *horse latitudes*), the cooled denser air descends, closing what is termed the *Hadley circulation* on both sides of the equator. The upward feature of the Hadley cell occurs at what is often called the *intertropical convergence zone* (ITCZ). The strong upward motion of air is characterized by heavy precipitation in convective thunderstorms and by relatively low surface pressures. The horse latitudes, on the other hand, are fair weather regions, with little precipitation, high surface pressures, and subsiding air.

The trajectory of an air parcel moving under the influence of a pressure gradient appears to be less constant than might be initially surmised, because of the *Coriolis force*. This fictitious force, associated with Earth's rotation, results in the deflection of all objects not at the equator to the right of the direction of motion in the northern hemisphere and to the left in the southern hemisphere; that is, orthogonal to the pressure gradient force. As a consequence, in the free troposphere away from the influence of frictional forces induced by Earth's surface, the winds will tend to rotate clockwise around high pressure systems (termed *anticyclonic circulation*) in the northern hemi-

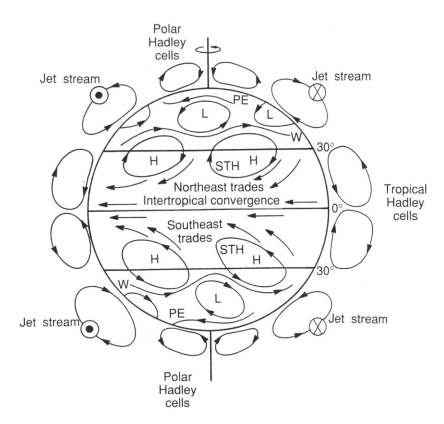

Figure 4.3 Principal features of the atmospheric circulation. The dots at the left side of the figure labeled "jet stream" indicate flow toward the reader; the crosses on the right indicate flow away from the reader. (This figure emphasizes the polar jet stream, which is variable in latitude. The weaker subtropical jet stream is not shown.) The highs (H) and lows (L) are primarily lower atmospheric features, extending only about 5 km into the atmosphere, whereas the Hadley cells can extend vertically as much as 10 to 20 km throughout the troposphere and partially into the stratosphere. W, westerlies; PE, polar easterlies; STH, subtropical highs. (Adapted with permission from S. H. Schneider and R. Londer, *The Coevolution of Climate and Life.* Copyright 1989 by Schneider and Londer, Sierra Club Books, San Francisco.)

sphere and counterclockwise in the southern hemisphere. Circulation around low pressure systems (*cyclonic circulation*) proceeds in the opposite directions. The Coriolis force arises because a moving object, such as an air parcel, tries to maintain its original direction with respect to an absolute coordinate system independent of Earth's rotation. On the Earth observing platform, however, the coordinate system is in continual rotation, and an object under observation undergoes an apparent deflection as a consequence of the interaction of the two coordinate systems, as we, the observers, are moving within the absolute coordinate system.

To make this situation more evident, consider Figure 4.4. Initially, wind is blowing from point P along vector V, and the coordinate system points north and east and rotates with Earth. Later, at point P′, the wind vector has maintained its direction, but has diverged from the moving coordinate system. The result is that the winds in the northern hemisphere appear to an observer on Earth to be deflected clockwise, i.e., to the right. In a similar fashion, the winds appear to be deflected to the left in the southern hemisphere.

The Coriolis force also acts on an air parcel moving at an angle to the equator: because angular momentum is conserved and the velocity of rotation decreases proportional to cos ϕ, the apparent change in angular momentum per unit time is proportional to the northward (southward) wind component v and the derivative of cos ϕ (i.e., sin ϕ). Note that the Coriolis force increases with latitude and vanishes at the equator.

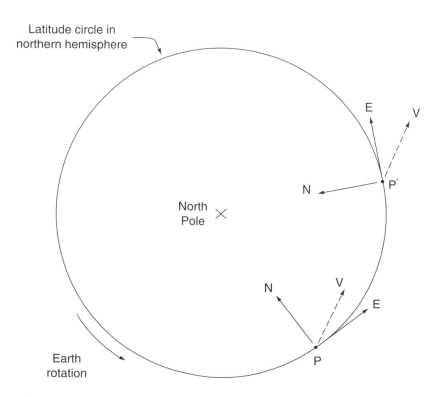

Figure 4.4 A depiction of the Coriolis force as seen by an observer above the North Pole (see text for discussion).

The Coriolis force, the atmospheric pressure distribution, and the vertical, temperature-driven flows create rather regular equatorward *trade winds* at the surface in the tropical latitudes, blowing from the northeast in the northern hemisphere and from the southeast in the southern hemisphere (Figure 4.5). In different seasons, the Sun-Earth inclination results in modifications of surface heating and thus of surface winds. In general, however, high pressure areas at the horse latitudes produce the trade winds, and also lead to prevalent southwesterly winds in the lower atmosphere in the northern hemisphere temperate latitudes and northwesterly winds in the southern hemisphere temperate latitudes. At higher latitudes, northeasterly winds dominate in the northern hemisphere and southeasterly winds in the southern hemisphere. Clouds have an important effect on this overall scheme. As the warm equatorial air rises, the condensation of water vapor to form clouds releases large amounts of energy (the *latent heat of condensation*). This heat enhances the tropical Hadley circulation.

The flows at temperate and high latitudes are far less regularly ordered than in the tropics because of strong influences by wandering high and low pressure systems. In the midlatitudes, much of the poleward heat transfer is accomplished by sporadic air mass exchanges across the latitudes: the movement of tropical and subtropical air masses toward the poles and of

Figure 4.5a Prevailing surface winds and the mean position of the intertropical convergence zone in January. (Reproduced with permission from A. H. Perry and J. M. Walker, *The Ocean-Atmosphere System*. Copyright 1977 by Longman Group UK Ltd.)

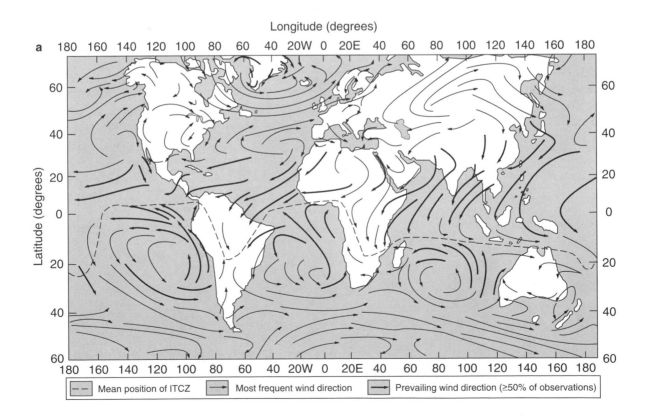

polar air masses toward the equator. The zones of contact between air masses are termed *warm fronts* and *cold fronts*, the designation depending on the nature of the air masses being advected.

The general westerly flow of the atmosphere at temperate latitudes is often disrupted near the ground by the complicating effects of surface cooling and heating and of orography (i.e., the presence of mountain ranges). Further aloft, the westerly flow is steadier in direction and speed, reaching a maximum in the *jet streams* at altitudes of 10 to 12 km. The jet streams vary with the seasons, being strongest in the winter, when their speeds often exceed 100 m s⁻¹. It is these streams that hasten the travel of many an eastward-moving airliner. Normally two jet streams exist in each hemisphere; the subtropical jet near 30° latitude and the polar jet, which can make large latitudinal excursions from 30° to 60°. Although stronger, the meandering polar jet stream does not show up as clearly on seasonal or annual average meridional cross sections of zonal winds, even though its speeds are normally greater than those of the subtropical jet.

Figure 4.5b Prevailing surface winds and the mean position of the intertropical convergence zone in July. (Reproduced with permission from A. H. Perry and J. M. Walker, *The Ocean-Atmosphere System.* Copyright 1977 by Longman Group UK Ltd.)

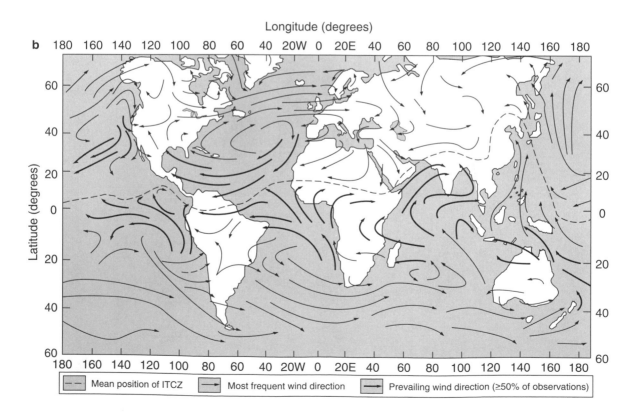

Patterns in the Stratosphere

In the stratosphere, the differential net heating sets up a meridional circulation pattern, which is shown in Figure 4.6. The pattern is characterized by generally upward motions above about 25 km in the summer hemisphere and downward motions in the winter hemisphere. This wind system joins that of the troposphere in the lower stratosphere.

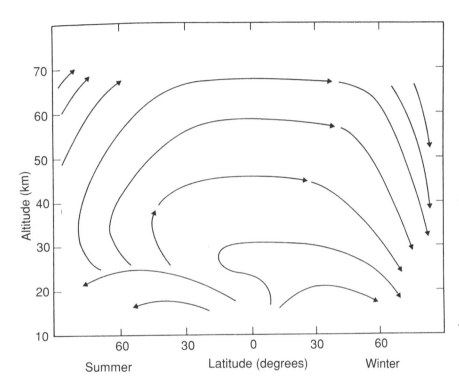

Figure 4.6 Lines of flow of the stratospheric circulation produced by uneven heating of the stratosphere and by the stratospheric extension of the Hadley circulation. (Reproduced with permission from T. Dunkerton, On the mean meridional mass motions of the stratosphere and mesosphere, *Journal of the Atmospheric Sciences,* 35, 2325–2333. Copyright 1978 by the American Meteorological Society.)

Pressure Gradients and Hydrostatics

The pressure, i.e., the force per unit area, varies in space and time, thus leading to flows of air from regions of high pressure to those of low pressure. It is easy to understand that the pressure force is proportional to the magnitude of the gradient, i.e., the space derivative of pressure. Let us consider an air parcel with dimensions dx, dy, and dz. With pressure p being a function of the space coordinates, the force exerted on the left-hand side of the parcel is given by

$$F_1 = p \, dy \, dz, \qquad (4.1)$$

and that exerted by the surrounding air from the right side to the left is given by

$$F_r = (p + \frac{\partial p}{\partial x}) \, dy \, dz. \qquad (4.2)$$

The difference force

$$\Delta F = F_r - F_l = \frac{\partial p}{\partial x} \, dx \, dy \, dz \qquad (4.3)$$

is the net pressure force, directed to the left on the parcel. Per unit mass of air, this balance would lead to a net pressure gradient force of

$$F = -\frac{1}{\rho} \frac{\partial p}{\partial x} \qquad (4.4)$$

in the x direction, ρ being the density of air (mass per unit volume). Similar expressions apply in directions other than the x direction. In the vertical direction z, the pressure force is largely balanced by the gravitational force g, so

$$\frac{\partial p}{\partial x} = -g\rho. \qquad (4.5)$$

Equation (4.5) is called the *hydrostatic equation*, which can be developed further by using the expression for ρ from the gas law:

$$p = \rho/RT \qquad (4.6)$$

so (R is the gas constant)

$$\frac{1}{p} \frac{\partial p}{\partial x} = -\frac{g}{RT}, \qquad (4.7)$$

and

$$p(z) = p(0) \exp\left(-\frac{g}{R} \int_0^z \frac{dz}{T}\right). \qquad (4.8)$$

The atmospheric pressure thus decreases with height by the factor e (≈ 2.72) for the case when

$$\frac{g}{R} \int_0^z \frac{dz}{T} = 1. \qquad (4.9)$$

This situation turns out to hold quite precisely for the upper troposphere and lower stratosphere. Inserting the average temperature profile of the atmosphere, it can be shown that

$$H = \frac{R\overline{T}}{g} \approx 8.3 \text{ km.} \qquad (4.10)$$

STABILITY AND VERTICAL MIXING

Patterns in the Troposphere

The vertical distribution of temperatures determines to a large degree the exchange of air in the vertical direction. Consider an air parcel that moves upward. Depending on the temperature of the air surrounding the parcel, stable or unstable conditions may occur. In reality, the atmosphere will consist of combinations of stable and unstable regimes. The situation is shown in Figure 4.7, where the dashed-dotted line indicates the *dry adiabatic lapse rate* Γ_α, i.e., the decrease of temperature with altitude when heat is neither gained nor lost. Near Earth's surface during daytime, strong radiation heating often establishes a temperature profile similar to that shown as a solid line. Since that lapse rate is greater than Γ_α, air parcels near the ground will rise to point B, where the air parcel will have reached the same temperature and density as the surrounding air. Height z_B defines the top of the turbulent planetary boundary layer, above which vertical mixing is strongly inhibited.

During daytime under anticyclonic conditions, the boundary layer can be easily identified visually from high mountains or aircraft, because that layer is often filled with pollutants from various anthropogenic activities and is less transparent. During nighttime, the cooling of Earth's surface will often create very stable conditions, as exemplified by point C in Figure 4.7. In this case, the air parcel cannot move upward, because it will always be cooler than the surrounding air. Thus, it has a higher density and experiences a downward force. Under such conditions, hardly any vertical mixing of boundary layer

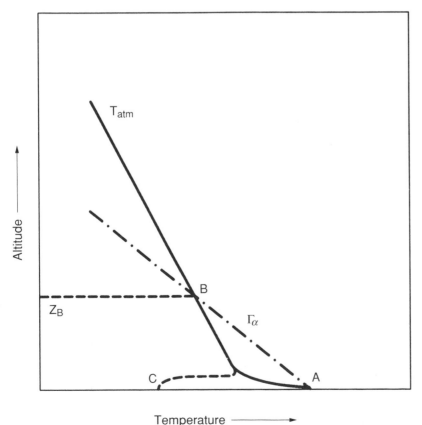

Figure 4.7 Types of lapse rates in the lower atmosphere. The symbols, points, and lines are discussed in the text.

air can occur, and intense buildups of air pollutants close to the surface can occur. The temperature profile CB in such cases is termed an *inversion*. During winter, inversions can be very persistent, because daytime heating of the surface is small and nocturnal cooling is large, especially if the ground is snow covered.

There are many possible stability combinations, each of which affects mixing of pollutants in the lower atmosphere. Several are illustrated in Figure 4.8. It is evident that local stability or instability can result in relatively prompt interaction of emission plumes with the ground, broad diffusion over a wide area, or efficient long-range transport. Assessments of the local environmental impact of a new

Figure 4.8 Patterns of smoke plumes produced by various combinations of atmospheric temperature conditions that vary with altitude. The dashed curves in the diagrams represent the adiabatic lapse rate; the solid curves indicate the vertical temperature variation. (Reproduced with permission from M. Neiburger, J. G. Edinger, and W. D. Bonner, *Understanding Our Atmospheric Environment.* Copyright 1973 by W. H. Freeman and Company.)

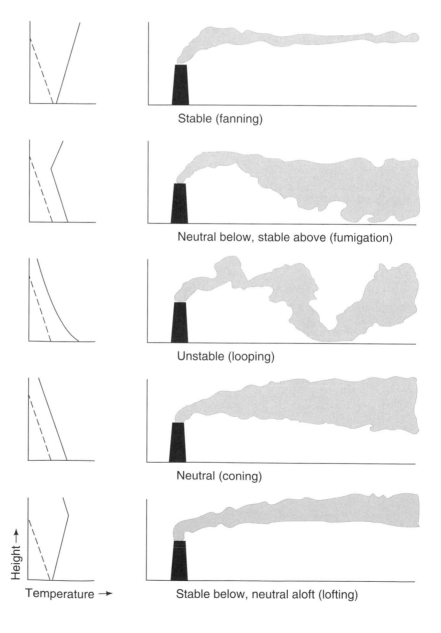

source thus require a thorough review of the typical meteorological characteristics of its region in addition to the engineering characteristics of the source itself.

Static Stability Criteria

An air parcel moving upward expands in an attempt to reach the same pressure as that of the surrounding air. The first law of thermodynamics, in combination with the gas law, requires that under adiabatic conditions (i.e., where no heat h is supplied or withdrawn), the rates of change of temperature T and pressure p are given by

$$dh = 0 = C_p \frac{dT}{T} - R \frac{dp}{p}, \qquad (4.11)$$

where C_p is the specific heat of air at constant pressure. By taking derivatives with respect to height and substituting from the hydrostatic equation (4.5), the rate of change of the temperature of the air parcel as a function of height can be derived:

$$\Gamma_\alpha = \frac{dT}{dz} = R \frac{1}{C_p \rho} \frac{\partial p}{\partial x} = \frac{1}{\rho C_p} \cdot g\rho$$

$$= \frac{g}{C_p} \approx -10\ ^\circ\text{C km}^{-1}. \qquad (4.12)$$

An air parcel moving upward will, therefore, experience a temperature drop; this is the *dry adiabatic lapse rate* Γ_α, which is valid as long as no condensation of water vapor occurs. Below a certain temperature, depending on the initial water vapor content of the parcel, condensation will occur, however, and the release of the energy of latent heat of condensation must be taken into account. In that case, the rate of change of temperature with height will, in general, follow the *moist adiabatic lapse rate*, which turns out to be $dT/dz \approx -6.5\ ^\circ\text{C km}^{-1}$.

Patterns in the Stratosphere

Unlike the fluctuating troposphere, the stratosphere is a region of continuous stability. In this region, temperatures are either constant or increase with height, thus producing a temperature inversion above the troposphere (see Figure 3.5). This temperature structure implies that material brought into the stratosphere by direct injection (aircraft emissions, volcanic eruptions, nuclear weapons testing, etc.) or as the products of chemical reactions will change altitude very slowly and can remain there for months or years.

Position of Tropopause

Because of stronger heating from below and a larger supply of latent heat in the form of water vapor, air parcels originating from near the surface in the tropics can penetrate higher against the stratospheric temperature inversion than can parcels originating at higher latitudes. As a consequence, the tropopause near the equator is located at a higher altitude (about 16–18 km) than it is over temperate and polar latitudes (8–12 km). In fact, a fraction of the upward branch of the tropical Hadley cell circulation can move into the stratosphere, and this is the natural pathway of transfer of tropospheric air into the stratosphere. The return flow from the stratosphere into the troposphere preferentially occurs near regions of the tropopause breaks, which occur near 20°–30° and 15 km between the high tropical tropopause and the lower midlatitude tropopause and also at 10–11 km between the midlatitude tropopause and the polar tropopause. (These latter breaks are variable in latitude and do not show up in monthly-averaged data.) Tropopause breaks are roughly conterminous with the locations of strong westerly jet streams.

As a consequence of the strong upward convection of air in the tropical regions, air parcels at the tropopause level in the tropics can reach very low temperatures of about −80 °C. The air entering the stratosphere is, therefore, very dry, typically with a water vapor mixing ratio of only about 3 ppmv. As a result, the stratosphere is largely devoid of clouds. Because the air masses that enter the stratosphere from the tropics were originally quite wet but passed through strong condensation and freezing processes, any soluble gases and particulate matter were efficiently removed by precipitation. Consequently, the air entering the stratosphere is very clean, containing negligible amounts of reactive chemical species. However, several industrially produced gases are so insoluble in water and chemically stable in the troposphere that they survive and reach the stratosphere, where they are broken down by the Sun's intense radiation and initiate many of the chemical reactions discussed later in this book.

Ozone Levels

The fact that the altitude of the tropopause in the tropics is 6–9 km higher than its altitude in higher latitudes significantly affects the distribution of the total ozone column. Although most ozone is created by photochemical reactions in the tropical stratosphere in the altitude range 25–30 km, transport to the lower stratosphere in the higher latitudes (Figure 4.6) and its accumulation there due to slow vertical mixing cause the largest total ozone column amounts to occur at high latitudes (see Figure 4.9). The lowest total ozone occurs in the tropics; consequently, by far the most ultraviolet radiation penetrates to Earth's surface in the tropical regions. In other words, the tropical biosphere is least protected from solar ultraviolet radiation, so the biosphere must have developed the strongest defense mechanisms against it (an example is the dark skin color of tropical natives). As can also be seen from Figure 4.9, total ozone peaks near springtime toward

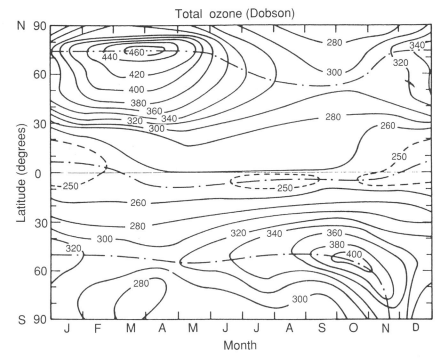

Total ozone (Dobson)

Figure 4.9 Column abundances of ozone as a function of latitude and season. The abundances are expressed in Dobson units; 1 DU = 2.69×10^{16} molecules O_3 cm^{-2}; 100 Dobson units corresponds to an ozone layer thickness of 1 mm if all ozone in a vertical column were at 1 atmosphere pressure and 0 °C. (Reproduced with permission from J. M. Wallace and P. V. Hobbs, *Atmospheric Science: An Introductory Survey*. Copyright 1977 by Academic Press, Inc.)

the higher latitudes in the northern hemisphere and near 55° latitude in the southern hemisphere. Minimum values occur in the autumn seasons. We should note, however, that Figure 4.9 represents the total ozone distribution as it was measured prior to the development of the springtime Antarctic "ozone hole" (see Chapter 8).

Most transfer of ozone from the stratosphere to the troposphere takes place during the springtime months, in both the northern and the southern hemispheres. Until about 1973, it was generally believed that this transfer was the main source of tropospheric ozone. This is, indeed, most likely still the case for the southern hemisphere. However, it is now clear that in the northern hemisphere even larger quantities of ozone can be formed by the reactions of chemicals emitted as a consequence of human activities. Considering the importance of tropospheric ozone for the overall photochemistry of the atmosphere (see Chapter 8), this issue is receiving much scientific attention.

REGIONAL CIRCULATION PATTERNS

Near the ground, influences other than temperature and pressure affect the direction and speed of air parcel motion. Friction between Earth's surface and the air slows the speed of the wind near the ground and the flow aligns itself more along the direction of the pressure gradient so as to enhance the flow outward from high pressure areas and inward toward low pressure areas.

The interactions of water and land with solar radiation cause significant modifications in wind and weather patterns. The simplest example is the land and sea breeze phenomenon, caused by the different heat capacities and conductivities of soil and water. For the former, the heat storage is limited to about the upper meter; for the latter, the upper hundred meters can store and exchange heat. The general situation is shown in Figure 4.10. During the day, as the land becomes warmer than the sea, the expanding and rising hot air creates pressure differentials, drawing sea air over the land near the surface. At night, the rapid cooling of the land creates a pressure differential in the opposite direction, and the ground-level flow is from land to sea.

The airflow demonstrated above for land–sea breezes on a day–night pattern is present in an extreme form in the monsoons that are so much a part of the climates of eastern and southern Asia and parts of Africa. In those regions, where large continental size and high mountains accentuate many of the features of air motions, the general flow of air during most of the winter months in the lower troposphere is seaward, because the continents become cooler than the seas as solar radiation fluxes are decreased. This pattern typically creates a dry season. In contrast, the large landmass readily heats up during the summer months, bringing water-laden airflows from the oceans over the land and depositing large amounts of precipitation upon it.

A final effect worth noting is the seasonal movement of the intertropical convergence zone in the equatorial regions (Figure 4.5; note the changing ITCZ location in South America, Africa, India, and southeast Asia), causing less precipitation in the northern hemisphere in the tropics and subtropics from December to April and in similar southern hemisphere regions from June to October. These periods form the dry seasons of the year in those regions, in contrast to the wet seasons some six months later. During the dry seasons, much burning of biomass, especially dry grass in the savannas, takes place, causing high levels of air pollution. We will discuss this important topic in Chapter 12.

EFFECTS OF ATMOSPHERIC MOTIONS ON ATMOSPHERIC CHEMISTRY

From the standpoint of atmospheric chemistry, the time scales for transport of airborne species between different atmospheric regions are of particular interest. One such scale is that for horizontal flow in the lower troposphere. The wind velocities in those cases are generally of the order of 5 m s^{-1}, which effectively moves chemical species or their precursors several hundred kilometers from the emission locations within a day or so.

On a longer time scale, the atmospheric winds move air parcels (and emissions) across entire continents in a few days. Species whose

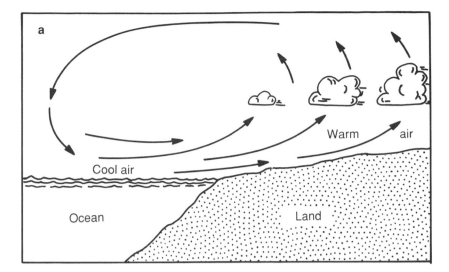

Figure 4.10 Air temperature distributions and airflows for (a) sea breezes in daytime and (b) land breezes in nighttime. (Reproduced with permission from G. T. Trewartha, *Introduction to Climate*, 4th Ed. Copyright 1968 by McGraw-Hill, Inc.)

low reactivity or poor solubility allows them to remain in the atmosphere as long or longer than that are therefore effectively moved from their place of emission to locations across national boundaries. A familiar example is that of the sulfur and nitrogen oxide emissions manifested in "acid rain," with industrial Europe and the United States exporting the pollution by air to Sweden and Canada, respectively.

On still larger spatial scales, the strong zonal winds in the middle to upper troposphere move trace atmospheric constituents with long lifetimes far from their geographical origins. It is these winds that are responsible for the deposition of dust from the deserts of China onto Hawaii or Alaska, from the Sahara onto Florida or Europe, and from central Russia onto the Arctic.

It is sometimes the case that atmospheric chemistry is concerned with the details of upper tropospheric motions such as the jet stream. If trace species are injected into the atmosphere at those altitudes, they can rapidly be transported from one part of Earth to another by the high winds. An example of this potential was provided in 1980 by the eruption of the Mount St. Helens volcano in Washington state. This large volcano supplied substantial amounts of dust and gases to the lower stratosphere. The stratospheric debris was carried eastward by upper atmospheric motions at an average zonal speed of approximately 25 m s⁻¹. As shown in Figure 4.11, the debris circled Earth in an undulating manner every 15–16 days and was deposited in some very localized areas in various parts of the northern hemisphere. Another example was the deposition of radioactive substances from the Chernobyl nuclear power plant accident in Russia in various parts of Europe. In northern Scandinavia the deposition was so heavy that large portions of the reindeer herds of the Lappish people had to be slaughtered.

Another important feature of the interaction of air motions with atmospheric chemistry concerns the time scale for transport of gases across the equator, a process that can allow species emitted in one hemisphere to affect the atmosphere of the other. At the equator, the converging flows from north and south tend to rise and move poleward

Figure 4.11 The estimated global trajectory at a pressure height of 200 millibars (about 12 km) of the ash from the Mount St. Helens eruption of 18 May 1980. Days lapsed are indicated on this polar plot, counting from May 18 (0). The North Pole is marked with a +; North America lies in the bottom half of the diagram and the Pacific Ocean is on the left. (R. Newell, ed., *NASA Conference on the Mount St. Helens Eruption*, NASA, 1982.)

at higher altitudes within the same hemisphere. Because cross-equatorial flow is slow, interhemispheric air exchange times of a year or so are typical. If the gas of interest is emitted predominantly in one hemisphere and has a long lifetime, the interplay of emissions and slow interhemispheric transport will produce a meridional concentration pattern of the form shown for methane in Figure 4.12. The more reactive the gas, the more dramatic will be the interhemispheric difference in concentrations. This is the situation for propane in Figure 4.12, the concentration decreasing from about 0.6 ppbv in the

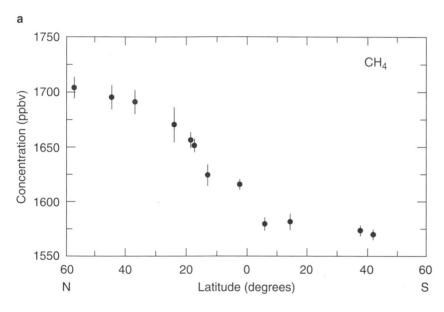

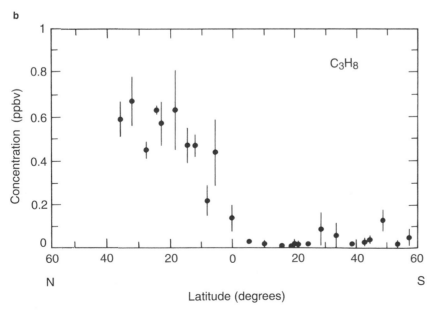

Figure 4.12 Interhemispheric profiles of the surface air concentrations of methane and propane. Methane is regularly measured at a number of sites around the world, and the data shown here are from annual mean measurements at a number of those sites. The error bars reflect the relative noisiness of the data at each site after seasonal and secular terms have been removed. (Reproduced by permission from L. P. Steele et al., The global distribution of methane in the troposphere, *Journal of Atmospheric Chemistry,* 5, 125–171. Copyright 1987 by Kluwer Academic Publishers.) The data for propane were obtained from measurements aboard the US Coast Guard vessel *Polar Star*, which sailed from Seattle, Washington to Punta Arenas, Chile, during the period November 15 to December 28, 1984. The concentration values represent daily measurements integrated over the latitude interval traveled by the vessel. The standard deviations of the daily measurements are indicated by the bars. (H. B. Singh, W. Viezee, and L. J. Salas, *Journal of Geophysical Research* 93, 15, 861–15, 878, 1988.)

northern midlatitudes to almost nothing across the ITCZ. If the emissions sources are not well known for a particular species, such concentration patterns can sometimes be used in connection with a knowledge of the meteorology to infer approximate source strengths.

The final time scale of interest involves transport between the troposphere and stratosphere, because that process controls the time scale on which ground-level trace species have influence on stratospheric chemistry and vice versa. Time periods for that transport of several years are dictated by the extremely inhibited mixing across the tropopause and in the lower stratosphere. In fact, most interregime transport probably results not by slow diffusive mixing but by the injection of tropical tropospheric air into the stratosphere by upward turbulent transport and by return flows from the stratosphere near jet stream tropopause folds at midlatitudes.

EXERCISES

4.1 What is the most efficient route and time of year for a sailing vessel to cross of the Atlantic Ocean from New York City, USA, to Le Havre, France? From Le Havre to New York City?

4.2 Denver and Kansas City are at about the same latitude, but their local air stability conditions are quite different. Using topography and weather information, explain why these differences might occur.

4.3 Resorts by the sea have always been welcome on hot summer days because of "ocean breezes." Which way do the ocean breezes blow at different times of day? Why?

4.4 If the atmospheric temperature gradient in a specific location is described by the dry adiabatic lapse rate, how will the pressure vary with height?

FURTHER READING

J. R. Holton, *An Introduction to Dynamic Meteorology*, 2nd Ed., Academic Press, New York, 1979.

J. M. Wallace and P. V. Hobbs, *Atmospheric Science: An Introductory Survey*, Academic Press, New York, 467 pp., 1977.

CHAPTER 5

Aerosols and Hydrosols

An *aerosol* is a stable suspension of solid or liquid particles or both in air. The atmospheric aerosol is therefore the combination of all the condensed-phase material present in the atmosphere as well as the air within which the suspended material resides. Similarly, a *hydrosol* is a stable suspension of solid particles in water. It most often refers to particles in fresh waters or in the ocean but is also appropriate for atmospheric water droplets containing large numbers of small particles—a common occurrence.

Atmospheric particles may be emitted or injected directly into the atmosphere, as happens, for example, by wind-driven erosion. Such particles are termed *primary particles*. Alternatively, if particles are created by accretion or nucleation from gas-phase molecules (*gas to particle conversion*) as happens particularly in the case of anthropogenic emissions, they are termed *secondary particles* or *accumulation mode particles*. Depending on the geographical location and the particle generation processes involved, both primary and secondary particle generation can be significant.

Hydrosol particles also have more than one source. For rivers and lakes, the most important is usually the influx of soil particles in runoff. For estuaries and oceans, particles in the river outflow are dominant. Also of significance for many chemical cycles are particles deposited from the air. Finally, detritus from living organisms is often important, especially in the larger particle sizes.

Aerosol and hydrosol particles play two distinctly different roles. As condensed-phase material, they have physical functions as absorbers, emitters, and scatterers of light. They also provide convenient surfaces for the deposition of molecules and/or ions from the fluid in which they are embedded. As chemical systems, they serve as media upon which reactions of interest can occur. Particles are involved in the final stages of many Earth system processes, so a complete understanding of chemical cycles and reaction chains often involves a detailed knowledge of the chemistry and physics of aerosols and hydrosols.

Figure 5.1 Ranges of equivalent diameters for some types of aerosol and hydrosol particles. For perspective, the diameters of molecules are also shown.

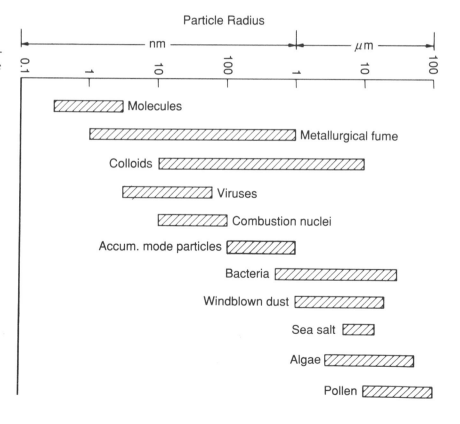

Size Characteristics

Although most particles are not spherical, it is convenient to designate the size of a particle by its *equivalent radius*, which is the radius of a sphere that experiences the same resistance to motion as the nonspherical particle. Under this definition or any other, atmospheric, marine, and freshwater particles cover a truly enormous size range, perhaps four or more orders of magnitude. The radii of some of the common types of particles are pictured in Figure 5.1. Primary atmospheric particles generated by natural processes, such as silt from wind-driven soil or salt particles from the sea, tend to be at the large end of the spectrum, whereas atmospheric particles resulting from industrial activity, such as combustion or ore processing, tend to be at the small end. Particles generated by the biosphere cover the entire range, pollen being of large size and organic aerosols formed by gas to particle conversion from emissions of low-volatility hydrocarbons being small (≈ 0.1 μm). Within fresh water and salt water, one finds viruses, bacteria, and algae and, covering a very wide range of sizes, colloidal organic particles formed as by-products of biological processes.

Tropospheric Aerosol Spectra

A common way to display the sizes of a group of particles is to plot the number distribution. Because of the wide size range, this display is generally accomplished by using a logarithmic scale for the abscissa and plotting normalized values, that is, the number of particles per unit size interval, on the ordinate. A typical spectrum for atmospheric particles in an urban area—Figure 5.2a—shows a high number concentration near an equivalent radius of about 5 nm (0.005 μm); this region is termed the *nuclei mode*. The second enhancement, near $r = 0.2$ μm, is termed the *accumulation mode*, because it is formed primarily from accumulation of nuclei mode particles and deposited gases. The figure, and especially associated plots of surface or volume distributions (see Exercise 5.1), show a peak also at about $r = 2–5$ μm; this size region is termed the *coarse mode*. In rural areas, sources of all particles, especially nuclei mode particles, are sparse and the overall particle number spectrum is reduced.

Figure 5.2a also shows a number spectrum for particles incorporated into rainwater. (This process, termed wet deposition, is discussed later in this chapter.) It can be seen that the particle content of rainwater reflects rather faithfully the spectrum of atmospheric aerosol particles.

Figure 5.2 Representative size spectra of aerosols and hydrosols. (a) Number distributions of average continental urban and rural aerosol particles and that of insoluble particles in rainwater. (L. Schütz and M. Krämer, Rainwater com-position over a rural area with special emphasis on the size distri-bution of insoluble particulate matter, *Journal of Atmospheric Chemistry, 5*, 173–184, 1987.) (b) Number distributions of stratospheric aerosol particles before (16 October 1981; dotted line) and after (23 September 1982; solid line and data) the eruption of the El Chichon volcano in Mexico. (J. M. Livingston and P.B. Russell, Retrieval of aerosol size distribution moments from multiwavelength particulate extinction measurements, *Journal of Geophysical Research, 94*, 8425–8433, 1989.) (c) Number distribution of hydrosol particles at a depth of 99 m in the Pacific Ocean. (A. Lerman, K. L. Calder, and P. R. Betzer, *Earth and Planetary Science Letters, 37*, 61–70, 1977.)

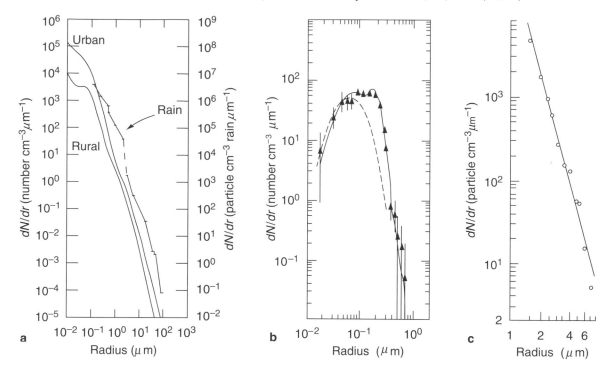

Particle Number, Surface Area, and Volume Distributions

If one knows the number of particles as a function of particle size, plus their average density, the particle surface area and volume as a function of size can be computed. The latter are useful because light scattering, deposition of gases, and chemical reactions within atmospheric droplets are better related to those characteristics. In addition, one can compute the number, surface area, and volume for the entire aerosol. For example, the total number of particles present is given by integrating the area below the number curve:

$$N = \int \frac{dn}{d \log D_p} \, d \log D_p, \qquad (5.1)$$

where D_p is the diameter of the particle. (Aerosol and hydrosol scientists use both equivalent radius and equivalent diameter, and equations for each are directly related.) In practice, data are available, not for a continuous distribution, but for the number of particles within a specific size range. Integration is thus performed numerically over discrete intervals, the limits of the integration being those of the measurements:

$$N = \sum_{D_p \text{(min)}}^{D_p \text{(max)}} \frac{\Delta n}{(\Delta \log D_p)} (\Delta \log D_p). \qquad (5.2)$$

Similarly, the total surface area and volume are given by

$$S = \sum_{D_p \text{(min)}}^{D_p \text{(max)}} \pi D_p^2 \frac{\Delta n}{(\Delta \log D_p)} (\Delta \log D_p) \qquad (5.3)$$

and

$$V = \sum_{D_p \text{(min)}}^{D_p \text{(max)}} \frac{\pi}{6} D_p^3 \frac{\Delta n}{(\Delta \log D_p)} (\Delta \log D_p). \qquad (5.4)$$

The Stratospheric Aerosol

Aerosol particles are present not only in the troposphere but also in the stratosphere. In fact, it has been determined that a layer of small particles is always located in the stratosphere at altitudes centered around 25 km at the equator and 17 km at the poles. Figure 5.3 shows a schematic diagram of that layer, indicating that except for the land-associated particle layer at very low altitudes in the northern latitudes, the highest concentrations of particles occur in the stratosphere.

The average size of the stratospheric particles tends to be much smaller than that of the particles in the lower atmosphere. This difference is seen in the typical size spectrum shown in Figure 5.2b, where the peak occurs at about $r = 0.06$ μm and virtually no particles above $r = 0.3$ μm are present. Regular observations of stratospheric particles by balloon-borne detectors demonstrate that the size spectrum is quite constant provided major volcanic activity is not occurring. At the time of a major eruption, however, particles larger than customary stratospheric sizes but too small to fall out promptly at the eruption site are injected into the stratosphere, as shown in Figure 5.2b.

Hydrosol Spectra

Hydrosol particles cover size ranges as wide as those for tropospheric aerosols, but size-specific measurements are fewer. Figure 5.2c shows a hydrosol spectrum at 99 m depth in the Pacific Ocean; the general pattern appears not to vary much with depth. (Note that it is similar

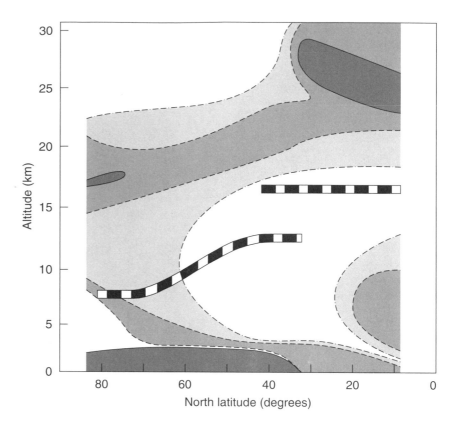

Figure 5.3 The observed spatial distribution of northern hemisphere aerosol particles of radius greater than 0.15 μm diameter in clean background air. The densely stippled regions indicate particle densities of 8–10 per cm^{-3}, moderately stippled regions those with densities of 5–8 particles per cm^{-3}, and lightly stippled regions those with 2–5 particles per cm^{-3}. The heavy dashed lines indicate the tropical and polar tropopause locations. (Reproduced by permission from G. Brasseur and S. Solomon, *Aeronomy of the Middle Atmosphere*, 2nd Ed. Copyright 1986 by Kluwer Academic Publishers.)

to the aerosol spectra of Figure 5.2a.) Not shown in the figure are the very small colloidal particles that are doubtless present, nor does the figure show the very large particles such as fecal pellets and disturbed sediment that contribute sporadically to the large particle sizes.

AEROSOL PARTICLE SOURCES

Wind-Driven Erosion

The most frequent method for the natural generation of aerosol particles is the erosion of soil by the wind. Soil particles are normally held in place by gravity and by attachment to neighboring particles. For each particle, there is a threshold wind velocity, determined by factors such as soil texture and composition, above which erosion occurs. Once the wind velocity passes that threshold, several motions can occur, as shown in Figure 5.4. First, the particle moves along the ground, its speed limited by the local wind motion. At higher wind velocities, particle levitation is possible. Subsequent collisions of particles with the soil produce levitation of smaller particles.

Although some wind-driven erosion occurs almost continually, most particle generation is the result of infrequent but powerful dust storms. The sporadic nature of these storms makes it difficult to predict storm frequency or storm particle generation. It is nonetheless clear that the effects of a few storms can overwhelm those of ongoing lower velocity wind events.

Anthropogenic Particles

In urban and industrial areas, most airborne particles are related to human activities such as transportation, fuel combustion, industrial processes, and solid waste. Construction activity and agricultural operations also produce significant amounts of airborne particles.

The sources of particles of different sizes tend to be distinct. The largest particles (those with diameters larger than a few micrometers) are generally the result of a mechanical process such as grinding or spraying. Smaller particles (those with diameters between a few tenths of a micrometer and a few micrometers) are usually related to incomplete combustion or a high-temperature industrial process. The smallest particles (those with diameters less than a few tenths of a micrometer) are the result of the condensation of vaporized material produced by a high-temperature process. Anthropogenic particles may coagulate with various degrees of efficiency to form larger particles but do not customarily fragment to form smaller ones.

Biomass Combustion

The combustion of biomass, whether natural (forest fires) or anthropogenic (fields burned following harvest), produces large amounts of airborne

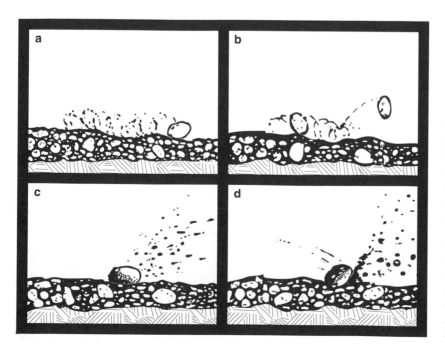

Figure 5.4 Motions of wind-driven soil particles. (a) Creeping motion of a particle moving as a consequence of a wind speed slightly greater than threshold; (b) A coarse particle lifted into the air by turbulent air fluctuations; (c) A particle collision with the surface, followed by breaking off of smaller particles that were encrusted on the colliding particle's surface; (d) A particle collision followed by "splashing" of the soil. (Adapted with permission from D. Gillette, Major contributions of natural primary continental aerosols: Source mechanisms, *Annals of the New York Academy of Sciences, 338*, 348–358. Copyright 1980 by New York Academy of Sciences.)

particles. Furthermore, these particles are emitted into buoyant plumes and may thus be carried substantial distances. The nature of the fuel and the efficiency of combustion are important variables in determining the output of particles from biomass burning.

Biomass-burning aerosols consist largely of partially oxygenated organic matter and constitute between a sixth and a half of all organic carbon aerosol released globally. A highly variable fraction of the remaining aerosol mass is black carbon (soot), which is abundant in intensely flaming fires. Accurate estimates of the emissions are difficult, but it is thought that biomass burning may be the principal source of atmospheric black carbon aerosol on a global basis. Black carbon aerosol particles are much smaller than organic carbon aerosol particles and affect the radiative properties of the atmosphere (as discussed later in this chapter) and the condensation efficiencies of clouds (by serving as condensation centers for cloud droplets).

Gas to Particle Conversion

An important source of very small particles in the atmosphere is the conversion of gases. This process is usually accomplished by accretion onto preexisting tiny particles but can sometimes occur by direct nucleation from the gas phase. A common example is the bluish haze occasionally seen over deciduous forests, which results at least in part from the condensation of terpenes and other vegetative emissions. A similar process occurs in urban areas, where anthropogenic organic emittants of low vapor pressure form oxygenated organic aerosols.

Sulfur-containing gases are also major participants in gas to particle conversion. The common sulfur species in the troposphere are sulfur dioxide (SO_2), dimethylsulfide (CH_3SCH_3), and hydrogen sulfide (H_2S). These reactive gases are rapidly converted to sulfuric acid. Because of its extremely low equilibrium vapor pressure, sulfuric acid condenses out onto cloud droplet and aerosol particle surfaces. The sulfate ion from the sulfuric acid and the ammonium and nitrate ions from gaseous ammonia and nitric acid are usually the most abundant species resulting from gas to particle conversion.

The stratospheric particle layer is also related to the gas to particle conversion process. In 1961, the German atmospheric chemistry pioneer Christian Junge, then working in the United States at the Air Force Research Laboratories in Cambridge, Massachusetts, discovered that this layer consisted largely of sulfuric acid particles. It did not seem possible for the layer to result from transported particles, because the air entering the stratosphere (mostly over the tropics) has been thoroughly washed during transport and is quite particle free. The alternative explanation was that one or more sulfur-containing gases were entering the stratosphere and being converted into particles after transport.

Because H_2S, SO_2, and CH_3SCH_3 react and are lost within the troposphere, another sulfur species was needed to explain the stratospheric particle data. That species turned out to be carbonyl sulfide (COS), which is chemically stable in the troposphere, has a low solubility in water, and is disintegrated by solar ultraviolet radiation to CO and S at heights of 20–25 km in the stratosphere. The resulting sulfur atoms are rapidly converted to SO_2 and H_2SO_4 by stratospheric chemical reactions and subsequently form the sulfate aerosol.

Volcanoes

Volcanologists report that it is rare for Earth to be without an active volcano, although the sizes of the eruptions vary greatly. Volcanic particles may be emitted into the lower atmosphere or, if the eruption is sufficiently energetic, into the stratosphere. The latter tends to occur once every year or so, and the abundance of stratospheric particles follows an oscillatory cycle from one eruption to the next, as shown in the 15-year series of observations pictured in Figure 5.5.

When the stratospheric sulfate layer is volcanically enhanced, so is the backscattering of solar radiation to space. This alteration of the radiation budget can be so severe that it can affect climate conditions near Earth's surface. In later chapters we will discuss examples of the influences of volcanic eruptions on climate.

AEROSOL PARTICLE TRANSPORT AND DEPOSITION

Particle Transport

Once particles are levitated into moving air currents, they can be transported over long distances. It has been realized over the past decade that the range of transport can be great indeed. An example is

Figure 5.5 Maximum stratospheric aerosol mixing ratios for particles with diameters greater than 0.30 μm for the period 1971–1986. All of the data are from instruments carried aloft by balloon, mostly from Laramie, Wyoming, USA. The dates of major volcanic eruptions that are believed to have perturbed the stratospheric aerosol concentration are indicated. (D. J. Hofmann, J. M. Rosen, J.W. Harder, and S. R. Rolf, Observations of the decay of the El Chichon stratospheric aerosol cloud in Antarctica, *Geophysical Research Letters, 14,* 614–617, 1987.)

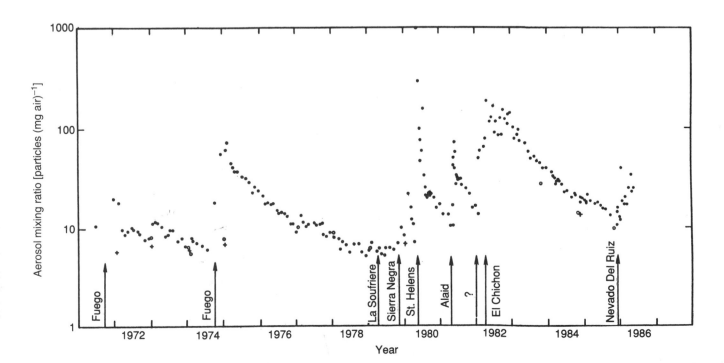

shown in Figure 5.6. The figure was constructed by identifying occurrences of high concentrations of dust at Enewetak (an island station in the Pacific) and using meteorological analyses to deduce where the air bringing that dust had been over the previous few days. The trajectories (there were several, because more than one air parcel arrived on that day) all originated from the vicinity of the China deserts and then traveled at pressure heights of 500–600 kPa (altitudes of about 3–6 km) to Enewetak. The picture is completed by reports of a widespread dust outbreak in China at the appropriate time. Similar diagrams have been constructed for several Pacific islands, and years of analyses over the South Atlantic have demonstrated that a fair fraction of Florida's airborne dust often originates in the African Sahara Desert.

Particle Deposition

Dry deposition results when an airborne particle comes into contact with a surface and is lost to it. The relationship between the concentration of a species i and the vertical flux Φ_i to a surface is often expressed by

$$\Phi_i = v_d [C_i], \qquad (5.5)$$

Figure 5.6 Air parcel trajectories arriving at Enewetak Atoll in the Pacific at noon universal time on November 26, 1982, during a period of heavy dust deposition. The bottom panel shows the horizontal components of the trajectories, the top panel the vertical components. The trajectories, carried back eight days, begin over China at a time when widespread dust storms were reported. (Courtesy of J. T. Merrill, University of New Hampshire.)

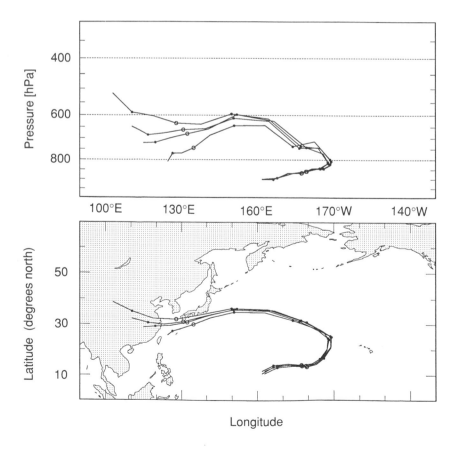

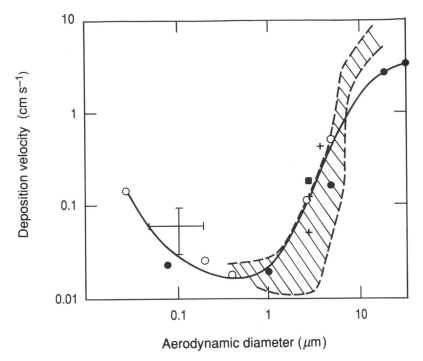

Figure 5.7 The deposition velocity of particles to grass. Field data for a number of experiments by different research groups are shown, some by data points and fitted lines, some by hatched areas. (Courtesy of J. A. Garland, Atomic Energy Research Establishment, UK.)

where $[C_i]$ is the concentration of the constituent at some reference height and v_d is a parameter with units of length divided by time called the *deposition velocity*. Besides being chemically influenced, deposition is size dependent, because gases or very small particles that possess high diffusion rates behave quite differently from larger particles, whose momentum constrains them to follow the motions of the atmospheric parcels in which they are embedded, and from still larger particles, which have enough momentum of their own to be relatively unaffected by air movements.

For airborne particles, the variation of deposition velocity with diameter can be as great as two orders of magnitude. Figure 5.7 illustrates this behavior. Particles larger than about 1 μm have too much momentum to follow the deflected air and forcefully impact on surfaces toward which they are heading. Much smaller particles are light enough to behave rather like gases, diffusing toward surfaces at a high speed. Near the deposition velocity minimum, particles have diameters of a few tenths of a micrometer, so both processes are inefficient and particle lifetimes are long. The same situation holds for particle filtration for clean rooms, an observation explaining why it is so difficult to achieve low indoor concentrations of submicrometer particles.

In *wet deposition*, interactions occur between gaseous molecules or airborne particles and atmospheric water droplets, and these interactions can be described in reasonable detail. Each individual particle is exposed repeatedly to raindrops, each time with some probability of inertial capture. By the same token, each raindrop has many opportunities to scavenge (i.e., to incorporate) particles, and individual raindrops reaching the ground may contain as many as 10,000 small particles. If the capture probabilities and raindrop characteristics remain constant during a particular rainfall, the change in concentration of aerosol particles in a given size range p over a small time interval is expressed by

$$\frac{dN_p}{dt} = -N_p \lambda_{p,j} , \qquad (5.6)$$

where N_p is the number of aerosol particles per unit volume in size range p and $\lambda_{p,j}$ is the *washout coefficient* for aerosol particles in size range p and at rain rate j, in units of fractional decrease per second. The washout coefficient incorporates within it much detailed physics and chemistry and can be expressed in great mathematical detail, but field data to justify these expressions are rather sparse. A selection of those data is shown in Figure 5.8.

A similar treatment applies to the study of scavenging of aerosol particles by snowflakes. For the same water content, snowflakes have much larger surface areas and fall much more slowly. As a consequence, they are 10 to 100 times more efficient as aerosol scavengers. This property can be seen in the data of Figure 5.8.

Figure 5.8 Typical washout coefficients for rain scavenging and snow scavenging of particles. The data for snow are shown as asterisks; all other data are for rain. Field data are given for a number of experiments by different research groups. Dashed lines through the rain data are provided to indicate trends in scavenging as a function of particle size.

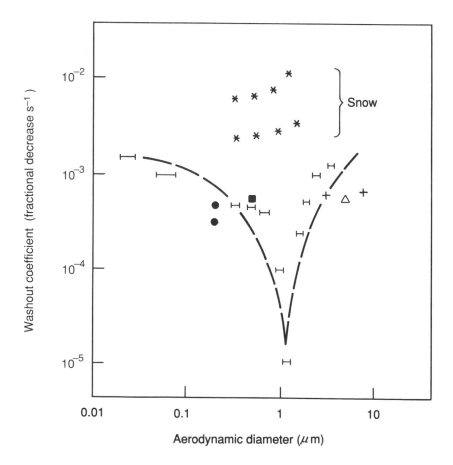

It is only recently that enough observations have been made over remote parts of continents and oceans to inaugurate the process of defining deposition patterns for airborne particles. The resulting data make it clear that the atmosphere is an important pathway for the transport of many natural and pollutant materials from the continents to the oceans and from one continent to another.

As with gases, one cannot simply consider the locations of the source regions but must combine information on sources, transport, and loss to arrive at a spatial assessment of deposition. Such an assessment has recently been performed for the deposition of particles into the world's oceans. We reproduce in Figure 5.9 one result of that study, a global assessment of the atmospheric deposition flux (the rate of deposition per unit area) of mineral aerosol. By far the highest inputs are into the Pacific Ocean just east of Asia, where the ocean receives windblown dust from the deserts of China, and into the Atlantic Ocean off the northwestern African coast, where the ocean receives windblown dust from the Sahara Desert. Other oceanic regions in the vicinity of continents receive perhaps a tenth of those amounts. Because of the paucity of land (especially deserts) in the southern hemisphere, the southern oceans receive much less deposition than do the northern oceans. Near Antarctica, the input drops by an additional factor of 10. Because the nutrient supply to many ocean regions is thought to be influenced by deposited aerosols, it is easy to see that with the atmospheric mineral input ranging over some three orders of magnitude, marine biota may be strongly influenced by their proximity to continents, especially continents undergoing major desertification.

Longitude (degrees)

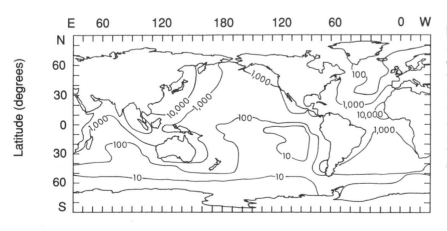

Figure 5.9 Estimated fluxes of mineral aerosol dust into the global oceans. The fluxes are highest off western Africa (as a consequence of the Sahara Desert) and off eastern Asia (as a consequence of the Mongolian Desert). The units are milligrams per square meter per year. (R. A. Duce, et al., The atmospheric input of trace species to the world ocean, *Global Biogeochemical Cycles, 5,* 193–259, 1991.)

Aerosol Particle Residence Times

The residence times of particles in the atmosphere are functions of the particle size and of the region of the atmosphere in which the particles are located. The predominant loss process for very small particles (those with diameters less than a few hundredths of a micrometer) is coagulation onto larger particles as a consequence of brownian motion. Coagulation is rapid, and the lifetimes of these small particles are a few hours or less. At the other end of the size spectrum are the very large particles, with diameters of perhaps 10 μm or greater. The gravitational settling of these particles or their rate of collection by hydrometeors is so rapid that their lifetimes are similar to those of the smallest particles. Particles of intermediate size are removed by collision with and diffusion to surfaces and by incorporation into precipitation. Depending on their chemical composition, particles in the 0.1–1.0 μm size range can also serve as nucleation sites for cloud droplets. This process of *nucleation scavenging* is the main mechanism for the removal of accumulation mode particles from the atmosphere. However, nucleation scavenging is not very efficient, especially at altitudes above about 4 km, and the lifetimes of accumulation mode particles can be quite long.

For particles in the lower and middle troposphere, residence times average about a week. In the stratosphere, residence times of months to years are more common. (These estimates are heavily dependent on the frequency of occurrence of clouds and rain, so they vary with location, season of the year, weather patterns, and the like.) These long residence times are a consequence primarily of the rapid decrease in the frequency of occurrence of clouds and precipitation as one ascends to higher altitudes.

Because of the dependence of aerosol particle lifetime on size, aerosol spectra change as particle aging occurs. At early times (that is, at locations close to the source or sources of the aerosol particles), the size spectrum resembles that of the particle source(s) plus small particles resulting from gas to particle conversion. The aerosol mass is often relatively high, the size spectrum is wide, and the temporal and spatial variation is great. Far from the sources, the aerosol particle mass has decreased, the particle size spectrum favors intermediate sizes, and variation is much smaller than was the case when the aerosol was young. It is possible, therefore, to infer something about the sources of an aerosol from measurements of the aerosol size spectrum.

HYDROSOL PARTICLE SOURCES AND DEPOSITION

Figure 5.9 showed that substantial amounts of mineral aerosol are deposited on Earth's surface. Are these amounts large enough to be really important? It is easy to demonstrate that they can dominate the hydrospheric cycles of some of the elements of interest. Figure 5.10 shows a vertical concentration profile for aluminum in the North Atlantic. The data show a high aluminum concentration near the surface and at the ocean bottom, and a lower concentration at intermediate depths. This profile is assumed to result from three different processes: atmospheric deposition at the ocean surface, aluminum scavenging by silica shells at middepth, and solubilization of aluminum minerals from the bottom sediments. The oceanic aluminum cycle (and those of fresh waters, not shown here) are thus the result of the combination of atmospheric transport and loss, marine *benthic* chemistry (benthic refers to the soil-water interface beneath a body of water), and marine biology.

A second major source of hydrosol particles is suspended sediment transported by runoff water, streams, and rivers. About 80% of the material transported by rivers and discharged into the oceans is in the form of suspended particles and only about 20% is in the form of dissolved species. It is, in fact, the reaction of solutes with solid surfaces that is thought to control the geochemistry of most trace elements, and these reactions take place in near-particle aqueous environments. In most cases the partitioning of solute species between soluble and particulate fractions is controlled not by the precipitation of pure solid phases but by adsorption on the surface of suspended solids of different chemical composition.

The third particle source for hydrosols is detritus from living organisms. This process is thought ultimately responsible for the sequestering of atmospheric carbon in the oceans. In regions of low biological production, particle generation is similarly low. When biological productivity is high, the particle input to deep water is high as well. It has been proposed that this whole process is enhanced in regions of high wind speed, as shown in Figure 5.11. When the wind velocity is high, ocean mixing is enhanced and dissolved nutrients are brought up into the surface ocean. Concomitantly, transport and deposition of atmospheric aerosol are enhanced. The incoming mineral matter interacts with colloidal organic carbon particles to form large particle aggregates, which have a high sinking velocity. As a result, high winds enhance productivity, large particle assemblage, and particle deposition to the bottom sediments. These processes restrict the lifetime of colloidal particles to an average of 1 to 2 months.

VISIBILITY

Visibility is commonly considered to be the greatest distance over which one can see and identify familiar objects with the unaided eye. The concept involves two quite different factors: (1) the degree to which light coming from the object is absorbed or scattered and (2) the visual threshold of perception.

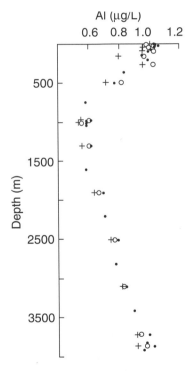

Figure 5.10 Concentration profiles of dissolved aluminum at a site in the North Atlantic Ocean. The three symbols identify three different measurement dates during a seven-month period in 1977 and 1978. (Adapted with permission from D. J. Hydes, Aluminum in seawater: Control by inorganic processes, *Science, 205,* 1260–1262. Copyright 1979 by the AAAS.)

In the atmosphere, solar radiation is absorbed and scattered by both gas molecules and particles. The scattering by gases in the clean atmosphere provides an upper limit of about 300 km to the visual range. At moderate or high aerosol loadings, however, radiation scattering by particles is the primary limitation to visibility. On the basis of field measurements, Robert Charlson of the University of Washington has shown that the total extinction coefficient ϵ_{ext} can be related to the total aerosol mass concentration (TPM) by

$$\frac{TPM}{\epsilon_{ext}} = 2.5 \times 10^{5}, \quad (5.7)$$

for TPM in units of $\mu g\ m^{-3}$ and ϵ_{ext} in units of m^{-1}. These particles, which are the accumulation mode particles that are products of human activity, are one of the principal factors in producing the brownish smoggy air now known throughout the world (see Figure 5.12 in the color insert).

Visual range is related to the extinction coefficient but is also a complicated function of the individual human eye. An approximate relationship for a black target against the sky is

$$R_{v} = \frac{3.91}{\epsilon_{ext}}. \quad (5.8)$$

Figure 5.11 A conceptual diagram of the deposition of particle matter in large water bodies. At low wind speeds (left), the mixed layer of the water is thin, nutrient supplies are low, and deposition of small biological particles is slow. At higher wind speeds (right), the mixed layer thickens, nutrients are brought up from deeper waters, and mineral dust from the atmosphere aggregates with biological particles to enhance deposition. (V. Ittekkot, Particle flux studies in the Indian Ocean, *Eos–Transactions of the American Geophysical Union, 52,* 527, 529, 1991.)

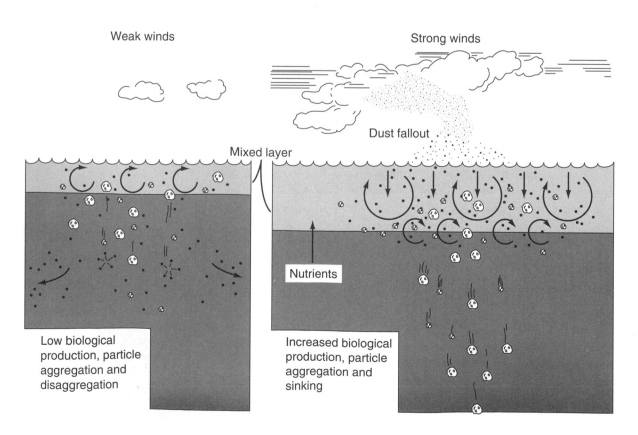

Weak winds

Strong winds

Dust fallout

Mixed layer

Nutrients

Low biological production, particle aggregation and disaggregation

Increased biological production, particle aggregation and sinking

This relationship is the basis for the international visibility code, given in Table 5.1, where codes 0–2 usually result from high concentrations of water droplets and the other codes may result from combinations of droplets and particles. In nonmeteorological situations, where visual range is often determined with nonblack objects, the constant in the numerator of Equation (5.8) is generally nearer 2 than 3.91, and the visual range is reduced.

TABLE 5.1. International Visibility Code

CODE NO.	WEATHER CONDITION	R_v (km)	ϵ_{ext} (km)
0	Dense fog	<0.05	>78.2
1	Thick fog	0.05–0.2	78.2–19.6
2	Moderate fog	0.2–0.5	19.6–7.8
3	Light fog	0.5–1.0	7.8–3.9
4	Thin fog	1–2	3.9–2.0
5	Haze	2–4	2.0–1.0
6	Light haze	4–10	1.0–0.39
7	Clear	10–20	0.39–0.20
8	Very clear	20–50	0.20–0.08
9	Exceptionally clear	>50	<0.08

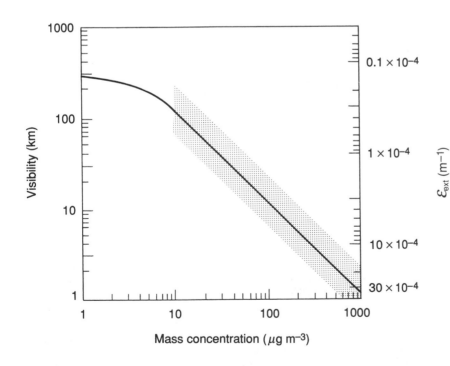

Figure 5.13 Visibility as a function of total airborne particulate mass concentration. More than 95 per cent of outdoor observations fall within the shaded band. Atmospheric total airborne particulate mass concentrations are rarely above 500 μg m⁻³ and rarely below 10 μg m⁻³. (Adapted with permission from R. J. Charlson, Atmospheric aerosol research at the University of Washington, *Journal of the Air Pollution Control Association, 18,* 652. Copyright 1968 by the Air Pollution Control Association.)

Equations (5.7) and (5.8) can be combined to relate visibility to total aerosol mass over the mass range generally encountered in the atmosphere. The result is shown in Figure 5.13, the leveling off at low mass concentrations being the visibility limit imposed by molecular gas scattering.

Scattering, Absorption, and Extinction Coefficients

Radiation intensity after passage through the atmosphere is given by Beer's law:

$$I = I_0 \, e^{-\epsilon_{ext} L}, \qquad (5.9)$$

where is I_0 the incident light intensity, I is the transmitted light intensity, L is the pathlength of the light, and ϵ_{ext} is the *extinction coefficient*. The extinction coefficient is the composite of interactions of light with particles and gases:

$$\epsilon_{ext} = \sigma_{ap} + \sigma_{sp} + \sigma_{ag} + \sigma_{sg}, \qquad (5.10)$$

where the first letter in each subscript indicates absorption or scattering, the second refers to gases or aerosol particles, and the terms may be defined either for individual or ensemble molecules or particles. Each of these four terms comes into play in the atmosphere in some conditions. In very clean air, where particles are few and gaseous molecules are relatively nonabsorbing, the term σ_{sg} is generally dominant. The parameter σ_{sg} is a function of wavelength, because the efficiency of the interaction of light with atmospheric constituents is a very sensitive function of the diameter D of the interacting body relative to the wavelength λ of the light. For the case $D_p \ll \lambda$ (this relation applies either to molecules or to particles with diameters less than about 0.03 μm), the interaction is termed *Rayleigh scattering*. Individual molecules have a scattering cross section σ_{sg} given in Rayleigh theory as

$$\sigma_{sg} = C_1 \lambda^{-4}, \qquad (5.11)$$

where the constant C_1 involves the index of refraction of the medium and the direction in which scattering occurs. A relationship with a different constant but the same wavelength dependence holds for small particles. The total Rayleigh extinction coefficient is essentially that due to molecular scattering and is given with satisfactory accuracy by multiplying Equation (5.11) by the number of scattering molecules per unit volume (N):

$$\epsilon_{ext} = N \sigma_{sg}. \qquad (5.12)$$

Because molecules have a higher scattering efficiency for light of shorter wavelengths, blue light is scattered more than red light. Hence, it follows from Equation (5.12) that the setting sun is red and the sky is blue.

At the opposite extreme from Rayleigh scattering constituents are the large particles, those with $D_p \gg \lambda$. Where these particles are numerous, they can scatter significant amounts of light, as can cloud and fog droplets. In these cases, the scattering cross section is about twice the geometrical cross section and shows little wavelength dependence, a characteristic explained by the fact that energy is lost from the incident light beam by both diffraction and scattering as a result of obstruction. Except near a large particle source such as a desert sandstorm, large particles are seldom abundant enough to play important roles in visibility.

Between these size extremes, where the diameter of the scatterer is close to the wavelength of the light being scattered, the scatterer sees a nonuniform electromagnetic field, and a complex process known as *Mie scattering* must be considered. We will not treat the process here in detail, but the important point is the wavelength dependence of the scattering cross section, which follows the relationship

$$\sigma_{sp} = C_2 \lambda^{-\alpha}. \quad (5.13)$$

The parameter C_2 is a complex function of the refractive index of the scattering particle and the scattering angle, and the parameter α can assume values between 1 and 2, being highest when the wavelength of light and the diameter of the particle are equal or nearly so.

EFFECTS OF AEROSOLS AND HYDROSOLS

What are the most significant roles played by the aerosols and hydrosols in chemistry or climate? One perspective is provided by comparing the mass of the atmospheric aerosol to the mass of reactive trace gases. The atmospheric aerosol mass is rarely lower than 1 μg m^{-3} at Earth's surface, often near 100 μg m^{-3} in urban areas, and occasionally as high as 30,000 μg m^{-3} in desert dust storms. The reactive trace gas concentrations may be as much as 10 ppmv total in urban areas but are significantly lower elsewhere. If one assumes an average molecular weight of 30 for the gases, their mass is of the order of 10^3–10^4 μg m^{-3}. Except in unusual circumstances, therefore, the mass of the reactive trace gas constituents in the urban atmosphere exceeds that of the aerosol by a factor of 10 to 100. In rural and background air, however, the reactive gas and aerosol masses are comparable.

Another role played by particles involves the hydrologic cycle (the global water cycle; see Chapter 6), where the particles may serve as condensation nuclei for cloud and fog droplets. Without the presence of the particles, vapor–liquid and liquid–solid phase changes would occur much less readily. Particles thus increase the efficiency of water recycling as precipitation. This effect also causes decreased visibility in regions with increased small particle formation or emissions, because the particles grow largely by incorporating water vapor; such decreases have been reported by several researchers.

Particles are important as well to the chemistry of the atmosphere and hydrosphere. They are responsible for the removal of many molecules from the gas phase, and a number of reactions of interest and importance occur on particle surfaces. These include oxidation of organic compounds to more toxic products, the catalysis of precursor molecules to inorganic and organic acids, and the sequestration of nutrients from fresh water and salt water.

For visibility in air or water, aerosol and hydrosol particles are generally the most important factors affecting this property. Together with clouds and fogs, which the smaller aerosol particles help to nucleate, particles normally set the limits to clarity.

To sum up, aerosol and hydrosol particles are ubiquitous constituents of the environment, and many questions of climate and chemistry cannot adequately be addressed without considering the effects of particles on processes of interest. In later chapters, we will discuss some of these processes in much greater detail.

EXERCISES

5.1 What factors would you expect to contribute to enhanced erosion and which to reduced erosion? Why are some parts of the world famous for frequent dust storms?

5.2 A parcel of air contains two sizes of particles and is located over a grassy surface. Some characteristics of the particles are given in the table below.

PARTICLE DIAM. μm	CONCEN- TRATION (No./cm³)	DENSITY (g/cm³)	WT. % Cl	DEP. VEL. cm/s
0.5	10⁴	1.6	2	0.03
2.0	10	1.8	5	1.50

Compute mass fluxes for the particles to the surface. Which size contributes the greater mass? Which contributes the greater amount of chlorine?

5.3 Average counts of atmospheric aerosol particles in Pasadena, California, during August and September 1969 (Whitby et al., *J. Colloid Interface Sci., 39*, 177, 1972) gave the following results (the number in the first and third columns is the center of the size interval and the upper and lower limits of the particle spectrum are at 0.09 μm and 5.8 μm):

D_p/μm	$\Delta N(D_p)$/no. cm⁻³	D_p/μm	$\Delta N(D_p)$/no. cm⁻³
0.11	1.8×10^3	1.05	5.2
0.14	1.4×10^3	1.27	2.5
0.18	1.6×10^3	1.48	1.4
0.25	1.5×10^3	1.82	1.4
0.35	3.6×10^2	2.22	0.67
0.44	1.8×10^2	2.75	0.26
0.55	5.4×10^1	3.30	0.11
0.66	2.2×10^1	4.12	0.24
0.77	9.4	5.22	0.02
0.88	5.3		

Plot $\Delta N(D_p)/\Delta \ln D_p$, $\Delta S(D_p)/\Delta \ln D_p$, and $\Delta V(D_p)/\Delta \ln D_p$. Compute N, S, and N. What particle size contributes the most to N? To S? To V? If the average particle density is 1.7 g/cm³, what is the aerosol mass concentration in μg m⁻³?

5.4 Aerosol particles are sometimes divided into "coarse mode" (those with diameters at or above 2 μg), and "fine mode" (those with smaller diameters). Given the aerosol chemical concentrations shown below and the size spectrum from Exercise 5.1, compute the mass of sulfate, nitrate, chloride, and ammonium in the two modes.

ION	COARSE MODE CONC. (WT %)	FINE MODE CONC. (WT %)
Sulfate	3	27
Nitrate	6	7
Chloride	5	1
Ammonium	1	9

5.5 Exercise 5.3 includes a computation of the mass of an urban aerosol distribution. Use this mass to estimate the visibility, assuming that light scattering from aerosol surfaces is the primary factor in visibility reduction.

FURTHER READING

R. A. Duce et al., The atmospheric input of trace species to the world ocean, *Global Biogeochemical Cycles, 5*, 193–259, 1991.

W. G. Dueser, P. G. Brewer, T. D. Jickells, and R. F. Commeau, Biological control of the removal of abiogenic particles from the surface ocean, *Science, 219*, 388–391, 1983.

E. J. McCartney, *Optics of the Atmosphere: Scattering by Molecules and Particles*, John Wiley, New York, 408 pp., 1976.

J. M. Prospero, R. J. Charlson, V. Mohnen, R. Jaenicke, A. C. Delany, J. Moyers, W. Zoller, and K. Rahn, The atmospheric aerosol system: An overview, *Reviews of Geophysics and Space Physics, 21*, 1607–1630, 1983.

CHAPTER 6

The Water Cycle and Climate

MAJOR WATER RESERVOIRS

The Distribution of Water on Earth

The presence of large amounts of liquid water is one of the most important characteristics of planet Earth, distinguishing it from all other planets in the solar system. The transport and distribution of water are, in addition, one of the most important features of Earth's climate. It is, in fact, the existence of varying temperatures and pressures in Earth's atmosphere and at its surface that causes water to constantly switch its solid, liquid, and gaseous states. This behavior allows life to exist on the planet, as Leonardo da Vinci recognized when he said, "Water is the driver of nature."

The distribution of water in its various reservoirs is given in Table 6.1. More than 97% of the water is in the oceans; nearly all of the remainder (the fresh water) is on the continents, predominantly in the polar ice caps. Despite recent "mining" of aquifers at far greater than replenishment rates, aquifers contain much more water than do rivers and lakes. The atmosphere contains only about one part in a hundred thousand of Earth's available water. The biosphere contains still less, but water is, of course, crucial in that regime, constituting about half of the biomass and being the transport medium for essential nutrients and waste products.

TABLE 6.1. Reservoirs of Available Water (liquid equivalent) on Earth		
RESERVOIR	VOLUME (10^6 km³)	PERCENTAGE OF TOTAL
Oceans	1350	97.3
Glaciers	29	2.1
Aquifers	8	0.6
Lakes and rivers	0.1	—
Soil moisture	0.1	—
Atmosphere	0.013	—
Biosphere	0.001	—

There is enormous variation in the amount of water present as vapor and condensed phases in the atmosphere. It can be as abundant as a few percent at low altitudes in the warm tropics and as scarce as a few parts per million in the cold lower stratosphere. In addition to the impact of atmospheric water on local and regional weather and climate, it has important consequences for atmospheric chemistry as well, as will be shown in subsequent chapters.

Characteristics of Oceans and Their Water

As residents on Earth's surface, we often find it hard to realize that the oceans and atmosphere are extremely thin layers on a very thick Earth, like the skin on an apple. Figure 6.1 uses data from Table 2.1 together with extensive height and depth measurements to plot the *hypsometry* (the pattern of heights as a function of area) of the solid, liquid, and gaseous Earth. For perspective, recall that the radius of Earth is about 6380 km. Thus, virtually the entire ocean depth and mountain and tropospheric height occur within about one half of one percent of the planet's radius. Among the characteristics of Figure 6.1 of particular interest are the small areas of regions that are extremely high or extremely deep, the rather large area of the (underwater) continental shelf, and the relatively small differences in the heights and depths of the continents and the ocean floor.

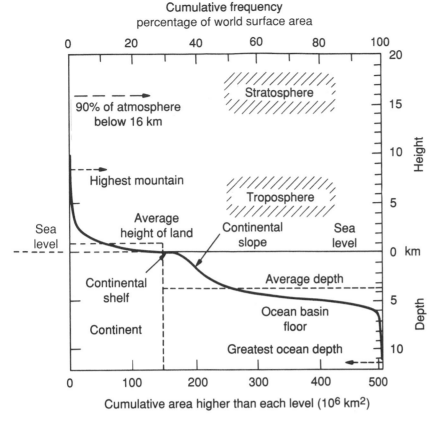

Figure 6.1 The cumulative hypsographic curve of Earth and its atmosphere. The atmospheric regimes are discussed in Chapter 3. 90% of the atmospheric gases are found below 16 km altitude and nearly all the ocean water is within 7 km of the surface. (Reproduced with permission from P.J. Wyllie, *The Dynamic Earth: Textbook in Geosciences.* Copyright 1971 by John Wiley & Sons.)

The water in the oceans is constantly in motion. Because of the importance of navigation throughout past centuries, ocean surface currents (those in the top few tens of meters) are well mapped and named, as shown in Figure 6.2. At and near the surface, the currents tend to follow the winds above them, and the flows approximately resemble those shown for the atmosphere in Figure 4.5, when the diversions of the currents introduced by the locations of the continents are taken into account.

A crucial parameter for atmosphere–ocean interactions is the temperature, which controls the rate at which gases dissolve in or are expelled from the ocean—cooler temperatures leading to greater solubility. A map of ocean temperatures is reproduced in Figure 6.3, in which strong winter–summer and interocean differences can be seen.

One of the most impressive wind-driven ocean surface currents is the Gulf Stream, which was first mapped by Captain Folger, a Nantucket whaleboat captain, at the request of Benjamin Franklin, then U.S. Postmaster General. The water flow in the Gulf Stream is enormous, reaching a peak of about 74 million cubic meters per second at 38° N latitude. (This is about 50 times larger than the combined river and groundwater flow to all the world's oceans!) In addition to aiding ocean shipping by providing a highway for more rapid passage across the Atlantic, the Gulf Stream is a major source of heat and humidity to western Europe, producing climate much warmer than is otherwise customary for those latitudes.

The trade wind systems of the tropics and subtropics are also drivers of important ocean surface currents. Because of the westward component of these winds, a strong westward current develops at and near the equator, carrying water from the eastern regions of ocean basins to the western parts. To compensate for these currents, an upwelling of deep ocean waters occurs, bringing to the surface deep, cold ocean waters rich in nutrients. The result in locations such as Peru is a cool climate characterized by frequent fogs and ocean waters teeming with marine life of all kinds. The cooler temperatures of the equatorial water near the western edges of the continents are readily seen in Figure 6.3.

For reasons not yet well understood, the normal patterns of temperature and trade wind circulation are sometimes strongly reduced or even reversed. The changes are strongest along the coast of South America, often resulting in massive fish death near the Christmas

Figure 6.2 A schematic diagram of the major ocean surface currents. (Reproduced with permission from T. J. Crowley and G. R. North, *Paleoclimatology*. Copyright 1991 by Oxford University Press, Inc.)

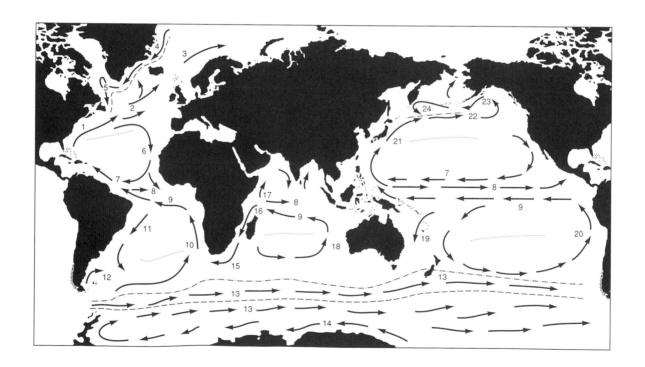

Figure 6.3 The average surface temperature (°C) of the oceans in February (top) and August (bottom). On the diagrams, shades of gray are used to indicate the loci of temperature and salinity bands. (Reproduced with permission from D. E. Ingmanson and W. J. Wallace, *Oceanography: An Introduction*, 4th Ed. Copyright 1989 by Wadsworth Publishing Co., Inc.)

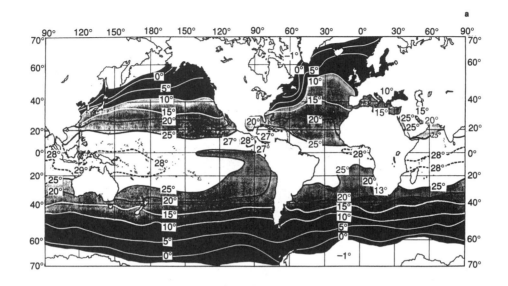

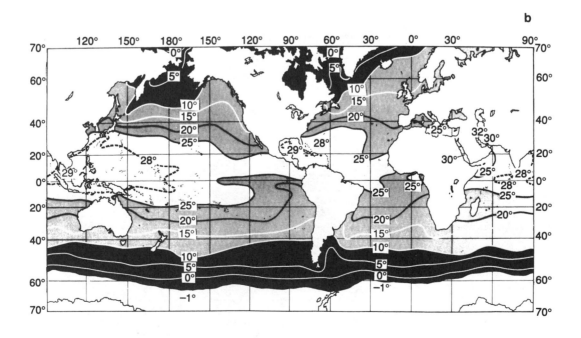

season. The phenomenon has come to be known as "El Niño" (Spanish for "The Child", a reference to Jesus Christ). During El Niño years, the entire circulation of the tropical atmosphere is highly disturbed, especially over the Pacific, but with worldwide repercussions for weather. Because of the interplay between meteorology, oceanography, and biology, the occurrence of an El Niño is clearly noticeable in the records of long-lived trace gases such as CO_2.

Ocean water is, of course, highly salty, a characteristic usually designated as the *salinity*, the total amount of dissolved material. Because the ratios of dissolved salts are very nearly constant in the world's oceans, the salinity can be quite accurately represented by measuring the chloride content and performing the following computation:

$$\text{Salinity} = 1.81 \times \text{chlorinity}, \quad (6.1)$$

where salinity and chlorinity are expressed in parts per thousand by weight (by convention, written without units). The salinity of ocean water is determined at the surface by the balance between evaporation (which increases salinity) and precipitation (which decreases it). Once the water parcel begins to sink, its salinity is changed only by mixing with dissimilar water. Figure 6.4 is a map of ocean surface salinity. Among the features of interest are the generally higher salinity of the waters at the warm equator, the generally lower salinity of the cool waters near the poles, the tongue of high salinity in the middle Atlantic resulting from outflow from the highly saline Mediterranean Sea, and the extreme salinity values of the shallow, rapidly evaporating waters of the Red Sea.

Figure 6.4 Average surface salinities of the oceans (⁰/₀₀) in August. (Annual variability is relatively modest.) On the diagram, shades of gray are used to indicate the loci of temperature and salinity bands. (Reproduced with permission from D. E. Ingmanson and W. J. Wallace, *Oceanography: An Introduction*, 4th Ed. Copyright 1989 by Wadsworth Publishing Co., Inc.)

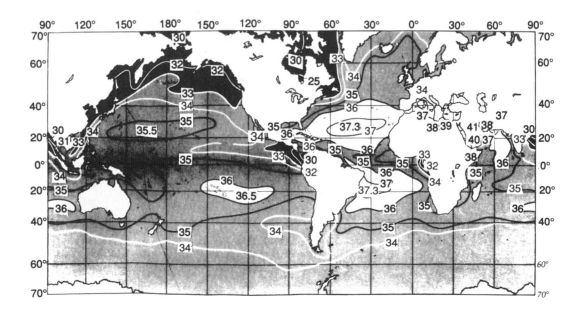

Ocean water far below the surface currents is also in motion, but the patterns are quite different from the flows at the surface. The vertical flow is termed the *thermohaline circulation*, because it arises from density variations. (Water is denser when colder or more saline.) High-density water sinks rather deeply and quasi-horizontal transport proceeds once an equilibrium depth has been reached. This flow at a few thousand meters, depth is balanced by shallower currents at a few hundred meters, depth moving toward the sites where the water is sinking.

An overview of the thermohaline and deep circulation is shown in Figure 6.5. The figure suggests that the transport system carries water and heat nearly around the whole Earth between the midlatitudes. The cycle can be pictured as beginning in the North Atlantic, where cold, dry Arctic air removes heat and promotes evaporation, thus increasing ocean salinity and density. This water sinks and flows south, forming the Atlantic deep water current at a few thousand meters, depth, with the magnitude of the flow being greater than 20 times the flows of all the world's rivers. This current turns east, flows past Africa and Australia, and rises from the deep ocean in the North Pacific, where it incorporates heat and fresh water transferred from the Asian continent. This relatively warm, less saline water works its way back to the North Atlantic at thermohaline depths of a few hundred meters.

The time scales for horizontal transport of chemical species by ocean surface currents are of the order of a few years. Trace constituents and heat can also be distributed vertically in the turbulent, wind-stirred mixed layer (about the upper 100 meters of sea water), a process

Figure 6.5 The large-scale salt transport system that operates in today's ocean, to compensate for the transport of water (as vapor) through the atmosphere from the Atlantic to the Pacific Ocean. Salty, deep water formed in the North Atlantic flows down the length of the Atlantic, around Africa, through the Indian Ocean, and finally northward into the deep Pacific Ocean. Some of this water upwells in the North Pacific, bringing with it the salt left behind in the Atlantic due to vapor transport. The atmospheric water vapor and ocean salt transport system is self-stabilizing. Records from ice and sediment tell us that this great conveyor system was somehow disrupted during glacial times and replaced by an alternative mode of operation. (Reproduced with permission from R. A. Kerr, Linking Earth, ocean, and air at the AGU, *Science, 239,* 259–260. Copyright 1988 by the AAAS.)

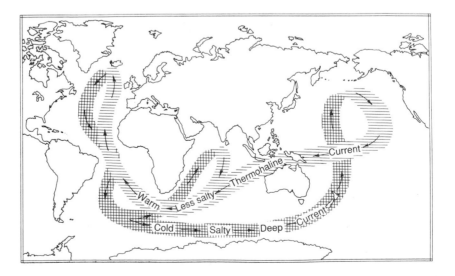

that occurs on about a yearly time scale. This vertical diffusion (a random process distinct from thermohaline flows) plays an important role in moderating seasonal atmospheric temperature fluctuations. The large capacity of the oceans to store heat through this mechanism stands in marked contrast to the more limited and transitory properties of the continents.

A final mixing time of interest is that for constituents of the deep ocean waters, which mix over hundreds or thousands of years. This stately pace guarantees that, once modified, the chemistry of ocean water will remain altered for a much longer period of time than will that of the atmosphere.

Rivers and Lakes

Table 6.1 shows that rivers and lakes hold only a tiny fraction of the water on Earth. As a reservoir, therefore, they are of little importance. As will be seen, however, their dynamic character compensates for their size, and their proximity to animals and humans makes them a vital source of water for household, industrial, and agricultural use. The chemical composition of natural fresh waters is characterized by wide variability, unlike the relatively stable chemical composition of the oceans.

The surface flows of rivers and lakes are largely determined by topography and the influx of runoff from precipitation; most flow velocities tend to be fairly rapid. Flow in the vertical direction in deep lakes is generally much slower, driven by slight seasonal variations in water temperature and water chemistry (hence differences in density). As a result, the chemistry of deep fresh waters can operate on time scales of seasons or longer. A dramatic example of episodic vertical mixing in lakes occurred in Lake Nyos in Cameroon. Around this lake—the only water source in a large geographical area—many people made their homes. Unknown to them, dissolved carbon dioxide from a submarine mineral-laden spring was accumulating at the lake bottom. The deep volcanic lake is usually stratified, with slightly denser CO_2-laden anoxic waters at the bottom underlying oxygenated waters in the surface layer. On 21 August 1986, some disturbance—perhaps a small volcanic eruption or a strong surface wind—caused a rapid overturning of the waters in the lake, releasing large amounts of CO_2 into the air. One thousand seven hundred people, unable to breath in the oxygen-poor air, died without knowing that they were victims of a natural but unpredictable hydrologic event.

WATER SOURCES, SINKS, AND FLUXES

The Hydrologic Cycle

Once the major water reservoirs are identified, the movements of water through them (termed the *hydrologic cycle*) can be investigated. Many processes are involved in these movements, as indicated in Figure 6.6. The basic cycle can be thought of as being inaugurated by evaporation from land and sea, maintained by condensation of the water vapor into clouds, and completed by the return of the water to the surface in the form of precipitation of various types. The hydrologic cycle is closely related to patterns of atmospheric circulation and temperature. Changes in either of these controlling variables can be expected to cause major changes in climate and biospheric conditions.

Sources and Transport of Atmospheric Water

Surface water will evaporate if the equilibrium vapor pressure at the temperature of the surface exceeds that in the atmosphere above. The rate of evaporation is highest when the evaporating surface is warm, the air dry, and vertical redistribution of the resulting water vapor is high; there is thus a correlation between evaporation and wind speed. The various factors favor evaporation in the warmer months and

Figure 6.6 Features of the overall hydrologic cycle of Earth. (Reproduced with permission from F. Press and R. Siever, *Earth*, 2nd Ed. Copyright 1978 by W. H. Freeman and Co.)

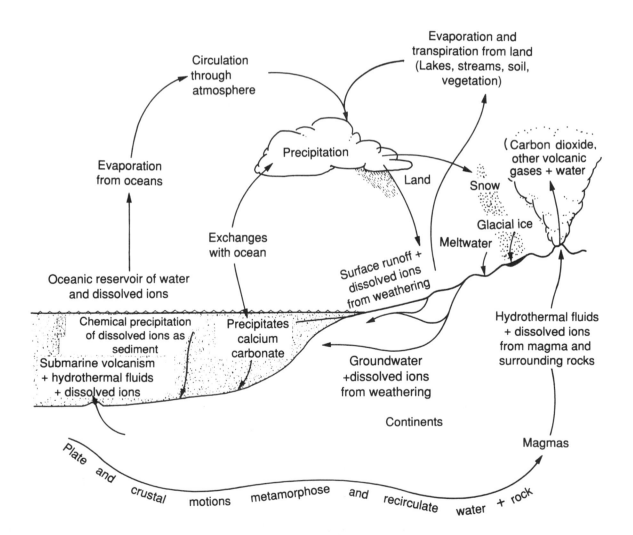

during the day. Most of the atmosphere's water is provided by the oceans, but contributions are also made by lakes and rivers, moist ground, and moist vegetation (the process of evaporation of water from moist plant surfaces is termed *evapotranspiration*).

Water vapor has an average lifetime of about 10 days in the atmosphere and can move thousands of kilometers before condensation occurs. The moisture transfer is characterized by two principal processes: exchanges among latitude bands (meridional moisture exchange) and exchanges between land and sea. Water evaporated from the subtropic ocean waters is transported to the equator by the dominant surface winds of the Hadley circulation. Near the equator, the surface winds in the northern and southern hemispheres converge in the intertropical convergence zone (ITCZ), thereby causing strong vertical convection and heavy precipitation. (The location of the ITCZ in winter and summer is indicated in Figure 4.5.) As a consequence, the tropical areas remove more water from the atmosphere than they supply. The land–sea moisture exchanges are in balance even though the sea loses more water by evaporation than it receives in precipitation and the land receives more by precipitation than it loses by evaporation. The balance is maintained by river and groundwater flows to the oceans.

Condensation and Loss from the Atmosphere

When unsaturated air is cooled, it eventually reaches the point of saturation. At that point (the *dew point*), any further cooling results in the deposition of water vapor onto convenient surfaces, generally the very small atmospheric particles that serve as condensation nuclei. The *hydrometeors* that are formed by this condensation process (a hydrometeor is any small condensed form of water falling through or suspended in the air) grow by further water vapor accretion. When sizes of the order of 10–20 μm are reached, the available water vapor is generally depleted and a stable cloud has formed.

An elementary fact of physics is that energy changes are involved in the transformation of water from one state to another. In the case of evaporation, this energy (the heat of vaporization) is 2260 J kg^{-1} at 25 °C. The energy is primarily derived from solar energy, and that retained in water vapor is known as *latent heat*. It is released to the atmosphere upon condensation and is a very important part of the atmospheric heat balance (see Figure 3.4). Because of the release of latent heat, clouds are warmer than the surrounding cloud-free air, and their buoyancy is enhanced, as will have been noticed by anyone who has flown through clouds in an airplane. Just as energy is required for water to evaporate, so it is given up when the water vapor condenses. To maintain the heat balance, a layer of water averaging 100 cm per year is evaporated from the surface and subsequently condensed in the atmosphere to produce precipitation.

The Classification of Clouds

Clouds come in many different forms, and their characteristics reveal much about the meteorological properties of the atmosphere. In the troposphere, four groups of cloud types are recognized: low-level, mid-level, high-level, and clouds with vertical development. The groups are subdivided into clouds with layered structure (*stratus*), those with clumped structure (*cumulus*), and those with filamentary structure (*cirrus*). The word *nimbus* denotes a cloud from which precipitation is falling. Cloud type characteristics are given in Table 6.2, and tropospheric clouds are illustrated in Figure 6.7.

TABLE 6.2. Types and Heights of Clouds	
TYPE	HEIGHT OF BASE (km)
Tropospheric clouds	
Low-level	
Stratocumulus (Sc)	0–2
Stratus (St)	0–2
Nimbostratus (Ns)	0–4
Mid-level	
Altocumulus (Ac)	2–7
Altostratus (As)	2–7
High-level	
Cirrus (Ci)	7–18
Cirrostratus (Cs)	7–18
Cirrocumulus (Cc)	7–18
Clouds with vertical development	
Cumulus (Cu)	0–3
Cumulonimbus (Cb)	0–3
Stratospheric and mesospheric clouds	
Polar stratospheric	15–25
Nacreous	20–30
Noctilucent	80–90

The layered tropospheric clouds generally occur in sequence during the passage of a cold or warm front. Figure 6.8 shows a common scenario, in which high-level cirrus and then cirrostratus clouds signal the front's approach, altostratus indicate the midpoint of frontal passage, and nimbostratus mark the frontal boundary and bring precipitation with them.

Vertically structured clouds are unrelated to frontal passage, forming instead as a consequence of intense solar heating of low-lying

a

Figure 6.7 Examples of tropospheric clouds. (a) stratocumulus; (b) stratus; (c) nimbostratus; (d) altocumulus; (e) altostratus; (f) cirrus; (g) cirrostratus; (h) cirrocumulus; (i) cumulus; (j) cumulonimbus. (M. Neiburger, J. G. Edinger, and W. D. Bonner, *Understanding Our Atmospheric Environment*, Copyright 1973 by W. H. Freeman and Co. Courtesy of the US National Oceanographic and Atmospheric Administration.)

b

c

d

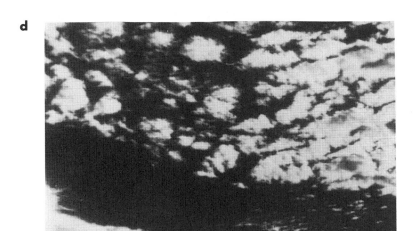

e

f

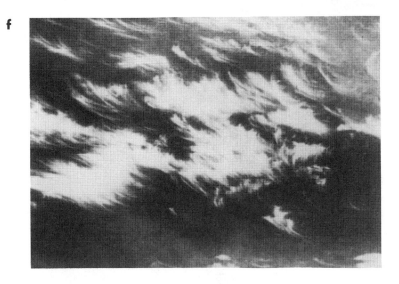

g

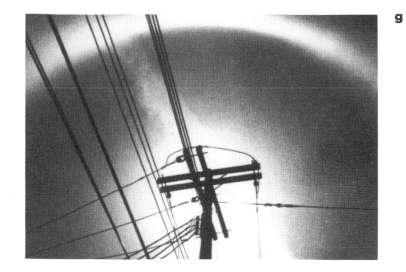

h

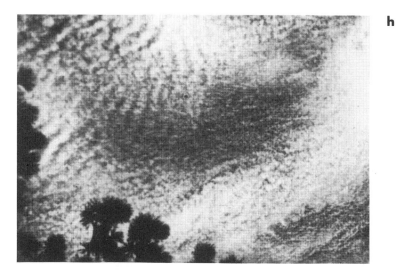

i

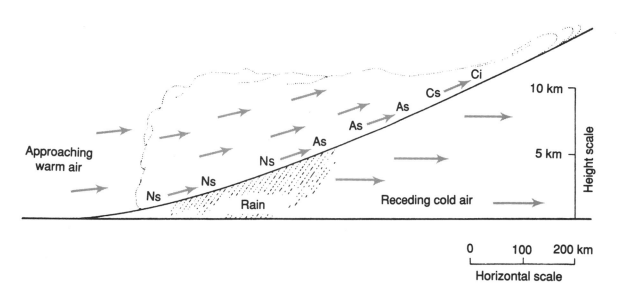

air, rapid ascent of the heated air, and water vapor condensation as the cooling air becomes supersaturated. This process is frequently followed by lightning, thunder, and heavy precipitation. Particularly large and vigorous cumulonimbus clouds occasionally penetrate the tropopause and enter the lower stratosphere.

The occurrence of clouds shows dramatic geographical variation. It turns out that the presence of liquid water clouds is largely restricted to the lowest 4 to 6 km at midlatitudes and in the tropics. Within those regions, clouds form and evaporate repeatedly. It is suspected that only a small fraction (perhaps as small as 10%) of the clouds that form ever precipitate; the remainder evaporate without ever generating the large droplets that transfer water from the tropospheric air to the ground. Even when precipitation develops, the raindrops often evaporate on their way through the cloud-free air below, so the drops never reach the surface.

Figure 6.8 A schematic representation of the sequence of clouds formed during the passage of a warm front. The cloud types are identified in Table 6.2. (Reproduced with permission from M. Neiburger, J.G. Edinger, and W.D. Bonner, *Understanding Our Atmospheric Environment*, Copyright 1973 by W.H. Freeman and Co.)

In contrast to the cloud-prone troposphere, the arid stratosphere is a place where clouds rarely form. However, stratospheric clouds are sometimes seen in the upward airflow over the Scandinavian mountains during the winter, when temperatures may be below –90 °C. Because of their colorful, iridescent appearance, they are called nacreous or "mother-of-pearl" clouds (see color insert Figure 6.9a). The highest clouds that are ever observed in the atmosphere occur during the summer at high latitudes, when vertical turbulence involving strong wave formation leads to temperatures as low as –150 °C near 90 km altitude. As these clouds are only visible from Earth's surface long after sunset, they are termed noctilucent. An example is shown in color insert Figure 6.9b.

A final cloud type is the polar stratospheric clouds (PSCs) that form in the middle stratosphere over the polar regions during the winter and early spring. These clouds often contain an equal weight mixture of nitric acid and water vapor and form at temperatures 5–10 °C warmer than would pure ice. As will be seen in Chapter 8, these clouds play important roles in stratospheric ozone depletion.

Air-Droplet Interactions and Precipitation

Clouds appear to be so substantial, and block sunlight so well, that it is truly surprising to find out how little water they contain. The water content of a typical cloud is only about 0.5 g m^{-3}, whereas liquid water has 10^6 g m^{-3}. The difference is that cloud water, in droplets with an average diameter of 20 mm, has about 5% of the surface area of liquid water with less than a millionth of the mass. This large surface area is important to the absorption and reflection of solar radiation as well as to the efficiency of cloud interactions with atmospheric particles and gases.

As air parcels move through clouds, the chemical constituents within the parcels are subjected to interactions with the droplets; this interaction frequently involves the transfer of soluble constituents to the liquid phase. In the lower half of the troposphere, cloud processing of air occurs about 15% of the time. On the basis of cloud coverage, typical heights and updraft velocities of air through convective clouds, and lifetime estimates for stratiform clouds, it appears that a typical air parcel has cloud encounter periods of a few hours or less, followed by cloud-free periods 5 to 20 times longer.

Continued condensation in rising air masses, as well as collisions and coalescence among the cloud droplets, leads to progressively larger droplets. Finally a few of the hydrometeors become so large that their tendency to settle overcomes the updrafts and frictional forces and they begin their fall to the ground. During this phase, they collect many smaller droplets and continue to enlarge. This mechanism for the development of precipitation is particularly common in the tropics. In cooler regions, precipitation often begins with the freezing of a few

Figure 6.10 The size distributions of hydrometeors in the atmosphere. Distributions vary widely under different conditions and those presented here are intended merely to be illustrative of typical observations. Note that the ordinates are logarithmic, except for cloud droplets.

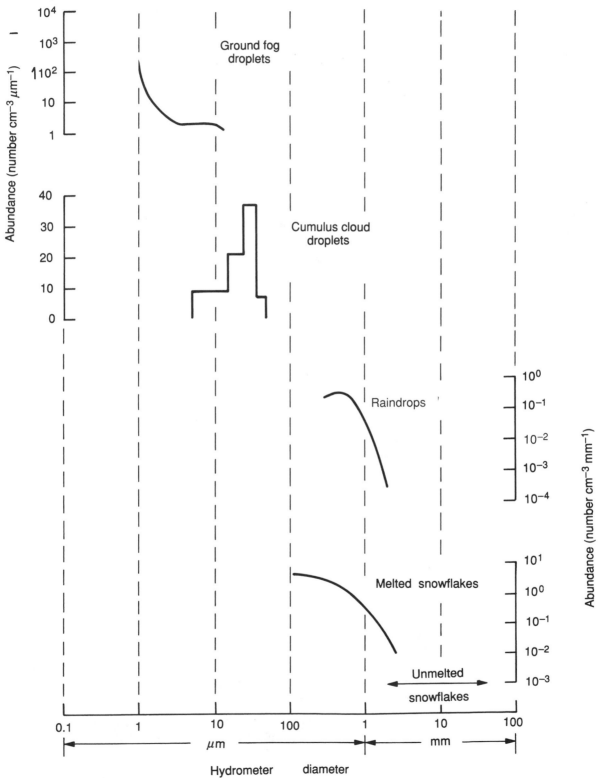

cloud droplets. Because the equilibrium vapor pressure of water over ice is substantially lower than it is over water, ice particles that form grow preferentially by deposition of water molecules from the gas phase or by transfer of water molecules from water droplets. The stage can easily be recognized in clouds by the replacement of the bright regions at the cloud tops by a diffuse, veillike boundary that is characteristic of the presence of ice crystals. When this stage in the cloud-forming process is reached, precipitation normally follows.

Atmospheric hydrometeors have a wide range of sizes. Typical hydrometeor size spectra for clouds, fog, rain, and snow are illustrated in Figure 6.10. Each type has a spread of more than an order of magnitude in size (and typically an order of magnitude in abundance as well, a variation not shown in the figure); the total size variation among the hydrometeors is four orders of magnitude. Fog droplets are very numerous but very small. Cloud droplets are fewer in number but larger. Raindrops and snowflakes are even larger and much less abundant than cloud and fog droplets.

Once fog forms near the ground or precipitation falls on the ground, the chemical constituents of the hydrometeors are transferred to the surface. The geographical variation of the occurrence of fog and of the volume of precipitation is great, as are the amounts of precipitation: the annual precipitation at different locations on the globe varies from almost nothing to more than 300 cm yr^{-1}.

The lifetimes of the hydrometeors also cover a wide range; raindrops typically live 10 minutes or so, cloud droplets a few minutes to a few hours. Given the wide spread of sizes, origins, and lifetimes for the hydrometeors, it would be surprising indeed if their chemical makeup and processes were not quite different as well. As we will see in Chapter 8, their chemical differences are at least as great as the differences already discussed.

The Water Budget

The overall assessment of the flows of water on and above Earth's surface constitute the *water budget*, which may be derived for regional, continental, or global systems. Flows for the individual continents are shown in Table 6.3. When the flows are normalized by the area to which they are delivered, the rates of precipitation are surprisingly uniform for most of the continental landmasses. (This does not mean, however, that all areas of the continents receive similar amounts of precipitation, simply that the *average* amount is similar.) The exception is South America, which receives twice the precipitation per unit area of any of the other continents.

A second interesting feature of Table 6.3 is the ratio of evaporative loss to that of gain by precipitation. The fraction is about 0.6 for Asia, Europe, and both the Americas; but it is higher for both Africa and Australia because of the large areas of desert that those continents

TABLE 6.3. Water Flow on the Continents[a]					
CONTINENT	PRECIP'N[b]	EVAP'N[b]	RUNOFF[b]	EVAP'N/ PRECIP'N[b]	NORM. PRECIP'N[c]
South America	29.4	19.0	10.4	0.6	1.6
Africa	20.8	16.6	4.2	0.8	0.7
Australia	6.4	4.4	2.0	0.7	0.8
North America	13.9	8.0	5.9	0.6	0.6
Asia	32.1	19.2	12.8	0.6	0.7
Europe	7.2	4.2	3.0	0.6	0.7

a. Adapted from M. L'Vovich and G. F. White, in *The Earth as Transformed by Human Action*, B. L. Turner and W. Meyer, eds., Cambridge University Press, Cambridge, UK, 1990.
b. 10^3 km^3 yr^{-1}.
c. m yr^{-1} over the continent.

contain and the high ambient temperatures that exist there during much of the year.

When the flows are considered on a global basis, the result is the global water budget shown in Figure 6.11. The budget contains several major pathways: precipitation, evaporation, evapotranspiration from vegetation, and vapor transport. All are highly sensitive to temperature, so they differ with geographical location, season, and climate. The oceans are primary factors in these pathways, receiving most of the precipitation and performing most of the evaporation.

One of the basic elements of the global water budget is water exchange between sea and land. As shown in Figure 6.11, more than

Figure 6.11 Earth's water budget. The units of the water flows are thousands of cubic kilometers per year. (Reproduced with permission from J.W.M. la Rivière, *Scientific American, 261* (3), 80–94. Copyright 1989 by Scientific American, Inc.)

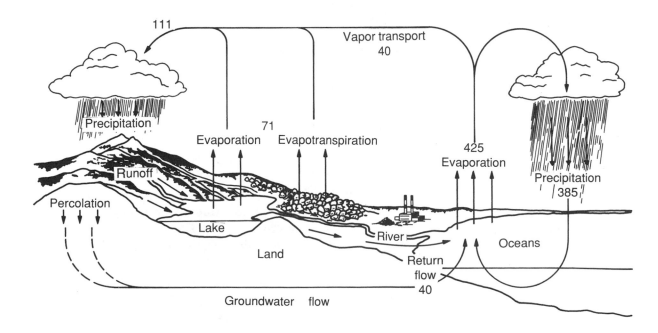

40,000 km³ of water per year are transferred through the air from the oceans to the continents. Part of this amount is retained in the form of groundwater and part is returned by runoff in the form of rivers. In the interim, it may be retained in reservoirs. Although freshwater reservoirs hold hundreds of times less water than do the oceans, much of it is at much shallower depths and its recycling over land is quite efficient. As a result, two-thirds of the precipitation falling on land is the result of evaporation from freshwater reservoirs.

The evapotranspiration process of vegetation plays a vital, if seldom appreciated, role in the water cycle over the continents and especially in the tropical forests. Figure 6.11 shows that precipitation over the continents amounts to about 111,000 km³ per year, some three times the amount of water transported from the oceans. The difference is mostly due to water supplied by evapotranspiration. The continental water supply is thus recycled about twice. For ecosystems and human habitation, these flows are crucial.

CLIMATE: THE SUM OF THE PARTS

Many of the characteristics of Earth that have been discussed in the present and preceding chapters combine to define Earth's *climate*. A simple definition of climate is that it is the long-term statistical behavior of temperature, solar radiation, cloudiness, wind flow, and precipitation in a particular geographical location or region. These variables are sometimes called the elements of climate. Climate is generally regarded in casual conversation as having seasonal or annual time scales and being some sort of average of the climate parameters over specified time periods.

Climate is influenced by the operations of five regimes. The *atmosphere* is the most important of the five, and the most rapidly varying. This envelope of air surrounding Earth responds quickly to external forces, such as the diurnal cycle of surface heating or changes in regional or global circulation. Closely linked to the atmosphere is the *biosphere*. Seasonal changes in vegetation affect the albedo of a geographical region, as well as its hydrologic cycle. Noting that humankind is also a part of the biosphere, we recognize that changes wrought by humanity such as deforestation, agriculture, and urbanization can also have profound effects on climate, locally and globally.

The third of the climate regimes, the *hydrosphere*, constitutes all the waters of Earth and tends to influence temperature and circulation on time scales of seasons to centuries. The oceans are crucial to climate because they absorb the bulk of the solar radiation that falls upon Earth; that energy vaporizes water, which moves up into the atmosphere and releases the absorbed energy as heat when the water vapor condenses into clouds. Ocean currents serve to transfer heat from the tropical regions, where the Sun is most intense, to the polar regions.

The fourth regime that influences climate is the *cryosphere*, that portion of Earth's surface with average temperatures below the freezing point of water. The bulk of the cryosphere is at or near the poles, but cryospheric regions occur atop high mountain ranges on all continents. Snow and ice are much better reflectors of solar radiation than are uncovered land and sea, and they cause a substantial decrease in surface heating. Cryospheric changes occur annually, but major variations in the cryosphere have time scales of centuries to millennia.

The fifth and slowest acting regime is the *pedosphere*, the solid portion of Earth's surface. The pedosphere rides on continental structures that evolve over time periods of millions of years as a consequence of the tectonic motions of the continents. Continents are more reflective of solar radiation than oceans, so geological periods when the continents were located primarily at high latitudes have been periods during which the planet's climate tended to be much cooler than average.

A technique used by atmospheric scientists to study the processes involved in climate is the *global climate model* (GCM), a computer representation of the forces and processes that constitute the climate system. GCMs are formidable mathematical tools, requiring the services of the largest computers and of highly competent mathematicians and computer scientists. In principle, each GCM must solve a set of equations describing the physical laws to which the atmosphere is subject. These laws fall into two categories. The first includes the basic thermodynamic and hydrodynamic laws that apply to all fluid systems. Included are expressions of the conservation of mass (matter can be neither created nor destroyed), Newton's second law of motion (momentum can be changed only by the application of a force), and the first law of thermodynamics (the internal energy of a system can be changed only by performing work on the system or by adding or removing heat).

The second category of physical laws are those that apply to the atmosphere and its environment. What is included in this category varies with the goals of the project involved, but in most cases one would include equations for the absorption and reflection of solar and infrared radiation, the evaporation and condensation of water, the transfer of heat by turbulent eddies, and the like.

The most common results of the GCMs—wind fields and temperatures, for example—are discussed in textbooks on meteorology. Climate models are also useful for anticipating changes in climate in response to changes in various forcing parameters such as the planetary heat budget or the geographical distribution of precipitation. Examples of results of this type will be presented in later chapters.

EXERCISES

6.1   A typical density for water in a dense cloud is 1 g m^{-3}. Given a cloud of this density and a volume of 0.7 km^3, compute the amount of energy released to the atmosphere when the cloud was condensed from the vapor. What happens to this energy if the cloud dissipates without producing any precipitation?

6.2   Using the cloud droplet spectrum in Figure 6.10, compute the water content (in g m^{-3}) of the cumulus cloud from which the spectrum was derived. From the same figure, extract the raindrop spectrum and assume that 10 mm of rain falls. Are the largest drops more important to the volume deposited than the smaller drops?

6.3   Given the Earth data in the appendix, convert the water fluxes in Figure 6.11 to equivalents in centimeters of precipitation or evaporation per year.

6.4   The current rate of water use by humanity amounts to some 44,000 km^3 each year (about 69% agricultural, 23% industrial, and 8% domestic). Some of this water is borrowed from surface flows and some from groundwater; only about a third is returned to the reservoirs from which it was taken. Given the present global water budget, are these numbers of concern?

6.5   Choose a geographical area no larger than 100 km by 100 km and no smaller than 10 km by 10 km. Construct for that area a water budget of the type shown in Figure 6.11. Present the sources of data. Assess the degree of uncertainty for each of the water flows in the budget. What is the impact of human water use in this area relative to the overall hydrologic budget of the area?

FURTHER READING

E. K. Berner and R. A. Berner, *The Global Water Cycle*, Prentice-Hall, Englewood Cliffs, NJ, 1987.

R. A. Freeze and J. Cherry, *Groundwater*, Prentice-Hall, Englewood Cliffs, NJ, 1979.

J. W. M. la Rivière, Threats to the world's water, *Scientific American, 261* (3), 80–94, 1989.

A. H. Perry and J. M. Walker, *The Ocean-Atmosphere System*, Longman, London, 1977.

Chemical Principles

MOLECULES, RADICALS, AND IONS

Earth is composed almost entirely of molecules, each made up of atoms bound together by chemical forces. Different combinations of atoms create molecules with widely differing reactivities. Molecules like carbon dioxide (CO_2) and dichlorodifluoromethane (CCl_2F_2) are practically unreactive with other atmospheric constituents. Molecules such as hydrogen peroxide (H_2O_2) or nitric oxide (NO), however, are quite reactive, and the hydroxyl radical (HO•) reacts with almost all atmospheric molecules.

An important example of a gas-phase reaction is

$$O_3 + NO \rightarrow O_2 + NO_2. \qquad (7.1)$$

In this reaction, as in any chemical reaction, the number of atoms of each type must be the same on each side of the equation. Reaction (7.1) is an *elementary reaction*, that is, it cannot be subdivided into several simpler reactions. There are certain rules concerning which species can react with each other, and how rapidly those reactions occur; these topics are the subject of this chapter.

Of great importance in initiating many chemical reactions are constituents known as *free radicals*. Free radicals are molecular fragments containing an odd number of electrons; thus bonding requirements are not satisfied. As a result, these fragments react vigorously in an attempt to achieve a more stable state. The formation of free radicals literally involves tearing molecules apart, a process that requires significant amounts of energy. In the atmosphere or hydrosphere, this process nearly always happens as a consequence of the absorption of solar radiation by a molecule or of a reaction with an energetic molecule or atom. For example, a molecule of hydrogen peroxide undergoes *photodissociation* (that is, fragmentation resulting from the absorption of a photon of light):

$$H_2O_2 + h\nu \ (\lambda < 550 \ nm) \rightarrow 2HO\cdot . \qquad (7.2)$$

Several characteristics of Equation (7.2) are worth noting. First, the symbol denoting a free radical is a centered dot following the chemical formula, the dot indicating that an unpaired electron is present. (This notation, however, is often not used by environmental chemists, particularly in the case of the only moderately reactive radicals NO and NO_2.) Second, the photodissociation reaction can take place only if the photon absorbed by the molecule carries enough energy to break a chemical bond that holds the molecule together. [In the H_2O_2 photodissociation reaction (7.2), the photon must have a wavelength shorter than 550 nm.] Finally, note that two free radicals are produced by the photodissociation process. This reaction fulfills the requirement that fragmentation of a molecule, not itself a radical, can only produce either two unsatisfied bonds or none, as in the reactions

$$HCHO + h\nu \ (\lambda < 330 \ nm) \rightarrow H\cdot + HCO\cdot \qquad (7.3)$$

$$HCHO + h\nu \ (\lambda < 360 \ nm) \rightarrow H_2 + CO . \qquad (7.4)$$

Once created, free radicals have a propensity to attack a large number of molecules (though not all). A typical reaction is that of the hydroxyl radical (HO•) with any organic compound containing hydrogen, such as methane:

$$CH_4 + HO\cdot \rightarrow CH_3\cdot + H_2O . \qquad (7.5)$$

Again, there are points about the reaction that are worth noting. One is that water is formed. Water is such a stable molecule that the driving force for its formation is strong, and the *abstraction* (pulling away) of a hydrogen atom from a hydrogen-containing molecule by the hydroxyl radical is energetically highly favored. A second point is that in a reaction of a molecule with a free radical, the products include another free radical (in this case, the methyl radical, $CH_3\bullet$), because no option exists in such a reaction for pairing of all the electrons involved. In turn, the $CH_3\bullet$ radical will react with molecular oxygen to create the new radical $CH_3O_2\bullet$, which reacts still further. In this way, a *chain reaction* is started, the first step being termed *initiation* and the subsequent steps *propagation*. Most of the important processes in Earth system chemistry are of this type.

Free radicals are lost only through reactions with other free radicals, a process termed *disproportionation* if two stable molecules result, or *recombination* if a single molecule is formed:

$$HO_2\bullet \ + \ HO_2\bullet \ \rightarrow \ H_2O_2 \ + \ O_2 \quad (7.6)$$

$$HO\bullet \ + \ NO_2 \ + \ M \ \rightarrow \ HNO_3 \ + \ M. \quad (7.7)$$

If either of the processes (7.6) or (7.7) is the last in a reaction chain, it can also be called *chain termination*.

Termolecular reactions like that shown in Equation (7.7) are quite important in atmospheric chemistry, because many significant atmospheric processes involve the union of two reactants. The formation of a chemical bond between the reactants releases energy that must be removed if the newly formed adduct (the combination of the two reactants) is to be stable. In the atmosphere, energy transfer is normally made to unreactive molecules, especially N_2 and O_2. M is the standard notation for any third body that removes excess energy without otherwise participating in the reaction.

Thus far, we have discussed chemical reactions taking place in the gas phase. Condensed phases are important as well. In the atmosphere, these include aerosol particles, raindrops, cloud droplets, and snowflakes. In the hydrosphere, lakes, rivers, groundwater, and the oceans are familiar examples.

The atmosphere's condensed phases often begin with cloud droplets, which form by the nucleation of water vapor on small preexisting particles known as *cloud condensation nuclei* (CCN). As the droplets grow by progressive condensation (accretion), they acquire chemical complexity because of the dissolution within them of particle constituents and of soluble atmospheric gases. Consequently, the atmosphere may be said to contain a large variety of suspended reaction vessels in which aqueous-phase chemistry can occur. In such systems, a characteristic not found in the gas phase is the tendency of many inorganic and some organic molecules to dissociate into ions. For sodium chloride, for example,

$$NaCl \ \rightleftharpoons \ Na^+ \ + \ Cl^-. \quad (7.8)$$

Note that in such an equation the sum of electrical charges on both sides of the equation must be equal, and the ionization process is realized by the transfer of an electron from the sodium atom to the chlorine atom.

If the compound undergoing dissolution is an acid, it will react with water to produce a hydrated proton (H_3O^+). [Although a proton in water exists in combination with a water molecule, this complexity is usually ignored when writing chemical reactions, as we do in this book except in Equation (7.9).] A molecule of nitric acid (HNO_3) in solution, for example, forms a hydrated proton and a nitrate ion:

$$HNO_3 + H_2O \rightleftharpoons H_3O^+ + NO_3^-. \qquad (7.9)$$

(Note that neither ion is a radical, as is the case for the uncharged species $H\cdot$ and NO_3.)

Many reactions of interest in liquid-phase environmental chemistry are acid–base reactions of the type shown in Equation (7.9). In these reactions, the *acid* is a substance that donates a proton (e.g., HNO_3) and the *base* is a substance that accepts a proton (e.g., H_2O). Consequently, by definition, proton transfers occur only in acid–base reactions.

The degree of acidity of a solution is defined by the concentration of protons present. Acidity is conveniently represented as the negative logarithm of that concentration (concentration being indicated by square brackets):

$$pH = -\log_{10}[H^+]. \qquad (7.10)$$

Ions of all kinds are relatively stable in aqueous solution because they are associated with several water molecules. These molecules move through solution with the ions and form a shield through which prospective reactants must pass. Because they have no unpaired electrons, and because of their hydration shield, ions are not particularly reactive. Free radicals can be present in solution, however, either because of solvation from the gas phase or because of a photodissociation process involving a molecule present in the solution. Free radicals in the aqueous phase, like those in the gas phase, can initiate many chemical reactions.

OXIDATION AND REDUCTION

Many chemical reactions involve the transfer of electrons from one reactant to another. An example that occurs in environmental chemistry is

$$Mn^{3+} + Fe^{2+} \rightarrow Mn^{2+} + Fe^{3+}. \qquad (7.11)$$

This reaction involves an oxidant (Mn^{3+}) that denotes an electron and a reductant (Fe^{2+}) that accepts an electron, and could also be represented as the sum of the two half-reactions

$$\text{Reduction: } Mn^{3+} + e^- \rightarrow Mn^{2+} \qquad (7.12)$$

$$\text{Oxidation: } Fe^{2+} \rightarrow Fe^{3+} + e^-. \qquad (7.13)$$

The facility with which electron transfer occurs in each reaction is expressed by a specific electrical potential, and thus each half-reaction can be related to others and to particular environments.

Just as it is convenient to characterize an aqueous solution by its acidity, a solution's *redox potential* is useful in gaining perspective on aqueous solution chemistry. By analogy to pH (a measure of the tendency of a solution to transfer protons), we can define pE (a measure of the tendency of a solution to transfer electrons) as

$$pE = -\log_{10}[e^-]. \quad (7.14)$$

The determination of pE is a routine electrochemical measurement that can be made in complex solutions such as those of waterlogged soils and sediments. If these environments contain high concentrations of one or more reducing agents, then large negative pE values result. If they contain high concentrations of one or more oxidizing agents, then large positive pE values will be measured.

CHEMICAL EQUILIBRIA

In some cases, particularly in solution, reactions do not proceed completely from reactants to products. Rather, they appear to stop while quantities of each of the constituents are still present. Such a system is said to have reached *chemical equilibrium*, a state that actually represents the point at which the rates of the forward and reverse processes are equal. The most common examples of equilibrium systems in environmental chemistry are those involving acids. For formic acid, for example, the ionization reaction is

$$HCOOH \rightarrow H^+ + COOH^- \quad (7.15)$$

and the recombination reaction is

$$H^+ + COOH^- \rightarrow HCOOH. \quad (7.16)$$

Formic acid is a *weak acid*, meaning it is only partly ionized in aqueous solution. The *equilibrium constant* for the paired reactions is defined as

$$K = \frac{[H^+][COOH^-]}{[HCOOH]}. \quad (7.17)$$

If we write a generalized equilibrium reaction as

$$\alpha A + \beta B \rightleftharpoons \gamma C + \delta D, \quad (7.18)$$

where α, β, γ, and δ are the *stoichiometric coefficients* indicating the number of molecules of each chemical species involved in the reaction, the equilibrium constant is defined by

$$K_{7.18} = \frac{[C]^\gamma [D]^\delta}{[A]^\alpha [B]^\beta}. \quad (7.19)$$

(It is conventional to write the equilibrium constant with the product concentrations, to the appropriate powers, in the numerator, and the reactant concentrations, to the appropriate powers, in the denominator.)

Because the distribution of species in the solution involves the concentrations of reactants and products, it can be perturbed by the addition of single constituents to the solution. For example, the addition of nitric acid (which almost completely ionizes to H^+ and NO_3^-) to a solution of formic acid would reduce the amount of formate ion, because the equilibrium of Reactions (7.15) and (7.16) would be shifted toward HCOOH due to the increase in the H^+ concentration. In environmental situations, nitric, sulfuric, and hydrochloric acids are essentially completely ionized (because of this property, they are referred to as *strong acids*), whereas carbonic, sulfurous, formic, and acetic acids are only slightly ionized (they are weak acids). The proportions of species in various ionization states as functions of pH for several common acids are shown in Figure 7.1. Ammonia, the most common base in atmospheric aqueous solutions, is shown as well.

Many reactions in aqueous solution can be influenced by the solution pH and thus by the acidity of any trace species in the solution. In raindrops, for example, the reaction

$$O_3 + OH^- \rightarrow HO_2{\cdot} + O_2^-{\cdot} . \qquad (7.20)$$

is directly dependent on pH, because the hydroxide ion is a reactant.

Figure 7.1 Equilibrium diagrams for several ion–molecule systems of particular interest in aqueous environ-mental chemistry.
(a) $CO_2{\cdot}H_2O$; (b) $NH_3{\cdot}H_2O$;
(c) $SO_2{\cdot}H_2O$; (d) H_2SO_4.

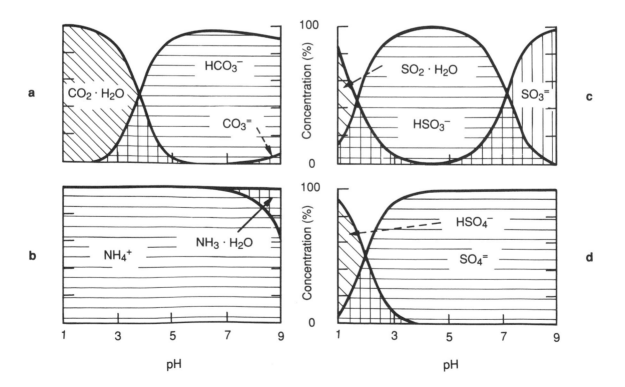

However, the reaction

$$HSO_3^- + HO\cdot \rightarrow SO_3^-\cdot + H_2O \qquad (7.21)$$

is indirectly dependent on pH, because, as shown in Figure 7.1, the solution pH controls the concentration of the HSO_3^- ion.

A final important point is the partition of a soluble or partially soluble species between the gas and aqueous phases. The relationship that defines the partitioning is known as *Henry's law* and is written in the simple case of species dissolving but not reacting in solution as

$$[C_{aq}] = H_c [C_g]. \qquad (7.22)$$

where H_c is the Henry's law constant and $[C_g]$ and $[C_{aq}]$ are the concentrations or partial pressures of species C in gas and aqueous solution, respectively. H_c may be regarded as a measure of the tendency for molecules in a gas to dissolve. The constant is temperature dependent in the sense that gases are more soluble at lower temperatures. An important demonstration of this concept is the incorporation of atmospheric CO_2 into seawater, a process whose efficiency is crucial to the atmospheric CO_2 budget. If the oceans warm, they will hold less dissolved CO_2 and will expel some to the atmosphere.

The atmosphere holds a number of examples of moderately or highly soluble species that are partitioned into atmospheric or surface water. Most gaseous compounds entering the atmosphere by natural or anthropogenic processes are rather insoluble, however, and remain predominantly in the gas phase until they are transformed into soluble compounds. The partitioning is determined principally by the affinity of the molecules with water; those molecules that dissolve readily are termed *hydrophilic*; those that dissolve with difficulty are termed *hydrophobic*.

ION PAIRING, COMPLEXATION, AND MINERALIZATION

In the usual formulation of the equations for predicting reactions, rates, and products, and in many laboratory experiments, the concentrations of reactants and other solution constituents are very dilute. At the high ionic concentrations often characteristic of environmental chemistry, however, interactions among the trace ions and with the polar water molecules (so-called *ion–ion* and *ion–solvent* interactions) play important roles. For example, oppositely charged ions undergo electrostatic attraction to form *ion pairs*, as in

$$Ca^{2+} + SO_4^{2-} \rightarrow CaSO_4. \qquad (7.23)$$

This equation has an equilibrium constant, just as did the acid-base equations above. In concentrated solutions, therefore, some fraction of the ions will be paired with ions of opposite charge while still remaining in solution.

Another aqueous process of interest is the formation of *complexes*, which are combinations of metal ions with anionic or neutral *ligands* such as sulfate, hydroxide, or ammonia. The distinction between ion pairs and complexes is that the former are created by electrostatic forces whereas the latter form by covalent bonding in which the metal atom shares a pair of electrons with the ligand. Complexes are common in environmental chemistry; for example, Fe(III) ions in solution are usually complexed to hydroxide ions (depending on the solution pH) and perhaps to sulfate ions as well.

Like the process of ion pairing, the complexation process can be assigned equilibrium (or *stability*) constants. To deduce what complexes are likely under specific conditions, consider that the metal complexes are formed in solution by stepwise reaction between metal ions (M) and ligands (L):

$$[M^{n+}] + [L] \rightleftharpoons [ML^{n+}], \qquad (7.24)$$

$$[ML^{n+}] + [L] \rightleftharpoons [ML_2^{n+}], \qquad (7.25)$$

and so forth. Stepwise stability constants can be written for each of these reactions, e.g.,

$$K_1 = \frac{[ML^{n+}]}{[M^{n+}][L]}, \qquad (7.26)$$

$$K_2 = \frac{[ML_2^{n+}]}{[M^{n+}][L]}, \qquad (7.27)$$

and so forth, higher values indicating more stable complexes. The distribution of metal ions into different complexes in concentrated solutions is given by considering all possible complexes that can form, together with their stability constants. A final equilibrium requiring discussion is that for a solid salt in equilibrium with a solution of its ions. In environmental chemistry, many of these salts have the composition of natural minerals. Solid calcite ($CaCO_3$), for example, is in equilibrium with its constituent ions in solution:

$$CaCO_{3(s)} \rightleftharpoons Ca^{2+} + CO_3^{2-}. \qquad (7.28)$$

the equilibrium constant being given by

$$K = \frac{[Ca^{2+}][CO_3^{2-}]}{[CaCO_3]_{(s)}} \qquad (7.29)$$

It is useful also to define the *ion product* as

$$IP = [Ca^{2+}][CO_3^{2-}], \qquad (7.30)$$

and the *solubility product*, which is the ion product at the point where saturation occurs:

$$K_{sp} = [Ca^{2+}]_{(sat)}[CO_3^{2-}]_{(sat)}, \qquad (7.31)$$

the saturation condition:

- If IP < K_{sp}, the solution is undersaturated with respect to the salt and the salt will dissolve in it.

- If IP = K_{sp}, the solution and solid phase are at equilibrium.

- If IP > K_{sp}, the solution is supersaturated with respect to the salt and the salt will precipitate.

Examples of each of these conditions are common in environmental chemistry and will be noted in future chapters.

CHEMICAL THERMODYNAMICS

What is the driving force for a chemical reaction? To put the question in the context of this book, why does ozone react with many other molecules in the lower atmosphere and carbon dioxide with none? The specialty that deals with such questions is *chemical thermodynamics*, a few principles of which we summarize here.

Any time a chemical reaction occurs, heat is absorbed or evolved. Most reactions of interest are exothermic, that is, they liberate heat. The heat flow Q for a chemical reaction is the difference between quantities termed the enthalpies of the reactants and products:

$$Q = \Delta H_R = \Delta H_f^\circ \text{ (products)} - \Delta H_f^\circ \text{ (reactants)} \qquad (7.32)$$

In this formula, ΔH_f°, the *enthalpy* of formation (or heat content) of a chemical compound, is the energy required to produce that compound from a set of basic components (N_2 and O_2, for example) at a given temperature and pressure. Tabulated data are generally for $T = 298.15$ K and $P = 1$ atm.

The change in enthalpy must be negative for an exothermic reaction. We can illustrate this principle for Reaction (7.5), in which the hydroxyl radical abstracts a hydrogen atom from methane. Using enthalpies of formation for each of the reactants and products involved (the numbers are available in many standard tabulations), we have

$$Q = (\Delta H_{CH_3\bullet} + \Delta H_{H_2O}) - (\Delta H_{CH_4} + \Delta H_{OH\bullet})$$

$$= (145.8 + -242.2) - (-75.0 + 39.0) \qquad (7.33)$$

$$= -60.4 \text{ kJ mol}^{-1}.$$

In this reaction the sum of the enthalpies of the products is less than that of the reactants, so heat is released by the reaction, i.e., the reaction is exothermic.

Molecules must collide if they are to have the opportunity to react. But only a small fraction of molecular collisions result in reactions, because the bonds holding the molecules together must be broken and new bonds established during the very short contact time (typically 10^{-12} s) if a reaction is to occur. As the reactants approach each other, they will proceed from a state in which they are independent entities to a temporary state in which they are chemically interacting. This *transition state* can be formed only if the reactants have sufficient energy to rearrange themselves into the transition state configuration. Figure 7.2 illustrates some of these concepts. If energy $E_a(r)$ is added to the system, as in the thermal motion of the reactants toward each other, the activation barrier may be overcome and a transition state (several different ones may be allowed) formed. If the reaction proceeds to completion, the system releases an amount of energy $E_a(p)$. The difference between the enthalpies of the reactants and products is the enthalpy change of the reaction ΔH; and $E_a(p) > E_a(r)$, if the reaction is exothermic. $E_a(r)$ is called the *activation energy* of the reaction. The reaction can be driven in the reverse direction as well as forward, but in that case the energy requirements are higher, the activation energy being $E_a(p)$, the sum of ΔH and $E_a(r)$. That reverse reaction is endo-thermic, i.e., heat must be added from an external source if the reaction is to proceed. In either case, the reactants must come together with enough kinetic, rotational, and vibrational energy to overcome the activation energy. For this reason, many bimolecular reactions are temperature dependent, in the sense that they become more probable at higher temperatures. In addition, if a molecule is

Figure 7.2 The concepts of activation energies of reactants and products and the change in enthalpy for a thermochemical reaction.

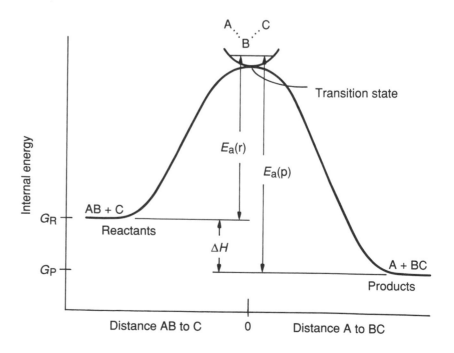

composed of both reactive and unreactive atoms or atom groups, the transition state can be achieved only if the colliding molecules are properly oriented. The *steric factor* for a specific reaction reflects this geometrical limitation on the reaction process.

Important differences exist between reaction probabilities in the gas phase and those in solution. A major one is that the diffusion rates in the two systems have very different magnitudes. In the liquid, diffusion is inhibited by the close contact of the molecules and by the attractive forces among them; each is contained within a *solvent cage* formed by loose bonding with half a dozen or so nearby water molecules. This circumstance limits opportunities for potential reactants to come into contact; but it ensures that, if they do diffuse into the same spatial region, their interaction time, and hence their probability of reaction, will be larger than in the gas phase.

SPEED OF REACTIONS

Even if a reaction of interest is thermodynamically favored, frequently the speed with which a reaction occurs is crucial to its importance, and thermodynamics says nothing about reaction speed. The science dealing with that topic is termed *chemical kinetics*. We restrict ourselves here to summarizing some basic principles of the kinetics of elementary reactions that are of special significance in atmospheric chemistry.

The speed of a chemical reaction (the *reaction rate*) is expressed as a rate of change in the concentration of a participating species with respect to time. The rate is positive if the species is created by the reaction and negative if it is being destroyed. Consider the elementary bimolecular reaction

$$A + B \rightarrow C + D. \quad (7.34)$$

For this process, the number of reactions per unit volume and unit time is given by

$$R = k[A][B], \quad (7.35)$$

where k is a constant of proportionality for a given reaction and is referred to as the *rate constant* or *rate coefficient*. Customary units for the rate are molecule cm^{-3}; thus for the bimolecular rate constants the units are cm^3 molecule^{-1} s^{-1}. The time rates of change of the participating species are then

$$\frac{d[A]}{dt} = -R; \quad \frac{d[B]}{dt} = -R;$$

$$(7.36)$$

$$\frac{d[C]}{dt} = +R; \quad \frac{d[D]}{dt} = +R.$$

As an example of the computation of the rate of a simple reaction, consider the oxidation of carbon monoxide in the atmosphere by the hydroxyl radical:

$$CO + HO\cdot \rightarrow CO_2 + H\cdot. \quad (7.37)$$

The rate of this reaction is given by

$$rate = k_{7.37}[CO][HO\cdot], \quad (7.38)$$

where the concentrations of the two reactants are known from measurements or calculated in models. The numerical value of the rate constant is determined in laboratory experiments by mixing known concentrations of the reactants and measuring their rates of decay and/or the rate of production of the products.

An example of a reaction in which the stoichiometric coefficients are not all unity (i.e., in which more than one identical molecule is involved) is Reaction (7.6), for which the rate is given by

$$rate = k_{7.6}[HO_2\cdot]^2, \quad (7.39)$$

and the species concentration changes by

$$\frac{d}{dt}[HO_2\cdot] = -2k_{7.6}[HO_2\cdot]^2 \quad (7.40)$$

and

$$\frac{d}{dt}[H_2O_2] = k_{7.6}[HO_2\cdot]^2. \quad (7.41)$$

A feature of chemical kinetics of vital importance to many considerations of atmospheric chemistry is that the thermodynamic properties of the reactants generally render reactions (and hence their rate constants) temperature dependent. The traditional form of this dependence was expressed a century ago by the Swedish chemist Svante Arrhenius as

$$k = A\,e^{-E_a(r)/kT}, \quad (7.42)$$

where $E_a(r)$ is the activation energy and A, the *preexponential factor*, is a characteristic of the reactants that reflects not only their collision rate but also certain molecular orientations that must be fulfilled for the reactants to gain access to a favorable transition state. The maximum value for A is thus given by the collision rate, which is of the order of 2–3×10^{-10} cm^3 molecule^{-1} s^{-1}. Such a high value rarely applies. For example, for the reaction between methane and the hydroxyl radical, laboratory studies give a value of 2.9×10^{-12} cm^3 molecule^{-1} s^{-1} for the preexponential factor. Because the collision rate alone would produce a rate constant about 100 times higher, the steric factor for the reaction is about 1%. A value of 1815 K has been determined for the quotient $E_a(r)/k$. We can therefore construct a table of rate constants for the atmospheric methane–hydroxyl radical reaction at the different temperatures that occur at different altitudes (Table 7.1).

TABLE 7.1. Rate Constants for the Hydroxyl Radical–Methane Reaction as a Function of Temperature				
REGION	ALTITUDE (km)	T (°C)	k^a	REACTION PROBABILITY
Boundary layer	0.1	25	6.6×10^{-15}	3×10^{-5}
Lower troposphere	1	9	4.6×10^{-15}	2×10^{-5}
Middle troposphere	6	−24	2.0×10^{-15}	1×10^{-5}
Lower stratosphere	20	−56	6.8×10^{-16}	3×10^{-6}

a. In cm^3 molecule^{-1} s^{-1}.

The reaction probability in the table is the fraction of collisions between reactants that actually result in a reaction; it is interesting that the fraction is so low. The table also shows that the rate constant differs by nearly a factor of 10 as one goes from ground level to stratospheric altitudes.

A few general rules applicable to gas-phase atmospheric reactions are worth listing here:

- Reactions are limited to those that are exothermic or very slightly endothermic.

- Reactions between two stable molecules, even if energetically favorable, tend to be very slow. (They may, however, readily take place on surfaces.)

- Reactions between free radicals and molecules are common in the atmosphere. Such reactions generally have positive activation energies and, as shown by Equation (7.42), their rates increase as temperature increases.

- Reactions between two free radicals are often very fast. Many of these reactions have negative activation energies, a finding indicating that they proceed faster at lower than at higher temperatures. At lower temperatures, the reactants have a greater probability of entering a suitable transition state if they are moving more slowly and thus are in contact longer.

- Termolecular reactions proceed faster at lower temperatures, because as the reaction partners approach each other with decreased kinetic energy, the transition state is more stable. Thus the third body M [see Reaction (7.7)] can more easily carry away the excess energy and stabilize the product.

- Radical formation and the initiation of chemical reaction chains often begin with the photodissociation of molecules by solar ultraviolet (or sometimes visible) radiation.

- Gas-phase ions are rare in the troposphere and stratosphere and do not play important chemical roles in those regions.

In solution, additional factors must be taken into account. If the reactants are ions with opposite charges, their electrostatic attraction will tend to enhance the probability of reaction. If they are similarly charged, their electrostatic repulsion will diminish it. Similar considerations come into play if the reactants are *polar molecules*, i.e., species with no net charge but with the charges distributed nonuniformly over the molecule. Many of these influences are summarized by a solution parameter known as the *ionic strength*, defined as

$$I = \frac{1}{2} \sum_i [C_i] \, Z_i^2, \qquad (7.43)$$

where C_i is the molar concentration of ion i and Z_i is the charge on the ion. For solutions at 25 °C, the relation between ionic strength and k_0, the rate constant for a reaction taking place in the solution at zero ionic strength, is given by

$$\log k = \log k_0 + m \left[\frac{I^{1/2}}{1 + I^{1/2}} \right] - 0.3I, \qquad (7.44)$$

where m is a constant whose value depends on the type of reaction and is often between 0.5 and 1.0. Equation (7.44) is derived from the theory of electrostatic interactions between ions and reflects the fact that high ionic strengths decrease the tendency for pairing of ions of opposite charge and slow the rates of reactions involving ions of opposite charges. Both effects are reversed for ions with like charge.

The rate constant adjustment of Equation (7.44) comes into play for solutions with ionic strengths greater than about 0.001. The ionic strength of rainwater is generally below this limit, but those for lake and river waters are often near the critical value; those for seawater, clouds, and fog and the water associated with atmospheric aerosol particles are generally above it. As a consequence, ionic strength considerations can be mandatory for considerations of condensed-phase chemistry.

The temperature dependence of reactions in solution is of interest in view of the wide temperature variations in aqueous solutions in the environment. For bimolecular reactions, a lower limit to the apparent activation energy is the temperature dependence of a diffusion-controlled reaction (an interaction in which every collision results in reaction), which corresponds to $E_a(r) \approx 12.8 \text{ kJ mol}^{-1}$ in the Arrhenius formulation of Equation (7.42). In the case of radical reactions, for which few experimental values are available, the activation energy can be estimated from the value of the rate constant at 298 K (often available) and a typical value for the preexponential factor in Equation (7.42) of $A = 1 \times 10^{10}$ L mol^{-1} s^{-1} (M^{-1} s^{-1}). Although rates increase as temperature increases, this effect is often more than counterbalanced by the fact that gases are less soluble in warm liquids than in cold ones.

General rules applicable to liquid-phase environmental chemistry are listed below.

- Reactions between two stable molecules in solution, even if they are energetically favorable, generally do not take place.

- Charge transfer reactions are extremely fast if they are exothermic.

- Reactions between free radicals and molecules in solution are often rapid if they are exothermic or very slightly endothermic.

- Reactions of ions in solution with other ions or with solution molecules are few and slow.

PHOTOCHEMICAL REACTION RATES

Photochemistry is the study of chemical reactions or processes initiated by the absorption of photons of radiation. In the atmosphere and in surface waters the only significant photon source is the Sun. When a photon is absorbed by a molecule, the photon energy increases the internal energy of the molecule. The energy may be used to increase the vibrational or rotational energy; normally the molecule will give up that energy eventually by collisional deexcitation to the surrounding medium. Alternatively, if the energy of the absorbed photon exceeds the bond strength of the molecule, the interaction will result in photodissociation, as was the case in Reaction (7.2). Some of the energy is then carried off by the molecular fragments as electronic, vibrational, rotational, or translational energy.

Computing Photodissociation Coefficients

Recall from Chapter 3 that the probability per second for radiation to be absorbed by a single molecule *i* is given by

$$\Psi_i = \int \sigma_i (\lambda) I(\lambda) \, d\lambda , \qquad (7.45)$$

where s$_i$ is the absorption cross section (often given in units of cm^2 molecule^{-1}), I(l)dl is the photon flux per square centimeter in the wavelength range l to l + dl, and the limits to integration are set by the wavelength limits at which molecule i absorbs radiation and the range of wavelengths emitted by the radiation source or transferred through the atmosphere. For a photodissociation reaction, only those photons with energies above the bond-breaking threshold are taken into account, and the probability for dissociation (the *photodissociation coefficient*) is given by

$$J_i = \int_{\lambda_{min}}^{\lambda_{max}} \sigma_i(\lambda)\, \phi_i(\lambda) I(\lambda)\, d\lambda , \qquad (7.46)$$

where the upper limit to integration is set by the photodissociation threshold wavelength, and the units of J_i are s^{-1}. The factor $\phi_i(\lambda)$ is the *quantum yield* of reaction, expressing the fact that the absorption of a photon may sometimes not lead to the reaction process of interest but to dissipation of the absorbed energy in some other way, such as radiation, collisional deactivation, or breaking of a different bond. If different dissociation pathways can be followed, the path weighting is given by fractional quantum yields. For example, $NO_3\cdot$ photodissociates in either of two ways:

$$NO_3\cdot \;+\; h\nu\,(\lambda < 700 \text{ nm}) \rightarrow NO + O_2 \qquad (7.47)$$

$$NO_3\cdot \;+\; h\nu\,(\lambda < 580 \text{ nm}) \rightarrow NO_2 + O . \qquad (7.48)$$

At each wavelength, there is a quantum yield appropriate to each reaction channel, and the quantum yields for all reaction and energy dissipation processes sum to unity.

Just as we can define a rate constant for a chemical reaction, so we can define a photodissociation coefficient J_i to express the probability that a molecule will participate in photochemical reactions. By analogy with reaction rates, a *photodissociation rate* is given by

$$\frac{d[C_i]}{dt} = -J_i\,[C_i] . \qquad (7.49)$$

The instantaneous rate of the photodissociation reaction (7.2), for example, is given by

$$R = J_{7.2}\,[H_2O_2] . \qquad (7.50)$$

It is important to note that the J values defined in this way are not constant but are functions of the incident radiation energy spectrum and must be recomputed, not only for each compound, but also for each new set of conditions such as a different altitude or a different time of day.

DEPOSITION TO SURFACES

For most gases, deposition to surfaces of various kinds occurs at rates comparable in significance to loss or change by chemical reaction. Like particle deposition (Chapter 5), gaseous deposition is customarily divided into two types. One is *dry deposition*, i.e., loss to a surface in the absence of active precipitation (the surface itself may be dry or wet), and *wet deposition*, in which gases are incorporated into fogs, clouds, rain, or snow and deposited as constituents of those hydrometeors.

Dry deposition results when a gaseous molecule comes into contact with a surface and is lost to it. The relationship between the concentration of a species i and the vertical flux Φ_i to a surface is expressed by the same formalism used for aerosol particle deposition:

$$\Phi_i = v_d [C_i], \quad (7.51)$$

where $[C_i]$ is the concentration of the constituent at some reference height and v_d is the deposition velocity. Much effort has been expended in attempting to predict deposition velocities on a theoretical basis. These efforts have progressed slowly, and environmental scientists generally rely on values of v_d derived from field and laboratory experiments. For gases, v_d shows a general dependence on aqueous solubility, especially when water is present on the surface of interest. Several other factors influence v_d determinations, however, including the reference height, the type of surface, and the atmospheric conditions. A selection of deposition velocities for gases is given in Table 7.2.

TABLE 7.2. Measured Deposition Velocities for Various Gases

GAS	SURFACE	v_d (cm s^{-1})
O_3	Soil, grass	0.5–1.8
SO_2	Pine forest	0.1–1.0
HNO_3	Grass	1.1–3.6

Under conditions of active precipitation, gases are incorporated into falling raindrops and lost to the ground in the precipitation. Wet deposition is an extremely efficient process and is quite effective at cleaning the air. As an example, we show in Figure 7.3 a time series of precipitation chemistry data from a single rain shower. The concentrations of four ions—H^+, NH_4^+, NO_3^-, and SO_4^{2-}—were measured as functions of time. From measurements of these ions at the ground, one cannot tell whether they came from gases or aerosol particles, although other evidence suggests that wet scavenging of both often occurs. The pattern in the figure is typical: high initial concentrations as the rainfall develops, low concentrations as heavier rain removes aerosol particles and soluble gases from the air, and high concentrations near the end of the storm as dilution of scavenged species is reduced and atmospheric concentrations recover.

The relative importance of wet and dry deposition depends markedly on local precipitation rates. During most seasons, the rates of the two processes are thought to be roughly comparable in the midlatitudes. In the humid tropics, wet deposition rates far exceed those of dry deposition. Deposition processes are crucial to the atmospheric cycles of soluble but relatively unreactive gases, of which ammonia is the best example.

Calculating Wet Deposition Fluxes of Gases

For a stagnant droplet, the flux of a gas i to the droplet surface is given by

$$\Phi_i = \frac{2\pi D_p \Delta_i}{1 + \beta \, Kn} [C_i] \qquad (7.52)$$

where

$$\beta = \gamma + \frac{4(1 - \alpha_i)}{3\alpha_i} \qquad (7.53)$$

$$\gamma = \frac{1.33 + 0.71 \, Kn^{-1}}{1 + Kn^{-1}} . \qquad (7.54)$$

Figure 7.3 A time sequence of precipitation chemistry data during a 13-hr rainstorm. The concentrations of all ions in the rainwater are seen to vary inversely with precipitation rate. (*Acidic Deposition: State of Science and Technology*, Vol. 1, National Acid Precipitation Assessment Program, Washington, DC, 1990.)

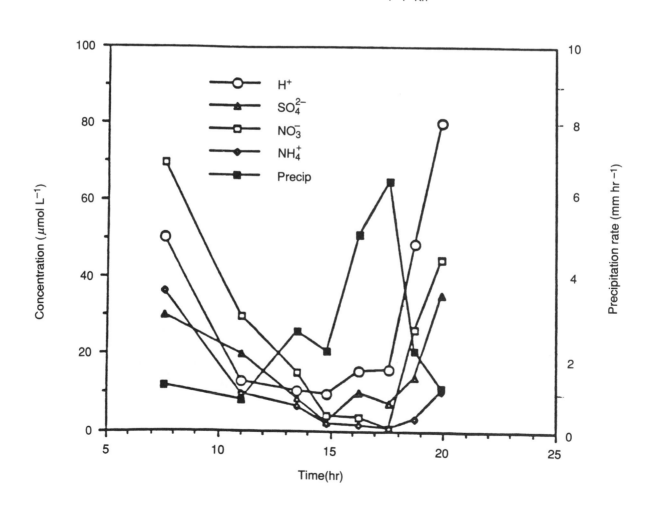

In these equations D_p is the droplet diameter, Kn is the Knudson number ($Kn = 2\lambda/D_p$), λ is the mean free path for gaseous molecules in the atmosphere, Δ_i is the gaseous diffusion constant in air, and α_i is the accommodation coefficient, i.e., the fraction of gas–droplet interactions that result in incorporation.

In the case of gaseous molecules encountering falling raindrops, the flux is increased by a ventilation coefficient f, which is a measure of the effect of turbulent fluid flow on the rate of diffusion of gas to the drop. The derivation of this term is beyond the scope of this book, but it typically takes a value of about 5. These considerations yield the rate of dissolution of gas i into a raindrop falling at terminal velocity, expressed as

$$(R_d)_i = f\,[C_i]\,\frac{2\pi D_p \Delta_i}{1 + \beta\,Kn}, \qquad (7.55)$$

where the units of $(R_d)_i$ are molecules or moles per unit time.

CHEMICAL SPECIES LIFETIMES

A useful concept for studies of environmental trace species is the species *lifetime*, which is defined by the process most effective in removing the species from the regime of interest. The most common limit is often a second-order chemical process, that is, a chemical reaction involving the species in question X and another reactant Y,

$$X + Y \xrightarrow{k} \text{products}. \qquad (7.56)$$

The lifetime of reactant X due to that process is defined as

$$\tau_X = 1/k[Y]. \qquad (7.57)$$

This example assumes that a second-order process provides the lifetime limit. It is also possible for first- or third-order processes to do so, as seen in part in Exercise 7.7.

Species X may have several possible reactions, any one of which may limit its lifetime. If so, the lifetime is computed by

$$\tau_X = \frac{1}{\sum_i k_i[Y_i]}. \qquad (7.58)$$

In some cases, especially for soluble atmospheric species that are relatively unreactive, the lifetime is set by absorption of the molecule onto atmospheric particles or Earth's surface.

EXERCISES

7.1 If rainfall has a pH of 4.0, what is the concentration of protons (H^+) in the rain?

7.2 Consider the reaction

$$HO_2\cdot\ +\ HO_2\cdot\ \rightarrow\ H_2O_2\ +\ O_2\ .$$

(a) Does the reaction as written satisfy the principle of conservation of mass? (b) Does it conserve bonding electrons? (c) What names are used by chemists to refer to this type of reaction?

7.3 At equilibrium, dissolved carbon dioxide from the atmosphere is present in rain at a concentration of about 2×10^{-5} M. The equilibrium constants for the acid dissociations of carbonic acid are $k_1 = 4.3 \times 10^{-7}$ M and $k_2 = 4.5 \times 10^{-11}$ M. Compute the concentrations of H^+, HCO_3^-, and CO_3^{2-} in the rain if no other ions are present.

7.4 Propene reacts in the atmosphere with both ozone and the hydroxyl radical. The rate constant for the reaction with ozone is 1.9×10^{-18} cm^3 $molecule^{-1}$ s^{-1}, and that for reaction with the hydroxyl radical is 2.9×10^{-11} cm^3 $molecule^{-1}$ s^{-1}. If the concentrations of propene, ozone, and the hydroxyl radical are 1 ppbv, 80 ppbv, and 8×10^5 radicals cm^{-3}, compute the rates of the reactions of propene with ozone and with the hydroxyl radical and state which reaction will cause the greatest rate of loss of propene.

7.5 The Henry's law constants for ethylene and hydrogen peroxide are 8.9×10^{-3} M atm^{-1} and 9.4×10^4 M atm^{-1}, respectively. If each molecule is present in the gas phase at a concentration of 5 ppbv, compute the liquid-phase concentrations.

7.6 What is the rate of dissociation of formaldehyde by the $HCO\cdot\ +\ H\cdot$ channel for the conditions given in Exercise 3.4?

7.7 Sulfur dioxide gas is present in an urban area at a concentration of 5 ppbv. Sulfur is also present on small particles having a total sulfate mass of 1.5 μg m^{-3} and a deposition velocity of 0.05 cm s^{-1}. If the v_d for SO_2 is 1.0 cm s^{-1}, does dry deposition of particles or SO_2 gas transfer more sulfur to ground surfaces?

7.8 Hydrogen peroxide gas in the atmosphere has many potential loss processes. It may photodissociate [Reaction (7.2)], it may react with the hydroxyl radical,

$$H_2O_2\ +\ HO\cdot\ \rightarrow\ HO_2\cdot\ +\ H_2O$$

with a rate constant of $k = 1.7 \times 10^{-12}$ molecule cm^{-3} s^{-1}, or it may be lost by adsorption onto surfaces. Because photodissociation and surface loss do not involve reaction with other molecules, they are examples of *first-order* loss processes. Photoprocess data are given below, and surface loss of H_2O_2 has been estimated to occur at midlatitudes with a deposition velocity of 0.5 cm s^{-1}, referenced to a fully mixed boundary layer 1 km thick. If the hydrogen peroxide concentration is 5 ppbv, the hydroxyl radical concentration is 2×10^6 radicals cm^{-3}, and the quantum yield for the photoprocess over the wavelength range is 1.0, determine whether the H_2O_2 lifetime is limited by photodissociation, hydroxyl radical reaction, or surface loss. How does this result change if the H_2O_2 concentration is only 1 ppbv?

Photoprocess Data for Hydrogen Peroxide		
WAVELENGTH[a]	$I(\lambda)$[b]	σ[c]
295–305	2.66×10^{13}	0.71×10^{-20}
305–315	4.20×10^{14}	0.42×10^{-20}
315–325	1.04×10^{15}	0.25×10^{-20}
325–335	1.77×10^{15}	0.14×10^{-20}
335–345	1.89×10^{15}	0.08×10^{-20}
345–355	2.09×10^{15}	0.05×10^{-20}

a. In nm.
b. In photon cm^{-2} s^{-1} $(10\ nm)^{-1}$.
c. In cm^2 $molecule^{-1}$.

FURTHER READING

B. J. Finlayson-Pitts and J. N. Pitts, Jr., *Atmospheric Chemistry: Fundamentals and Experimental Techniques*, John Wiley, New York, 1098 pp., 1986.

R. M. Harrison, S. J. deMora, S. Rapsomanikis, and W. R. Johnston, *Introductory Chemistry for the Environmental Sciences*, Cambridge University Press, Cambridge, UK, 1991.

B. B. Hicks, Measuring dry deposition: A reassessment of the state of the art, *Water, Air, and Soil Pollution, 30*, 75–90, 1986.

C H A P T E R 8

The Chemistry of the Atmosphere

ATMOSPHERIC SPECIES: DIVERSITY AND ABUNDANCE

At the time of the Second World War, barely two dozen different chemical species were known to be present in Earth's atmosphere. A decade later, as scientists began investigating the chemical and biological implications of atmospheric composition, that number was still less than 100. Today nearly 3000 species have been identified. Some are found in the atmospheric gas, some in aerosol particles, large and small, and some in hydrometeors (rain, snow, fog, clouds, and dew). A number of them appear in several of these reservoirs. Many of the compounds have both natural and anthropogenic sources and thus were present in the preindustrial atmosphere as well as that of today.

Of the atmosphere's reactive compounds, methane is by far the most abundant, at a northern hemisphere concentration of about 1.7 ppmv. When all reactive molecules are considered, tropospheric air of minimum chemical complexity contains more than four times as many different reactive organic gases as inorganic ones. The organic species are usually emitted to the atmosphere as pure hydrocarbons, such as butane, propane, and a very great variety of solvents, industrial chemicals, and combustion products. Their reactions in the atmosphere lead to partly oxygenated compounds, which can be more readily removed by precipitation or taken up at Earth's surface than can their precursor molecules.

Tropospheric aerosol particles of minimum complexity contain about a dozen major constituents, mostly the common inorganic ions and inorganic oxides found in Earth's crust. Carbon in the form of soot is frequently present in urban areas as a product of combustion. Organic material is ubiquitous and chemically diverse.

In atmospheric water droplets, the ionic species present are similar to those found in aerosol particles, the inorganic anions nitrate and sulfate being most important. The aqueous droplets are nearly always acidic. It is likely that soil-derived oxides are universal in precipitation, but the data are still too sparse for certainty. Hydrated formaldehyde and formic acid appear always to be present.

The stratosphere is chemically simpler than the troposphere but still possesses much chemical diversity. Several tens to a hundred different species are always present, including a number of free radicals, thus creating the conditions for a rich and vigorous chemistry. Much is still unknown about the stratospheric aerosol, because it can only be studied by the difficult experimental feat of capturing individual particles in situ and returning them to Earth for analysis, but oxides of aluminum and iron, sodium and potassium from meteoric infall, sulfate from oxidized COS or SO_2 of volcanic origin, and liquid or frozen water appear always to be present, as well as several of the more common inorganic ions.

Because of the great variability of the atmosphere, there is no such thing as a typical trace species composition. Nonetheless, of the several thousand trace constituents that have been detected in tropospheric air, about 170 distinct interacting chemical species, many of them of natural origin, are commonly present. The stratosphere is less complex but is nonetheless known to commonly contain at least 40 different species. Atmospheric chemistry is thus constantly performing a myriad of chemical reaction experiments in an uncontrolled and moving "reaction vessel."

Catalytic Ozone Chemistry

Ozone is the central species in stratospheric chemistry. Nowhere are there more than about 10 ozone molecules per million molecules of air. It is present, in fact, in such minute quantities that all the ozone in the atmosphere would only make up a layer about 3 mm thick if it could be segregated under standard temperature and pressure conditions at Earth's surface. Nonetheless, ozone's interactions with many stratospheric constituents and its strong absorption of biologically harmful ultraviolet radiation ensure its central role. As we will see, it is also of fundamental importance to tropospheric chemistry.

Ozone in the stratosphere is produced by the photodissociation of molecular oxygen by solar ultraviolet radiation. This process produces two oxygen atoms, each of which combines with molecular oxygen to yield ozone:

$$O_2 + h\nu\,(\lambda \leq 242\,\text{nm}) \rightarrow O + O \qquad (8.1)$$

$$2[O + O_2 + M \rightarrow O_3 + M] \qquad (8.2)$$

$$\text{Net:}\ 3O_2 + h\nu \rightarrow 2O_3\,. \qquad (8.2a)$$

Because most oxygen in the atmosphere is present as O_2, it is clear that there must also be processes that reconvert O_3 into O_2. The initial proposal by pioneer atmospheric scientist Sidney Chapman of Oxford University in 1930 was that this recycling process mainly occurred through the reaction sequence:

$$O_3 + h\nu\,(\lambda \leq 1140\,\text{nm}) \rightarrow O + O_2 \qquad (8.3)$$

$$O + O_3 \rightarrow 2O_2 \qquad (8.4)$$

$$\text{Net:}\ 2O_3 + h\nu \rightarrow 3O_2\,. \qquad (8.4a)$$

For about 40 years, it was generally accepted that this sequence explained the cycle of stratospheric ozone. However, research over the past two decades has shown that several minor constituents play an essential role in stratospheric ozone destruction due to catalytic reaction cycles that may be summarized as

$$X\cdot + O_3 \rightarrow XO\cdot + O_2 \qquad (8.5)$$

$$O_3 + h\nu \rightarrow O + O_2 \qquad (8.5a)$$

$$O + XO\cdot \rightarrow X\cdot + O_2 \qquad (8.6)$$

$$\text{Net:}\ 2O_3 + h\nu \rightarrow 3O_2\,. \qquad (8.6a)$$

In this reaction set, $X\cdot$ and $XO\cdot$ are radicals that catalyze the conversion of O_3 to O_2. As proposed by Paul Crutzen in 1970, in the stratosphere the most important catalysts are the oxides of nitrogen

NO and NO_2 (X = NO), present as a consequence of ground-level emissions of N_2O. Nitrous oxide emissions are largely from microbiological processes in soils, especially from soils that have been heavily fertilized. N_2O is quite unreactive in the troposphere, with a lifetime between 150 and 200 years. In the stratosphere, it is broken down by several processes, such as reaction with the energetic oxygen atoms [in an electronic energy state denoted as $O(^1D)$] produced by solar ultraviolet radiation:

$$N_2O + O(^1D) \rightarrow 2\,NO . \qquad (8.7)$$

This process leads to NO_x production that allows the catalytic chain of Reactions (8.5) to (8.6) to be initiated. NO can also be directly injected into the stratosphere by aircraft emissions, by nuclear explosions, or by solar proton bombardment. Although NO production by these latter means is much smaller than that in the troposphere by soil microbial activity, fossil fuel and biomass burning, and lightning, the oxidation of NO to soluble HNO_3 strongly limits the likelihood of its transfer to the stratosphere. Limitations of this type on the transfer of reactive tropospheric species to the stratosphere are an important and characteristic atmospheric property.

A second natural chain involves HO• and HO_2• [HO in this case is the X of Equations (8.5) to (8.6).] In the stratosphere, these species arise primarily as a consequence of the reactions of water vapor and other species with excited oxygen atoms:

$$H_2O + O(^1D) \rightarrow 2\,HO• . \qquad (8.8)$$

Despite the very low water vapor content of the stratosphere, enough of these radical species are produced to be of substantial chemical importance.

Another catalytic cycle for ozone destruction in the stratosphere is that involving chlorine. It was proposed in 1974 by Richard Stolarski and Ralph Cicerone, then of the University of Michigan, and linked to chlorofluorocarbons (CFCs) in that same year by Mario Molina and Sherwood Rowland, then of the University of California at Irvine. The CFCs are not broken down in the troposphere by reactions with the hydroxyl radical or any other potential oxidizing constituent. Of particular importance for the stratospheric ozone layer are the chemically very stable gases $CFCl_3$ and CF_2Cl_2. Over a period of a few years, these gases move from their release points at the surface up into the stratosphere. Above about 20–25 km, the available solar radiation is energetic enough to destroy them, thereby releasing chlorine atoms and chlorine monoxide molecules that are even more powerful ozone-destroying catalysts than are NO and NO_2 on a molecule-for-molecule basis. The relevant reactions (illustrated for $CFCl_3$ and involving several steps) are

$$CFCl_3 + h\nu\,(\lambda < 260\,\text{nm}) \xrightarrow{nO_2} CO_2 + HF + 3\,(Cl• \text{ or } ClO•) , \qquad (8.9)$$

followed by the catalytic ozone destruction cycle:

$$Cl\cdot + O_3 \rightarrow ClO\cdot + O_2 \quad (8.10)$$

$$O_3 + h\nu\,(\lambda \leq 1140\,nm) \rightarrow O + O_2 \quad (8.10a)$$

$$ClO\cdot + O \rightarrow Cl\cdot + O_2 \quad (8.11)$$

$$Net:\ 2\,O_3 + h\nu \rightarrow 3\,O_2, \quad (8.11a)$$

where Cl• and ClO• are the X• and XO• of Reactions (8.5) and (8.6). Chlorine is also supplied to the stratosphere by occasional volcanic eruptions and by tropospheric to stratospheric transport of methyl chloride (CH_3Cl), a product of seaweed. The natural chlorine content of the stratosphere as a consequence of these sources is about 0.6 ppbv. The quantity currently supplied by CFCs and other industrial organic chlorine compounds is about 3 ppbv, some five times the natural background.

An additional chlorine destruction chain for stratospheric ozone, originally proposed by Mario and Luisa Molina of the Massachusetts Institute of Technology, is as follows:

$$2[Cl\cdot + O_3 \rightarrow ClO\cdot + O_2] \quad (8.11b)$$

$$ClO\cdot + ClO\cdot + M \rightarrow Cl_2O_2 + M \quad (8.12)$$

$$Cl_2O_2 + h\nu\,(\lambda \leq 400\,nm) \rightarrow 2\,Cl\cdot + O_2 \quad (8.13)$$

$$Net:\ 2\,O_3 + h\nu \rightarrow 3\,O_2. \quad (8.13a)$$

A natural constraint to stratospheric ozone depletion is provided by the reactions between the catalysts, as in

$$HO\cdot + NO_2 + M \rightarrow HNO_3 + M \quad (8.14)$$

$$ClO\cdot + NO_2 + M \rightarrow ClONO_2 + M \quad (8.15)$$

leading to products that do not destroy ozone. In this way, two "ozone killers" neutralize each other. The reservoirs for the reactive radicals are temporary, however, because the molecules can photodissociate:

$$HNO_3 + h\nu\,(\lambda \leq 330\,nm) \rightarrow HO\cdot + NO_2 \quad (8.16)$$

$$ClONO_2 + h\nu\,(\lambda \leq 450\,nm) \rightarrow ClO\cdot + NO_2. \quad (8.17)$$

Computer model calculations incorporating these reactions indicate that the steady increase of $CFCl_3$, CF_2Cl_2, and other CFCs leads to substantial reductions in ozone concentrations in the stratosphere, mainly above about 25 km.

The three catalytic cycles that dominate stratospheric ozone chemistry are shown with their linkages, sources, and sinks in Figure 8.1. The NX, HX, and ClX boxes summarize several of the stratosphere's catalytic cycles leading to the destruction of ozone. Some of the more important reactions connecting the cycles are also shown, as are some of the source and sink reactions for each cycle. A full understanding of stratospheric chemistry involves the simultaneous consideration of all of these reactions.

Polar Ozone Chemistry

Among the more dramatic of the recent discoveries concerning atmospheric chemistry has been the Antarctic "ozone hole," a rapid and accelerating decrease in the ozone over Antarctica each September and October. The decrease in total ozone content was notable during the 1980s, as shown in Figure 8.2 (in the color insert). Attention was immediately focused on the possibility of an enhanced chlorine destruction cycle, a proposal that received eventual experimental confirmation from data such as those shown in Figure 8.3. This figure shows the development of the ozone hole polewards of 70°S at about 16 km altitude between 30 August (a) and 21 September 1987 (b). Although in the cold regions south of 70°S high volume ratios of ClO• are already established by the end of August, the photochemical breakdown of ozone, which occurs mostly during daytime, has not yet begun. A few weeks later, about half the ozone has been destroyed.

The vertical distribution of ozone over Antarctica in 1987 was measured by balloon flights from McMurdo Station (78°S). The

Figure 8.1 The catalytic cycles of stratospheric ozone chemistry. PSC, polar stratospheric clouds. Contributions from the ionosphere, a highly ionized section of the mesosphere, are small, but are included here for completeness.

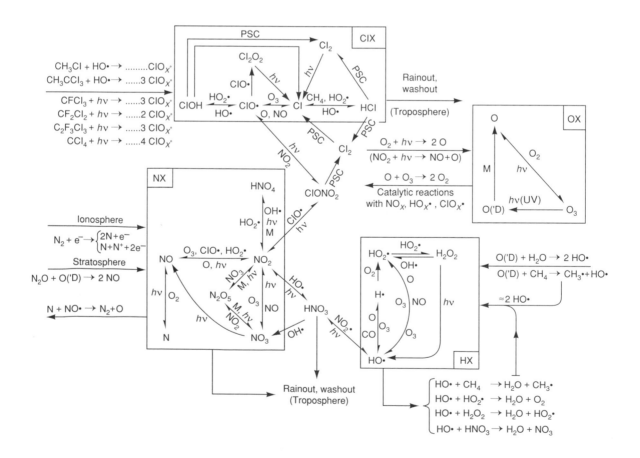

Figure 8.3 Concentrations of ozone (dotted lines) and the chlorine monoxide radical (solid lines), as measured on ER-2 aircraft flights from South America to Antarctica. (a) 30 August 1987; (b) 21 September 1987. (J. G. Anderson, W. H. Brune, and M. H. Proffitt, Ozone destruction by chlorine radicals within the Antarctic vortex. The spatial and temporal evolution of ClO–O$_3$ anticorrelation based on in situ ER-2 data, *Journal of Geophysical Research, 94,* 11,465–11,479, 1989.) Ozone concentration profiles (c) and temperature profiles (d) over McMurdo Station near the time of the ER-2 flights. (D. J. Hofmann, J. W. Harder, J. M. Rosen, J. V. Hereford, and J. R. Carpenter, Ozone profile measurements at McMurdo Station, Antarctica, during the spring of 1987, *Journal of Geophysical Research, 94,* 16,527–16,536, 1989.)

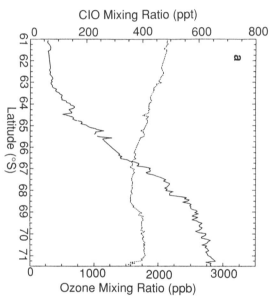

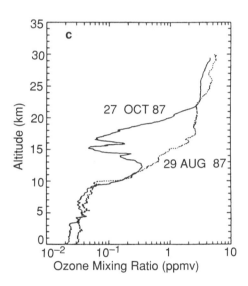

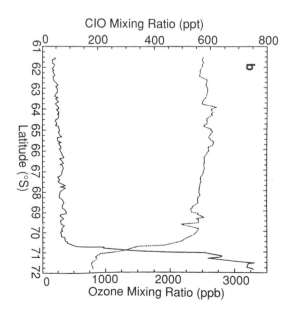

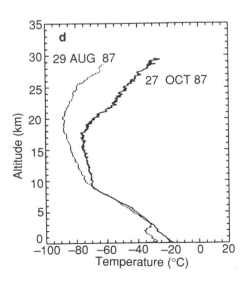

pattern for late August (Figure 8.3c) represented a total column ozone concentration of 274 Dobson Units. By late October, the striking losses in the 12–22 km region had reduced that concentration to 148. The warming of the stratosphere that occurs during the austral spring is shown in Figure 8.3d.

The ozone hole mechanism was eventually shown to be caused by chemical reactions occurring on frozen particles consisting of mixtures of nitric acid and water molecules. The formation and freezing out of HNO_3 implies a loss of NO and NO_2 from the stratosphere under cold conditions. Furthermore, the surfaces of the polar stratospheric clouds serve as reaction sites (see Figure 8.1). Two common catalyst sequestration molecules are hydrogen chloride and chlorine nitrate. In principle, these species can react with each other as follows:

$$ClONO_2 + HCl \rightarrow Cl_2 + HNO_3 , \qquad (8.18)$$

with chlorine radicals being subsequently produced when the Cl_2 molecules are photodissociated in sunlight:

$$Cl_2 + h\nu \, (\lambda \leq 450 \, nm) \rightarrow 2Cl\cdot . \qquad (8.19)$$

Reaction (8.18) in the gas phase turns out to be so slow that it is unimportant. Both chlorine nitrate and hydrogen chloride readily attach to the ice surface and react with each other, however, and in that medium Reaction (8.18) proceeds very much faster than in the gas phase. Once the products of the reaction are formed, the molecular chlorine vaporizes into the surrounding air, whereas the nitric acid is strongly bound into the ice matrix. The reaction may thus be written

$$ClONO_2 + HCl \xrightarrow{ice} Cl_2\uparrow + HNO_3(ice) . \qquad (8.20)$$

As a consequence of Reaction (8.20), the nitrogen dioxide that might otherwise be regenerated from the chlorine nitrate and be capable of interfering with the ozone loss cycle is not available to do so. Ozone is thus unprotected against destruction by the powerful Cl• and ClO• radicals.

In its polar location, Antarctica receives no solar radiation for several months each year and stratospheric temperatures cool to −80 °C or colder, sufficient to form the nitric acid–ice or pure ice particles and condense the sequestering species on them. As the Sun makes its appearance in late September, the chlorine photodissociation of Reaction (8.19) begins and ozone depletion occurs. The situation is exacerbated by the atmospheric motions, which over Antarctica in the winter are dominated by an almost circumpolar circulation system, the polar vortex, that retains the chlorine oxide-rich air over Antarctica for many months.

Solar Proton Effects on Stratospheric Chemistry

From time to time, large fluxes of protons and smaller quantities of helium and other heavy nuclei are emitted in bursts from the Sun. The main societal interest in solar proton events is that the ionization they cause leads to the disturbance of radio wave propagation in the upper atmosphere and thus to communications interruptions. While the protons and other particles are too energetic to be efficient in dissociating nitrogen, large numbers of energetic secondary electrons are produced. Collisions of these electrons with N_2 produce ions, atoms, and electrons, e.g.,

$$N_2 + e^- \rightarrow N^+ + N + 2e^-. \qquad (8.21)$$

Subsequent reactions lead to the generation of NO via, e.g.,

$$N^+ + O_2 \rightarrow N + O_2^+ \qquad (8.22)$$

$$NO^+ + e^- \rightarrow N + O \qquad (8.23)$$

$$N + O_2 \rightarrow NO + O. \qquad (8.24)$$

As the solar protons are guided by Earth's magnetic field into the polar caps at geomagnetic latitudes greater than 65°, the NO production is expected to occur in high latitude regions during the days following a solar proton event.

With the availability of satellite instrumentation, the fluxes of energetic protons have been monitored for the past 30 years. One recorded solar proton event, the largest to have been monitored by satellite, occurred intermittently during the period 2–10 August 1972. Computer models of the stratospheric chemistry that resulted

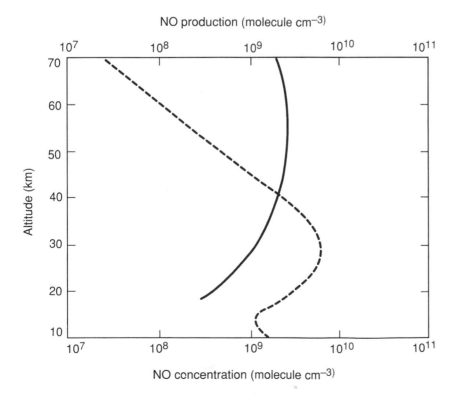

Figure 8.4 The background concentration of NO in the absence of solar proton events (dashed line) and the production of NO at high geomagnetic latitudes (>55°) as a function of altitude, computed for the solar proton event of August 1972 (solid line).

indicated that NO production should have been substantial, especially above 25 km, as shown in Figure 8.4. As a consequence, the stratospheric ozone concentrations should have been markedly reduced.

More than a year after the model results were published, satellite observations of ozone concentrations during the period of the solar proton event were analyzed in detail. As shown in Figure 8.5, the data demonstrated a strong decrease in stratospheric ozone concentrations, in reasonable agreement with that predicted by the model. (In fact, later corrections to the measured profiles made the correspondence even better.) In a very convincing way, the correctness of the theory that NO_x catalytic reactions could destroy ozone, which was still only a hypothesis in the 1970s, was confirmed in an impressive way, and stratospheric ozone models had been tested by a completely independent technique.

Figure 8.5 (a) Zonally averaged total ozone above 40 km at high northern latitudes dur-ing July and August 1972. (b) The percentage decrease of ozone as a function of altitude. The reference profile (not shown) is that of the seven days that preceded the main solar proton flux on August 4. Data are shown for two 7-day periods centered 8 and 19 days after the event. The dashed line is a model calculation of the ozone reduction for 1 September 1972 resulting from the catalytic effect of solar proton-produced NO.

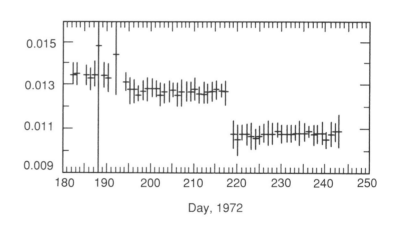

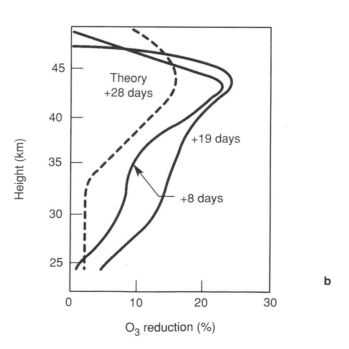

Urban Smog Chemistry

It is clear that for most of the gases emitted at Earth's surface by natural and anthropogenic processes there must exist removal mechanisms of comparable magnitude, because the gases are not observed to accumulate at rates proportional to their rates of emission. Only in the last two decades have atmospheric scientists had a fundamental understanding of the processes involved in this removal. As will be seen, most of these processes are initiated by reactions with the hydroxyl radical, which itself is formed by the action of solar ultraviolet radiation on ozone and water vapor.

In the early part of this century, ground-based measurements and in situ balloon-based observations made it apparent that most of the atmosphere's ozone is located in the stratosphere, the peak concentration occurring at altitudes between 15 and 30 km. For a long time, it was believed that tropospheric ozone originated from the stratosphere and that most of it was destroyed by contact with Earth's surface. Ozone was known to be produced by the photodissociation of O_2, a process that can only occur at wavelengths shorter than 242 nm. Because such short-wavelength radiation is present only in the stratosphere, no tropospheric ozone production is possible by this mechanism. In the mid-1940s, however, it became obvious that production of ozone was also taking place in the troposphere. After heavy injury to vegetable crops had occurred repeatedly in the Los Angeles area, it was shown that plant damage could be produced by ozone, which was known to be a prominent constituent of photochemical smog. The overall reaction mechanism was eventually identified by Arie Haagen-Smit of the California Institute of Technology as

$$NMHC + NO + h\nu \rightarrow NO_2 + \text{other products} , \qquad (8.25)$$

where NMHC denotes various reactive nonmethane hydrocarbons (ethylene, butane, etc.), the catalyst is NO_x ($NO + NO_2$), and $h\nu$ indicates a quantum of solar radiation of wavelength less than about 410 nm. Ozone formation by this mechanism is possible because solar radiation dissociates the NO_2 formed in Reaction (8.25):

$$NO_2 + h\nu \, (\lambda \leq 410 \text{ nm}) \rightarrow NO + O , \qquad (8.26)$$

and the recombination of O with molecular oxygen then produces ozone:

$$O + O_2 + M \rightarrow O_3 + M . \qquad (8.26a)$$

Reactions (8.26) and (8.2) are two reactions in a chemical triad that links NO_x and O_3. The third reaction of the group is

$$NO + O_3 \rightarrow NO_2 + O_2 . \qquad (8.27)$$

These three rapid reactions establish a *photostationary state* among the concentrations of the principal reactants, often written as

$$\frac{[NO_2]}{[NO]} = \frac{k_{8.27}[O_3]}{J_{8.26}} . \qquad (8.28)$$

However, the cyclic nature of Reactions (8.26), (8.26a), and (8.27) does not result in the net production of ozone (as can be seen by adding the reactants and products of the three reactions). Ozone production is only possible if, as occurs in Reaction (8.25), NO is converted to NO_2 by reacting with gases other than O_3 itself. Some examples of how this occurs in the atmosphere are shown below.

In 1971, a fundamental aspect of the atmosphere's reactivity was discovered by Hiram Levy, then of Harvard University, who pointed out that ozone photolysis by solar radiation at wavelengths shorter than about 310 nm leads to the production of the hydroxyl radical through the reactions:

$$O_3 + h\nu \, (\lambda \leq 310 \, nm) \rightarrow O(^1D) + O_2 \qquad (8.29)$$

and

$$O(^1D) + H_2O \rightarrow 2 HO\cdot . \qquad (8.30)$$

Levy's observation was crucial, because it was known that the hydroxyl radical was capable of reacting with a very large number of molecules. The exact role of hydrocarbons in Reaction (8.25) was then revealed to be

$$RH + HO\cdot \rightarrow R\cdot + H_2O , \qquad (8.31)$$

where R is a chemist's notation for any organic fragment consisting solely of carbon and hydrogen, such as C_2H_5 or C_3H_7. This reaction is followed by

$$R\cdot + O_2 + M \rightarrow RO_2\cdot + M \qquad (8.32)$$

and

$$RO_2\cdot + NO \rightarrow NO_2 + RO\cdot . \qquad (8.33)$$

This bypassing of the ozone consumption reaction (8.27) enables nitric oxide to be oxidized to nitrogen dioxide without destroying ozone, the result being sharply enhanced ozone concentrations when suitable amounts of both NO_x and NMHC are present.

The Crucial Role NO_x Concentrations

Throughout the 1950s and 1960s, it was thought that significant ozone formation through the mechanism described above could only take place in atmospheric environments that were heavily contaminated with automobile exhaust emissions and strongly illuminated by sunlight under stable meteorological conditions. It was only in the 1970s that the occurrence of photochemical smog was also discovered in Europe. At about the same time, the first suggestions were made that significant in situ photochemical production and destruction of tropospheric ozone could also take place on global scales.

In 1973 Paul Crutzen, then of the University of Stockholm, showed how ozone can be produced by methane and/or carbon monoxide oxidation cycles, depending on the presence of NO and NO_2 as catalysts. This sequence of reactions begins with the attack of the hydroxyl radical on methane:

$$CH_4 + HO\cdot \rightarrow CH_3\cdot + H_2O, \qquad (8.34)$$

followed by the reaction sequence

$$CH_3\cdot + O_2 \rightarrow CH_3O_2\cdot \qquad (8.35)$$

$$CH_3O_2\cdot + NO \rightarrow CH_3O\cdot + NO_2 \qquad (8.36)$$

$$CH_3O\cdot + O_2 \rightarrow HCHO + HO_2\cdot \qquad (8.37)$$

$$HCHO + h\nu\,(\lambda \leq 330\,nm) \rightarrow HCO\cdot + H\cdot \qquad (8.38)$$

$$HCO\cdot + O_2 \rightarrow CO + HO_2\cdot \qquad (8.39)$$

$$H\cdot + O_2 + M \rightarrow HO_2\cdot + M \qquad (8.40)$$

$$3[HO_2\cdot + NO \rightarrow HO\cdot + NO_2] \qquad (8.41)$$

$$4[NO_2 + h\nu\,(\lambda \leq 410\,nm) \rightarrow NO + O] \qquad (8.41a)$$

$$3[O + O_2 + M \rightarrow O_3 + M] \qquad (8.41b)$$

$$\text{Net:}\ \ CH_4 + 8O_2 + 5\,h\nu \rightarrow CO + 4O_3 + 2HO\cdot + H_2O, \qquad (8.41c)$$

leading to a net production through formaldehyde (HCHO) of four ozone molecules and two hydroxyl radicals for each CH_4 molecule that is oxidized to CO.

The carbon monoxide produced through the methane reaction chain or emitted directly from fossil fuel combustion can then itself interact with the hydroxyl radical. The dependence of this oxidation cycle and its net result on the local abundance of NO_x can be shown as follows. To begin, note that the carbon monoxide oxidation can proceed either through reactions that produce ozone in the presence of the catalyst molecules NO and NO_2:

$$CO + HO\cdot \rightarrow H\cdot + CO_2 \qquad (8.42)$$

$$H\cdot + O_2 \rightarrow HO_2\cdot + M. \qquad (8.42a)$$

$$HO_2\cdot + NO \rightarrow HO\cdot + NO_2 \qquad (8.43)$$

$$NO_2 + h\nu\,(\lambda \leq 410\,nm) \rightarrow NO + O \qquad (8.43a)$$

$$O + O_2 + M \rightarrow O_3 + M \qquad (8.43b)$$

$$\text{Net:}\ \ CO + 2O_2 + h\nu \rightarrow CO_2 + O_3, \qquad (8.43c)$$

or, if NO_x concentrations are low, through reactions leading to ozone destruction:

$$CO + HO\cdot \rightarrow H\cdot + CO_2 \quad (8.43d)$$

$$H\cdot + O_2 + M \rightarrow HO_2\cdot + M \quad (8.43e)$$

$$HO_2\cdot + O_3 \rightarrow HO\cdot + 2O_2 \quad (8.44)$$

$$\text{Net: } CO + O_3 \rightarrow CO_2 + O_2. \quad (8.44a)$$

In either case, no net destruction of radicals occurs. Because the rate coefficient for Reaction (8.43) is about 4000 times larger than that for Reaction (8.44), the ozone-producing carbon monoxide oxidation branch is more important than the ozone destruction branch when NO to O_3 concentration ratios exceed 1:4000, i.e., for NO volume mixing ratios larger than about $5-10 \times 10^{-12}$ (5–10 pptv) in the lower troposphere.

Similar but more complex reaction cycles also take place during the atmospheric oxidation of the hydrocarbon gases, with methane being the most important example in extensive portions of the atmosphere, especially in remote marine environments. Methane is so abundant, in fact, that its oxidation, in addition to that of CO, plays a large role in the tropospheric ozone and hydroxyl radical balances. Again the availability of NO determines the results. To briefly summarize the detailed reaction sequences, the oxidation of one methane molecule to carbon monoxide and then to carbon dioxide yields (depending on the various species abundances):

- In NO-poor environments: a net loss of 3–4 odd hydrogen ($HO\cdot$ and $HO_2\cdot$) radicals.

- In NO-rich environments: a net gain of 0.5–1 odd hydrogen radicals and 3–4 ozone molecules.

These results are very important for the photochemistry of the "background" troposphere, because NO, CO, and CH_4 emissions to the atmosphere are now strongly influenced by anthropogenic activities and because CH_4 and CO are the principal reaction partners of $HO\cdot$.

If no chemical reactions took place involving NO and NO_2 other than their interchange by reactions such as those given above, the atmospheric chemical lifetimes of the NO_x molecules would be infinite. However, conversion of NO_x to nitric acid (HNO_3) occurs within a few days, largely by Reaction (8.14), i.e., NO_2 with hydroxyl radicals. Nitric acid is rather unreactive in the gas phase, but it is highly soluble in water. As a result, it is readily removed from the troposphere by rain (leading in part to "acid rain," typically defined as rain having a pH reading less than about 5), so that nitric acid formation effectively removes NO_x from the troposphere.

The NO_x atmospheric lifetimes of only a few days are quite short for a chemical compound and lead to a drop in concentration by approximately a factor of 10 for each 1000-km travel distance from the source region. Because the principal sources of the nitrogen oxides are anthropogenic in nature, highly urbanized regions have NO_x concentrations up to several tens of ppbv. Away from these regions, however, the NO_x concentrations drop rapidly to very low values. For instance, over oceanic regions the concentrations can be as low as a few parts per trillion. Because the crossover from the "high-NO_x" oxidation of CO to the "low-NO_x" oxidation of CO occurs at 5–10 pptv, Earth's atmosphere is host to both processes at different times and places.

For the future, it is important to realize that the potential for ozone formation in the troposphere is large and is limited only by the availability of NO and NO_2 as catalysts. It can be roughly estimated that if all worldwide hydrocarbon and carbon monoxide emissions would result in ozone formation, then the tropospheric ozone concentration could be more than 10 times larger than that produced by the principal natural source for tropospheric ozone, downward transport from the stratosphere. Ozone is a phytotoxic and poisonous gas, so if its concentrations were to grow to such high values, there would be serious environmental consequences.

Hydroxyl Radical: The Atmosphere's Detergent

Although only about 10% of all atmospheric ozone is located in the troposphere, its presence there is of fundamental importance for the chemical composition of the entire atmosphere because of its role in the generation of the hydroxyl radical by Reactions (8.29) and (8.30). The central role of HO• occurs because it is a fragment of the very stable water molecule, to which it can revert by abstracting a hydrogen atom from a nearby molecular target. Because a very great percentage of emitted molecules contain one or more hydrogen atoms, the abstraction-initiated sequence (8.31) and (8.32) is a common one. Its crucial importance is that once the initial hydroxyl attack occurs the molecule that has been attacked is headed for increased oxidation, increased aqueous solubility, and rapid scavenging from the atmosphere. Without the hydroxyl radical, the composition of the atmosphere would be totally different and hazardous to many of the present forms of life on Earth, because there would be large accumulations in the atmosphere of most trace gases, many of which would act as air pollutants and greenhouse gases. It is truly remarkable to note that the cleansing of the atmosphere is primarily accomplished by a very tiny concentration of hydroxyl radicals, typically present at mixing ratios of a $1–4 \times 10^{-14}$, despite the fact that the atmosphere contains nearly 21% molecular oxygen by volume. However, the reactions with O_2 of all gases that are emitted into the atmosphere are far too slow to be of any significance. Furthermore, the uptake of emitted gases into cloudwater is, in most cases, very low. Consequently, reaction with

HO• is needed to prepare the gases for removal. Besides the many hydrocarbon gases scrubbed out in this way, hydroxyl radicals oxidize NO_x to nitric acid and H_2S and SO_2 to sulfuric acid. These acids are, in most locations, the main ingredients of acid precipitation.

THE CHEMISTRY OF PRECIPITATION

The Major Constituents

An important part of the atmospheric cycle of many of the trace species is the condensed water chemistry in which they participate. In recent years, as chemical analyses of rain, fog, clouds, and dew have been successfully accomplished, it has become obvious that the aqueous chemistry of the atmosphere is fully as complicated as the gas-phase chemistry but qualitatively different in rates, processes, and implications. The understanding of the aqueous-phase chemistry is still evolving, but many of its controlling processes and interesting reactions appear sufficiently well established that we can confidently describe them here.

The single most important species in clouds and precipitation is probably the hydrogen ion, whose concentrations can be indicated by specifying the solution acidity or pH value. A few typical values for precipitation are shown in Figure 8.6, together with pH values for a few common liquids. The presence of atmospheric CO_2 assures that nearly all atmospheric water droplets will be acidic, and natural and anthropogenic nitrogen and (especially) sulfur species increase the acidity, i.e., lower the pH value, to at least pH 5.0. Most rain near

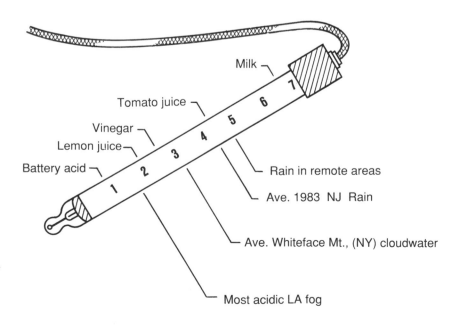

Figure 8.6 pH values in atmospheric water of various types, compared with pH values for several common liquids. (T. E. Graedel, Corrosion-related aspects of the chemistry and frequency of occurrence of precipitation, *Journal of the Electrochemical Society, 133,* 2476–2482, 1986.)

urban areas has pH levels nearer 4.0. Cloud and fog droplets are nearly always more acidic than rain, apparently because longer lifetimes and smaller drop sizes inhibit dilution of the acidic constituents. In some fogs, the pH of the droplets has been measured to be as low as 1.7, close to that of battery acid! It is no wonder that materials exposed to such fogs deteriorate rapidly.

The chemical composition of the various forms of water in the atmosphere varies from event to event as well as from cloud to fog to rain to dew. Some idea of this variation is given in Figure 8.7 in which are plotted the ranges of species that have been measured in precipitation at various sites throughout the world. In general, ionic concentrations and acidities tend to be higher in clouds and fogs than in rain, snow, and dew, though the overlap is substantial. The ions may arise from dissolved gases or from aerosol particles incorporated into the hydrometeors. In the case of chloride, dissolved sea-salt particles are important.

A few organic constituents have received detailed measurement attention in atmospheric water droplets. Formaldehyde (HCHO) is an atmospheric species with high aqueous solubility and many anthropogenic and natural sources. As a consequence, it is very common in droplets, where it assumes the hydrated form $CH_2(OH)_2$. Another organic constituent commonly measured in precipitation is the total organic acid (RCOOH) content, largely consisting of formic and acetic acids. Gas-phase organic acid concentrations can reach several parts per billion, so dissolved gas-phase species doubtless provide some of the liquid-phase source. The remainder is thought to come from oxidation of dissolved aldehydes. In remote regions, such as tropical forests, organic acids often result in precipitation acidity below pH 5.0.

As a result of the presence of its ionic constituents, the typical atmospheric water droplet is best described as an electrolyte. The ionic strengths of the types of precipitation vary widely, from about 1×10^{-5} for rain to about 1×10^{-2} for evaporating fogs. In fact, fog droplets have ion concentrations high enough to produce significant ion pairing effects. In addition, soil particles, organic detritus, soot, and other constituents must be considered. The result is that atmospheric solution chemistry encompasses the full spectrum of liquid-phase chemical processes.

Inorganic Chemistry

The atmosphere's aqueous-phase chemistry, like that of the gas phase, is modified by reactions among its constituents. For many of these reactions, the solution acidity is an important controlling factor. The initial acidity of atmospheric water droplets is established by the composition of the cloud condensation nuclei and the trace gases in the air parcel within which nucleation occurs. The counterion is often ammonia, but it can be alkaline soil dust. The presence of HCO_3^- in

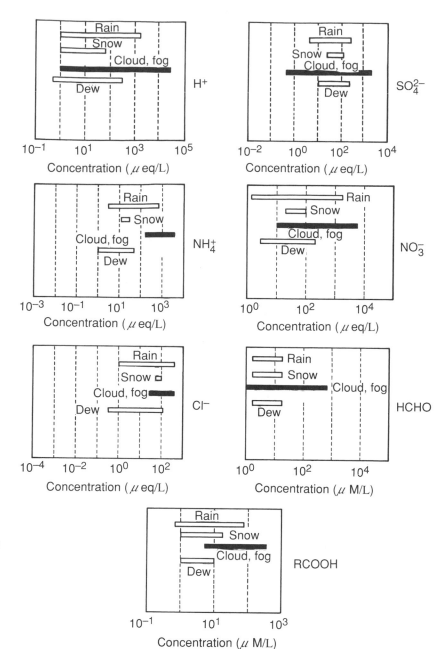

Figure 8.7 Concentration ranges for ionic and molecular constituents of precipitation in different atmospheric regimes. (Adapted from T. E. Graedel, Corrosion-related aspects of the chemistry and frequency of occurrence of precipitation, *Journal of the Electrochemical Society, 133,* 2476–2482, 1986.)

solution does not appear to affect solution chemistry in any major way. Other common but relatively uninteresting trace components are the chloride salts. The conversion of chloride ions to chlorine atoms can lead to the generation of HCl, among other products, but no efficient path for the generation of chlorine atoms has been identified. Chloride's greatest influence is likely to be as an electrolyte constituent, i.e., as a contributor to the overall ionic strength of the solution.

Ammonia [and the ammonium ion (NH_4^+)] and nitric acid [and the nitrate ion (NO_3^-)] are the most important inorganic nitrogen compounds in atmospheric water droplets. Ammonia reacts with strong acid anions, as evidenced by the large concentrations of ammonium salts found in aerosol particles. Ammonia principal sources are decaying natural organic matter, livestock wastes, fertilizers, and some industrial activities, so it is most common near large population and agricultural centers. The NO_x compounds (NO and NO_2) have low solubility in aqueous solutions; as a consequence, they play no significant role in atmospheric droplet chemistry. However, the higher oxides NO_3 and N_2O_5 can interact strongly with the aqueous phase.

Sulfur compounds are important constituents. The most abundant are the sulfates, including sulfuric acid, ammonium hydrogen sulfate (NH_4HSO_4), and ammonium sulfate [$(NH_4)_2SO_4$]. The solution concentration of SO_2 is the highest of any of the sulfur gases. Its principal reactions (in the customary bisulfite form) are usually with H_2O_2 and O_3:

$$HSO_3^- + H_2O_2 \rightarrow HSO_4^- + H_2O \quad (8.45)$$

$$HSO_3^- + O_3 \rightarrow HSO_4^- + O_2. \quad (8.46)$$

A central concern in aqueous atmospheric chemistry is the formation of the two inorganic acids: sulfuric (H_2SO_4) and nitric (HNO_3). In solution, each acid is in equilibrium with its ions,

$$H_2SO_4 \rightleftharpoons H^+ + HSO_4^- \quad (8.47)$$

$$HSO_4^- \rightleftharpoons H^+ + SO_4^{2-} \quad (8.48)$$

$$HNO_3 \rightleftharpoons H^+ + NO_3^- \quad (8.49)$$

It appears from a variety of evidence that nitric acid is formed primarily in the gas phase and dissolves in precipitation, whereas sulfuric acid is usually formed within cloud droplets by oxidation of dissolved sulfur dioxide.

Photochemistry

Hydrogen peroxide turns out to be an important oxidizer in water droplets, despite relatively low concentrations of the order of 1–10 μM. It also serves as a source of solution hydroxyl radicals when it is photolyzed:

$$H_2O_{2(aq)} + h\nu\,(\lambda \leq 380\,nm) \rightarrow 2HO\cdot. \quad (8.50)$$

A second source of free radicals in aqueous droplets is related to the ability of anions to form complexes with trace metal ions in concentrated solution. The most important example is that of iron ions in aerosol particles or raindrops, present as a consequence of the high abundance of iron in windblown soil particles and in particles from industrial processes. In the resulting solutions, iron forms complexes in which doubly or triply charged iron ions [termed Fe(II) or Fe(III)] are complexed with solution constituents such as hydroxide ions. The crucial characteristic of the mono- and dihydroxy complexes of Fe (III), which are common, is their ability to absorb visible and ultraviolet radiation over the 290- to 400-nm band. Upon absorption of a photon, the Fe(III) complexes undergo charge transfer from the complexed anion (the ligand) to the metal. The net result is reduction of the metal coupled with the oxidation of the ligand, a process that transforms the ligand into a free radical. The redox process is followed by the escape of the free radical into the bulk solution and solvation of the resulting Fe(II) complex.

For the mono- and dihydroxy complexes, the equations are

$$[Fe^{III}(H_2O)_5(OH)]^{2+} + H_2O + h\nu\,(\lambda \leq 400\ nm) \rightarrow$$
$$[Fe^{II}(H_2O)_6]^{2+} + HO\cdot \quad (8.51)$$

$$[Fe^{III}(H_2O)_4(OH)_2]^+ + H_2O + h\nu\,(\lambda \leq 400\ nm) \rightarrow$$
$$[Fe^{II}(H_2O)_5(OH)]^+ + HO\cdot. \quad (8.52)$$

Note that because hydroxide ions are involved in the predominant complexes, the relative abundances of the iron complexes are pH dependent.

It turns out that the most common fate of the HO• in solution is oxidation of formaldehyde to formic acid:

$$HCHO_{(aq)} + HO\cdot + O_2 \rightarrow HCOOH_{(aq)} + HO_2\cdot, \quad (8.53)$$

and the resulting HO$_2$• radical can oxidize the Fe(II) created by Reactions (8.51) and (8.52):

$$Fe^{II} + HO_2\cdot + H_2O \rightarrow Fe^{III} + H_2O_2 + OH^-. \quad (8.54)$$

The overall result of these reactions is a catalytic cycle in which iron in solution is the agent for a suite of organic oxidation reactions, as shown in Figure 8.8.

Organic Chemistry

Aldehydes (i.e., organic compounds containing the –CHO group) are among the important organic constituents present in atmospheric droplets. They are readily generated in the gas phase and are highly soluble in water. Once present, they are oxidized to carboxylic acids by reaction chains such as that with iron involving hydroxyl radicals:

$$RCHO_{(aq)} + HO\cdot \rightarrow \cdots \rightarrow RC(O)OH_{(aq)}. \quad (8.55)$$

The organic acids, like inorganic acids, ionize in solution:

$$RC(O)OH \rightleftharpoons H^+ + RC(O)O^- \quad (8.56)$$

but the equilibria are such that the organic acids are only partially

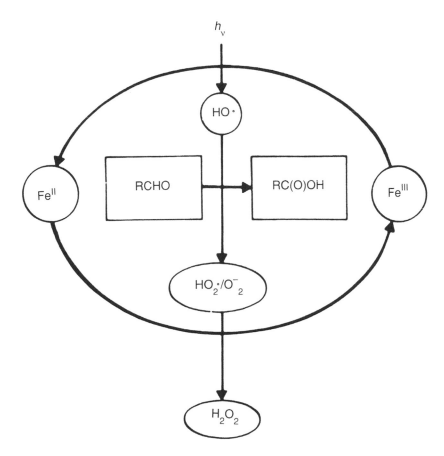

Figure 8.8 The iron catalytic cycle for the production of organic acids from aldehydes in atmospheric droplets. (T. E. Graedel, M. L. Mandich, and C. J. Weschler, Kinetic model studies of atmospheric droplet chemistry. 2. Homogeneous transition metal chemistry in raindrops, *Journal of Geophysical Research, 91,* 5205–5221, 1986.)

ionized and are thereby termed "weak." Nonetheless, organic acids can be the major acid constituents of atmospheric water droplets if the precursors of the inorganic acids are present only at very low concentrations and if large amounts of organic precursors are available. Such conditions exist in tropical forested environments as a consequence of the oxidation to organic acids of the isoprene (C_5H_8) emitted in those areas in great abundance.

Chemical Processing of Air by Clouds

It has been estimated that only about 10% of all clouds produce precipitation, the others evaporating as surrounding conditions change. Nonetheless, all clouds play important roles in atmospheric chemistry by virtue of their interactions with air passing through them. As soon as air enters a cloud, its chemistry changes in major ways. One of the reasons is a change in the chemically reactive ultraviolet radiation that is received. The main reason, however, is that compounds that are strongly water soluble, such as the free radicals HO• and HO_2• and the oxidizing molecules HCHO and H_2O_2 are incorporated into the liquid phase. This process leaves behind in the gas phase the less water-soluble components, such as NO, CO, and CH_4. Therefore, reactions that are important in the gas phase, such as the ozone formation cycle

$$HO\cdot + CO + O_2 + M \rightarrow HO_2\cdot + CO_2 + M \quad (8.56a)$$

$$HO_2\cdot + NO \rightarrow HO\cdot + NO_2 \quad (8.46b)$$

$$NO_2 + h\nu \, (\lambda \leq 410\,nm) \rightarrow NO + O \quad (8.56c)$$

$$O + O_2 + M \rightarrow O_3 + M \quad (8.56d)$$

$$Net: \ CO + 2O_2 + h\nu \rightarrow CO_2 + O_3 \quad (8.56e)$$

are strongly inhibited. Instead, a rapid ozone destruction cycle occurs in the liquid phase, leading to the formation of formic acid:

$$HO_2\cdot \rightleftharpoons H^+ + O_2^- \quad (8.57)$$

$$O_2^- + O_3 + H^+ \rightarrow HO\cdot + 2O_2 \quad (8.58)$$

$$HCHO_{(aq)} + HO\cdot + O_2 \rightarrow HCOOH_{(aq)} + HO_2\cdot \quad (8.58a)$$

$$Net: \ HCHO_{(aq)} + O_3 \rightarrow HCOOH_{(aq)} + O_2 . \quad (8.58b)$$

Further reactions involving formic acid in the aqueous phase lead to additional ozone loss:

$$HCOOH_{(aq)} + HO\cdot + O_2 \rightarrow CO_2 + HO_2\cdot + H_2O \quad (8.59)$$

$$HO_2\cdot \rightleftharpoons H^+ + O_2^- \quad (8.59a)$$

$$O_2^- + O_3 + H^+ \rightarrow HO\cdot + 2O_2 \quad (8.59b)$$

$$Net: \ HCOOH_{(aq)} + O_3 \rightarrow CO_2 + O_2 + H_2O . \quad (8.59c)$$

This sequence leads to the direct production of CO_2 without the formation of CO as an intermediate product, as is always the case in the gas phase following the photochemical breakdown of formaldehyde [Equations (8.38) and (8.39)].

Jos Lelieveld and Paul Crutzen, both of the Max-Planck-Institute for Chemistry in Mainz, Germany, have estimated that air spends an average of 13–20 hr in a cloud-free environment followed by 3–4 hr inside clouds. Taking account of both types of environments in a photochemical model, Lelieveld and Crutzen calculated that the resulting inhibition of gas-phase chemistry leads to reductions in ozone production in the lower half of the troposphere. For NO_x-rich regions, the reduction is about 40%. In NO_x-poor regions, net ozone destruction rates are enhanced by factors ranging from 1.7 to 3.7. As a consequence of lower ozone production, the calculated concentrations of HO• are substantially smaller than those derived from calculations that do not take cloud chemical reactions into account. An associated result is the significantly smaller production of CO from the oxidation of CH_4 and other hydrocarbons.

Atmospheric gas-phase chemistry thus cannot be properly assessed without a comprehensive knowledge of liquid-phase processes.

Chemical Properties

Unlike atmospheric gases, in which every like molecule has the same chemical options, each aerosol particle is an individual, an agglomerate from many different sources. Despite this uniqueness, some similarities exist in the chemical makeup of fine and coarse particles (the division between the two types being set at about 2.5 mm diameter; see Chapter 5).

Because fine particles are produced by combustion processes of various kinds and by condensation (the deposition of gaseous molecules onto preexisting small nuclei), their constituents tend to be elemental carbon and simple inorganic and organic species: ammonium, sulfate, and nitrate ions, small oxidized organic molecules, and the like. Trace metals (especially from the combustion of coal) are also common.

Coarse particles reflect their origins more than do fine particles, and they are changed to a much smaller degree by their residence in the atmosphere. Those originating as windblown dust are largely crustal oxides of iron, aluminum, silicon, and titanium. Those whose origins are from vegetation are rich in waxes, long-chain fatty acids, and related compounds. Over the oceans, sea-salt aerosol particles reflect the chemistry of the oceans from which they came.

During their lifetime in the atmosphere—several days at least for fine particles, several hours for coarse particles—the particles are exposed to atmospheric water vapor, which they readily adsorb. Several quite different studies have demonstrated that at relative humidities above 40% the aerosol water content can often be at least 30% of the total particle weight. Figure 8.9 shows such data over a 14-hr period, the aqueous fraction being more than 50% at its peak. The significance of this result can be appreciated by considering an individual particle of diameter 0.6 μm. If the weight of the water shell is 30% of the total weight, the shell will be approximately 130 monolayers thick. If the weight is only 10% of the total, the shell will still be 40 monolayers thick and water-soluble salts will dissolve to form a concentrated aqueous solution in contact with an undissolved or partially dissolved core. As the ambient humidity changes, this shell will become thicker or thinner but will always be present at moderate thicknesses under most atmospheric conditions. As well as providing a dynamic chemical environment at the aerosol surface, the varying water content affects aerosol microphysical properties such as the instantaneous settling velocity and agglomeration rate.

In addition to its evolution in size as a consequence of agglomeration and condensation, the aerosol particle evolves chemically. This process is illustrated by Figure 8.10, which shows aerosol chemical analyses at different hours during the day. All of the measured components (ammonium, nitrate, and sulfate ions and total organics) show marked changes from hour to hour.

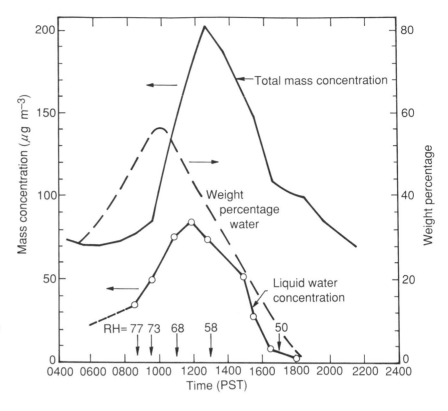

Figure 8.9 The diurnal variation in liquid water concentration in atmospheric aerosols in Pasadena, California, on 15 September 1972. Note the dependence of the fractional water content on relative humidity (RH), indicated at the bottom of the figure. (Reproduced with permission from W. Ho, G. M. Hidy, and R. M. Govan, Micro-wave measurements of the liquid water content of atmospheric aero-sols, *Journal of Applied Meteorology, 13*, 871–879. Copyright 1974 by American Meteorological Society.)

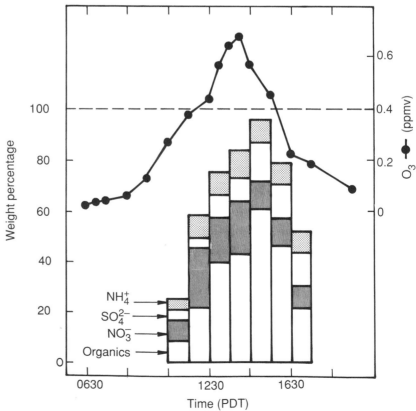

Figure 8.10 Diurnal concentration patterns of nitrate, sulfate, and ammonium ions, and total aerosol organics (as weight percentage of the total dry aerosol) in Pasadena, California, on 25 July 1973. Concurrent ozone concentrations are also shown, because the generation of aerosol ions and molecules as well as the ozone concentrations are related to the vigor of the gas-phase chemistry. (Adapted with permission from D. Grosjean and S. K. Friedlander, Gas-particle distribution factors for organic and other pollutants in the Los Angeles atmosphere, *Journal of the Air Pollution Control Association, 25*, 1038–1044. Copyright 1975 by Air Pollution Control Association.)

When an aerosol particle sorbs water, especially if the water subsequently becomes acidified, many constituents in the solid state will dissolve. Some, however, resist dissolution even under highly acidic conditions. It is possible to say a good deal about the insoluble core of a typical aerosol particle. In urban areas the most common cores are primarily soot or fly ash. Fly ash is largely composed of inorganic oxides with chemically reactive transition metal atoms present on the exterior. Soluble inorganic salts such as ammonium nitrate are also present. In less urban environments, soil constituents, both inorganic and organic, are common. The organic material contains a variety of compounds derived from vegetation. About 5–10% of the atmosphere's reactive carbon and 20–30% of its reactive sulfur and nitrogen reside on particles rather than in the gas phase. A portion of this material is dissolved or dissolvable; a smaller portion is insoluble.

A final point is that each particle is an individual chemical system, even though most analyses are performed on bulk collections of particles. Individual particles may differ in amount of soluble material, mineral type and content, and many other characteristics. Partly because chemical analyses of individual aerosol particles are so difficult, atmospheric chemists have yet to address these complications satisfactorily.

Chemical Reactions

The liquid water shell provides a concentrated chemical environment for reaction processes. Because the water shell is thin and because anions and cations are abundant, ionic strengths are very high, perhaps of the order of 5 to 20. A few measurements of aerosol pH give values of well under 2, a further indication of a dynamic chemical environment. Finally, evidence suggests the frequent presence of organic surface films on aerosol particles. If such films occur, they may partially or totally inhibit exchange between atmospheric gas and condensed phases, thus creating small isolated reaction vessels in the atmosphere.

The evidence for chemical reactions within atmospheric aerosol particles is persuasive. One might first consider the case of sulfate ions. They are ubiquitous in aerosol particles, their concentration evolves with time, SO_2 (the precursor) is moderately soluble in the water shell, and oxidation of dissolved to SO_2 sulfate occurs by reaction with H_2O_2 and O_3 and is catalyzed by both soot and transition metals. These facts do not preclude formation of sulfuric acid in the gas phase followed by incorporation into particles but suggest that oxidation within particles is likely to be very efficient and thus to play an important role.

An alternative and very important function for sulfur-containing aerosol particles is that they act as nucleation sites for cloud droplets. This role renders sulfate important for determining rates of cloud formation and of cloud optical properties as a consequence of gas to particle conversion processes. The possibility that sulfur emissions are linked in this way to climate change is discussed in Chapter 19.

Chemical Perspective

With the perspective of the above discussion, we can now proceed to sketch a typical physical and chemical cycle for atmospheric aerosol particles and their associated adsorbed water. The cycle is illustrated in Figure 8.11. Stage 1 is the emission of a condensable gas (SO_2 is a common example), emission of a precursor to a condensable gas (e.g., H_2S), or emission of a small particle. The precursor gas or condensable gas undergoes chemical reactions that result in condensation; we term the product the sulfate aerosol. Direct sources for particles are of three types. The first is sea spray, which produces a droplet reminiscent of seawater, but often with enhanced trace metals and organics (such as fulvic or humic acid) from the sea surface microlayer. The resulting aerosol we term the chloride aerosol. A second source of particles is mechanical processing of solid surfaces, leaf litter and windblown dust being good examples. Leaf litter is almost entirely organic, with C_{16}–C_{30} lipids predominating. Windblown dust is primarily composed of inorganic oxides, some humus or other organic material being common as well. We term these emittants the organic aerosol. A third set of particle sources includes those involving combustion of fossil fuels or biomass to generate soot, the carbon aerosol.

Regardless of source, the smallest aerosol particles agglomerate with each other and incorporate water vapor to arrive at a form typified by the diagram labeled 2 in Figure 8.11. Here the particle has a solid core covered by an aqueous shell making up perhaps 30% of the total particle mass. Atmospheric gases that are highly water soluble will then dissolve in this shell to form a complex and chemically reactive solution. We have indicated six gases whose dissolution appears particularly likely.

In stage 3 of the particle cycle, species that are relatively insoluble begin to accumulate on the particle surface. Such a process is often favored when a relatively large molecule has both hydrophilic (water-loving) and hydrophobic (water-hating) parts; fatty acids such as *n*-decanol [$CH_3(CH_2)_8CH_2OH$] are good examples.

In stage 4 in Figure 8.11, the organic layer rather than a water surface presents itself to gaseous molecules. If the organic layer does not react with the gaseous molecule (with NH_3, say), the flux of the gaseous molecules into the particle will be depleted. If reactions do occur, however, some of the molecules in the organic layer are transformed. Two processes whose occurrence is reasonably well established are the formation of nitrated arenes by NO_2 reaction with PNAH and the formation of ketones from arenes. This process produces a particle with the general characteristics shown in 5.

The final step in the cycle is the deposition of the particle on a surface at the ground. (Deposition may, of course, occur at any of the stages.) With only a few exceptions, it is these deposited particles that are available for analysis. The complete cycle thus involves particle generation, growth by agglomeration and/or condensation, incorporation and chemical processing of atmospheric gases, and, finally, deposition to Earth's surface.

VARIATIONS ON LARGE SPATIAL SCALES

Variations in the concentrations of chemical species are seen over different parts of Earth because of the locations of major emissions sources, the diffusion and dispersion produced by motions of air parcels, and the removal of species from the atmosphere by reaction or deposition. A particularly nice demonstration of such variations was provided by a gas filter radiometer experiment aboard one of the flights of the NASA space shuttle. The radiometer looked down upon Earth from space and made determinations of the carbon monoxide concentrations averaged throughout the middle troposphere (about 6 km altitude). A sample result is shown in Figure 8.12 (in the color

Figure 8.11 A schematic cycle for atmospheric aerosol-particle chemistry. The individual processes are discussed in the text. (PNAH is an abbreviation for polynuclear aromatic hydrocarbons such as pyrene or anthracene.) Particle deposition can occur at any stage but, for simplicity, is shown only for the last. (T. E. Graedel, D. T. Hawkins, and L. R. Claxton, *Atmospheric Chemical Compounds: Sources, Occurrence, and Bioassay*, Academic Press, Orlando, 1986.)

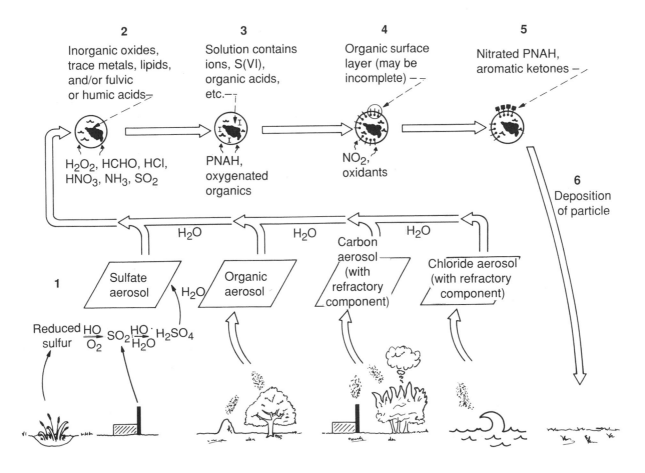

insert). The data show strong peaks in CO concentrations downwind of major source regions, particularly downwind of northern Europe and in northern South America and southern Africa, where large regions were undergoing their annual biomass burning cycle. Away from these regions, concentrations are lower by factors of 3 or 4. Clearly, the variations are sufficiently large that the information of Figure 8.12 can only be regarded as providing some perspective on typical transient concentrations, rather than giving definite values, but the potential for systematic monitoring of air quality by satellite is very great.

A more extended view of air quality with time is provided by long-term measurements from Earth-orbiting satellites. Figure 8.13 (in the color insert) illustrates such data for tropospheric ozone in 2-month blocks for the period 1979–1989. Distinctive features of these displays include the high ozone values in summer in the northern hemisphere, presumably due to smog reactions from industrial emissions, and high values toward the end of the year off the coasts of Africa and South America, thought to be a consequence of extensive biomass burning. Concentrations are low throughout the year over the Pacific Ocean, but not the Atlantic. Ozone levels over the Indian Ocean tend toward the low side, but some parts show moderate ozone levels in the spring of the year. Some surprising patterns are revealed by careful study; for example, in May–June the ozone levels over the entire Atlantic Ocean exceed those in either the western United States or central Asia.

VARIATIONS WITHIN AND BETWEEN URBAN AREAS

Variations in atmospheric quality are perhaps greatest within and between urban areas because most anthropogenic emissions to the atmosphere occur there. As with the subcontinent distance scale discussed above, this point is perhaps best made by example. A convenient data set for that purpose is that of the Global Environment Monitoring System (GEMS), established by the World Health Organization. At present, 14 countries have contributed data to the program, which has concentrated on sulfur dioxide and suspended particulate matter.

Comparisons of trace atmospheric species among cities require that data be collected over the long term and with substantial completeness of measurements at several sites within each city. For cities meeting these criteria, GEMS has averaged the data over the period 1980–1984. The results for SO_2 are shown in Figure 8.14. They indicate that the concentration of a single trace gas can differ by a factor of as much as 3 among different sites within the same urban area and as much as 30 between different urban areas, even when long-term averages are used.

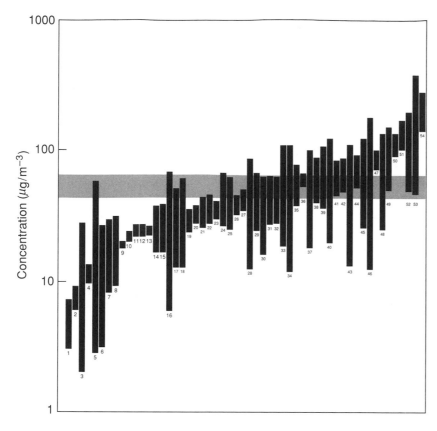

Figure 8.14 The range of annual averages of sulfur dioxide concentrations measured at multiple sites within cities throughout the world for the period 1980–1984. Each bar represents a city, coded as follows: 1, Craiova; 2, Melbourne; 3, Auckland; 4, Cali; 5, Tel Aviv; 6, Bucharest; 7, Vancouver; 8, Toronto; 9, Bangkok; 10, Chicago; 11, Houston; 12, Kuala Lumpur; 13, Munich; 14, Helsinki; 15, Lisbon; 16, Sydney; 17, Christchurch; 18, Bombay; 19, Copenhagen; 20, Amsterdam; 21, Hamilton; 22, Osaka; 23, Caracas; 24, Tokyo; 25, Wroclaw; 26, Athens; 27, Warsaw; 28, New Delhi; 29, Montreal; 30, Medellin; 31, St. Louis; 32, Dublin; 33, Hong Kong; 34, Shanghai; 35, New York; 36, London; 37, Calcutta; 38, Brussels; 39, Santiago; 40, Zagreb; 41, Frankfurt; 42, Glasgow; 43, Guangzhou; 44, Manila; 45, Madrid; 46, Beijing; 47, Paris; 48, Xi'an; 49, São Paulo; 50, Rio de Janeiro; 51, Seoul; 52, Tehran; 53, Shenyang; 54, Milan. Few cities in the developing world have been extensively monitored and therefore do not appear in this figure; many would be expected to have high levels of sulfur dioxide, particularly if they make extensive use of high-sulfur coal for heating and energy generation. The gray shading indicates the concentration range recommended by the United Nations Environment Program as a reasonable target for preserving human health. (Global Environment Monitoring System, *Assessment of Urban Air Quality*, United Nations Environment Program, Geneva, 1988.)

SUMMARY

In this chapter, we have discussed several different chemical regimes, separating them largely for convenience in presentation, because the atmosphere does not allow them to be studied independently. The level of our understanding of these systems, all important for atmospheric chemistry, is variable. In the case of tropospheric gas-phase chemistry, very large computer models are now available to handle hundreds of chemical reactions among scores of different species. The important oxidizer is clearly identified as HO•. Provided the models are given accurate information concerning emissions, and provided they have done a good job of simulating atmospheric motions, the results are reasonably accurate; certainly accurate enough to be useful for scientific investigations or for policy planning. Because gas-phase molecules and radicals dissolve on precipitation and aerosols, gas-phase chemistry has a strong impact on precipitation chemistry and tropospheric aerosol chemistry.

The level of detailed understanding of precipitation chemistry is less satisfactory. The principal oxidizer is H_2O_2, probably followed in importance by HO• and O_3. Transition metal ion catalysis is present but needs to be more fully explored. Water droplets form around aerosol particles and dissolve molecules and radicals from the gas phase, so they reflect both gas-phase and aerosol-phase chemistry. Condensed-phase material also contributes to the gas phase, especially by forming and vaporizing HNO_3 and HCl.

The general thrust of tropospheric aerosol chemistry is outlined by chemical analyses of collected particles, although the details are relatively unexplored in theoretical models. It is apparent that aerosol and precipitation chemistry will have strong similarities when local humidity is high, at least, for under such conditions the aerosol particles are covered with a substantial shell of water. Interactions of aerosol chemistry with that of other regimes are poorly understood, but it appears that aerosols are receptors of gas-phase species and suppliers of species to precipitation when the aerosol particles serve as condensation nuclei.

The chemistry of gases in the stratosphere appears to be understood better than that of any other atmospheric chemical system, in part because not too many species are present, and those that do reside in the stratosphere are relatively small and more easily studied in the laboratory. Interactions between the gas and aerosol (or ice) phases are important and are beginning to be understood. The effects of surface reactions, however, were long neglected and underestimated, as evidenced by the failure to predict the Antarctic ozone hole or its chemistry. The ozone hole demonstrates, however, that interactions between condensed and gas phases can dominate the chemistry under some conditions.

Atmospheric chemistry as a scientific discipline is less than a quarter-century old, and much has been learned. It is clear, however, that the incredibly complex array of chemical species, emissions sources, and reactions, combined with the multiplicity of phases and the interactions among them, leave much yet to understand.

EXERCISES

8.1 The hydroxyl radical reacts with many atmospheric species, but only one (or possibly two) can control the atmospheric lifetime of the radical. The following table gives atmospheric concentrations and hydroxyl radical rate constants for several important hydroxyl reactants. All values are given at 25 °C, and all reactions are known or assumed to be second order, that is, no other reactants are involved. Compute the hydroxyl radical lifetime set by each of these reactants and state which one controls the atmospheric lifetime of the hydroxyl radical.

Hydroxyl Radical Reactants and Related Data		
REACTANT	CONCENTRATION (ppbv)	k (cm³ molecule⁻¹ s⁻¹)
O_3	80	6.8×10^{-14}
CO	150	2.3×10^{-13}
CH_4	1700	7.7×10^{-15}
SO_2	10	1.5×10^{-12}

8.2 Assuming hydrochloric acid is the only acidic species present, its concentration is 4 pptv, and it has come to equilibrium with surrounding water droplets, find the droplet pH that would result. (The Henry's law constant for HCl is 2.0 M atm^{-1}.)

8.3 Air containing 1 ppbv HNO$_3$ and 5 ppbv SO$_2$ undergoes cooling and nucleation to produce a cloud with water content 1 g m^{-3}. The Henry's law constants for the two gases are 9.1 $\times$ 10^4 M atm^{-1} and 1.4 M atm^{-1}. Compute the average pH of the cloudwater, neglecting any effects caused by gas transport to droplets of different sizes. (Acid dissociation constants are available in any chemical handbook.)

8.4 Apart from SO$_2$, H$_2$S and COS are generally the most abundant sulfur gases in the urban atmosphere. Typical concentrations and Henry's law constants for the three species are given in the table at the upper right. Compute their relative concentrations in cloud droplets of pH 5.0 if the droplets live long enough to come to equilibrium with the surrounding gas.

Urban Sulfur Gas Parameters		
GAS	CONCENTRATION (ppbv)	H$_c$ (M atm^{-1})
H$_2$S	1.5	1.0 $\times$ 10^{-1}
COS	0.6	2.1 $\times$ 10^{-2}
SO$_2$	5.0	1.4

8.5 The table below gives typical ionic concentrations in seawater, lakewater, rainwater, and fog droplets. Compute the ionic strength of each. Consider the solution reaction

$$NO_2^- + O_3 \rightarrow NO_3^- + O_2,$$

which has a rate constant at infinite dilution and 25 °C of $k = 1.6 \times 10^5$ M^{-1} s^{-1} and an activation energy of 13.8 kcal mol^{-1}. Compute the rate constant at 25 °C for each of the four regimes. Repeat the calculation for 10 °C.

Characteristic Ion Concentrations in Natural Waters				
	CONCENTRATION, M			
Ion	**Seawater**	**Freshwater**	**Rainwater**	**Fog droplets**
H$^+$	1 $\times$ 10^{-8}	2 $\times$ 10^{-8}	1 $\times$ 10^{-4}	2 $\times$ 10^{-3}
Mg^{2+}	5 $\times$ 10^{-2}	4 $\times$ 10^{-4}	4 $\times$ 10^{-6}	8 $\times$ 10^{-5}
Na$^+$	0.48	5 $\times$ 10^{-4}	2 $\times$ 10^{-6}	3 $\times$ 10^{-4}
Ca^{2+}	1 $\times$ 10^{-2}	1 $\times$ 10^{-3}	5 $\times$ 10^{-7}	2 $\times$ 10^{-4}
Cl$^-$	0.56	3 $\times$ 10^{-4}	2 $\times$ 10^{-5}	2 $\times$ 10^{-4}
SO$_4^{2-}$	3 $\times$ 10^{-2}	2 $\times$ 10^{-4}	5 $\times$ 10^{-5}	3 $\times$ 10^{-4}
NO$_3^-$	3 $\times$ 10^{-5}	6 $\times$ 10^{-6}	1 $\times$ 10^{-4}	1 $\times$ 10^{-3}

FURTHER READING

G. Brasseur and S. Solomon, *Aeronomy of the Middle Atmosphere*, Reidel, Dordrecht, 1984.

J. Lelieveld and P. J. Crutzen, Influence of cloud photochemical processes on tropospheric ozone, *Nature, 343,* 227–233, 1990.

S. E. Schwartz, Acid deposition: Unraveling a regional phenomenon, *Science, 243*, 753–763, 1989.

J. H. Seinfeld, Urban air pollution: State of the science, *Science, 243*, 745–752, 1989.

R. P. Wayne, *Chemistry of Atmospheres*, 2nd Ed., Oxford Univ. Press, Oxford, UK, 1990.

CHAPTER 9

Aquatic Chemistry

Just as the atmosphere receives water, gases, and particles from Earth's surface and performs physical and chemical transformations on them, so the deposition of the atmosphere's water, gases, and particles drives many physical and chemical processes on the surface. The surface is predominantly water in one form or another. As we saw in Chapter 6, nearly three-fourths is ocean, and lakes and rivers are prominent features of the land. Beneath the surface, groundwater moves and changes. Both groundwater and sea bottom water are sometimes altered radically in the vicinity of the volcanic and magmatic activity discussed in Chapter 2.

From a scientific standpoint, the atmosphere is the simplest to study of Earth's chemical systems, perhaps because the atmosphere permits relative isolation of the chemical constituents from one another, so the actions of each can be determined with precision. Such is seldom the case with aquatic chemistry—the chemistry of water at or beneath Earth's surface. In these systems, solids, liquids, and gases are inextricably combined, biological systems are generally present to add their own chemical complexity, and sampling is often a formidable task. Nonetheless, much is known of them, and several of the more common systems are discussed below.

LAKE AND RIVER CHEMISTRY

The surface water that people most frequently observe is that in lakes and rivers. The water in lakes and rivers comes from precipitation and is modified by interaction with soil and rock. In most parts of the world, the impacts of humanity are also evident in added flows to lakes and rivers from manufacturing, water and sewage treatment plants, agricultural drainage, and the like. As a consequence, the chemical composition of surface fresh waters is characterized by enormous variability. A good picture is provided by data for several hundred river sampling stations in the United States, given in Table 9.1. In most cases, the waters are slightly alkaline (that is, they have pH values higher than 7). Because substantial precipitation falls near coastal regions, seasalt aerosol incorporated into precipitation has a strong influence on river and lake chemistry, contributing perhaps a third or more of the sodium, chloride, and sulfate on a global basis. Given this and other influences, relatively high concentrations of many common ions are present, and ionic strengths are generally of the order of $I = 0.003$. Flowing rivers naturally incorporate air, so they have a substantial oxygen content.

The passage of water through lakes is a much more stately process than in the faster flowing rivers. In most lakes, modest water runoff feeds into a contained basin drained by a restricted outflow. The result is that the aqueous solution often has a long time to interact with the local environment, especially the biota and the rock or sediment at the banks and bottom of the lake. The interaction time is not constant but is driven by the sporadic nature of precipitation; therefore the resulting outflow is a series of events rather than a continuous process. Figure 9.1b shows the rate of discharge from a hydrologic catchment system in southern Norway throughout one annual cycle. The outflow pattern was characterized by a highly active May, an active June–July period, and small events throughout the fall.

Figure 9.1a shows measured sulfate concentrations in the outflow throughout the year. The pattern is one in which sulfate levels build slowly during periods of low flow, with sulfur being gradually added to the catchment by dry deposition. During periods of significant precipitation, the outflow rate increases sharply, and the rapid

TABLE 9.1. Concentrations of Major Chemical Parameters in U.S. Rivers, 1974–1981

CONSTITUENT[a]	Conc.[b] (mg L^{-1})
Na^+	7–69
Mg^{2+}	4–22
Ca^{2+}	16–67
K^+	2–5
Cl^-	7–53
SO_4^{2-}	11–117
NO_3^-	0.2–0.9
O_2	8.7–10.5

a. pH = 7.3–8.1.
b. These data are the 25th and 75th percentiles for mean concentrations at some 300 sampling stations on U.S. rivers, as reported in R. A. Smith, R. B. Alexander, and M. G. Wolman, Water-quality trends in the nation's rivers, *Science, 235,* 1607–1615, 1987.

passage of the rainwater through the catchment results in a decrease of sulfate concentrations in the outflow relative to those during periods of light or negligible precipitation.

An important development in environmental chemistry in the last decade or two has been the discovery of the high acidity of precipitation (see Chapter 8) and of strongly acidified lakes. Figure 9.2 (in the color insert) illustrates the measured acidities of lakes in Sweden. Note that the most acidified are those in the southwestern corner of the country. It is this area that receives air currents from the direction of England and Germany, air that over the past decades has been strongly loaded with oxides of sulfur and nitrogen. The deposition of these species, both during dry periods and in precipitation, is regarded as the principal factor causing the high acidity.

The chemical nature of the surfaces on which the acidic precipitation falls or on which acidic gases and particles are deposited by the wind determines the degree to which the acidic species are neutralized. If the water is primarily in contact with unweathered rocks, little neutralization will occur and the pH of the lake may be well below 5. Conversely, if the lake is in contact with well-developed soil, much of the acid will be neutralized and little damage will occur.

The acid-neutralizing capacity of soil is related in complex ways to the soil's organic constituents, particularly humic acid. Humic acid

Figure 9.1 (a) The observed sulfate concentrations in stream water at Storgama, southern Norway, for the year 1978. (b) The observed stream water discharge rate. (Adapted with permission from N. Christophersen, L. H. Dymbe, M. Johannssen, and H. M. Seip, A model for sulphate in streamwater at Storgama, Southern Norway, *Ecological Modelling, 21*, 35–61. Copyright 1983/84 by Elsevier Science Publishers BV.)

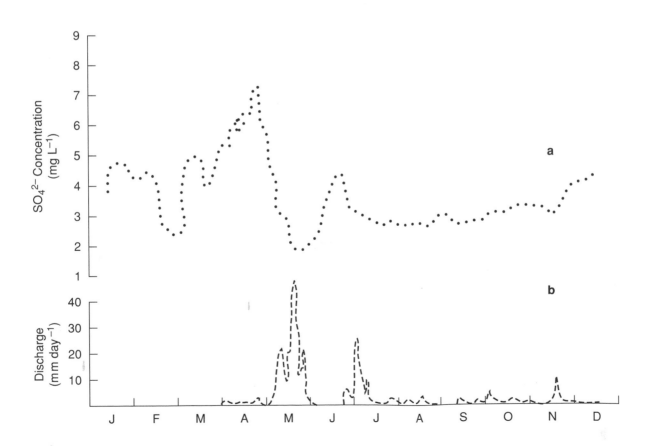

occurs in rotting vegetable matter and, as a consequence of soil runoff and of freshwater biology, is abundant as a constituent of fresh waters. Humic acid is not a simple chemical species but a designation for a group of very large molecules derived from lignin and other natural materials. It is important to the cycling of trace metals, free radical interactions, and many other chemical properties of fresh waters but is so chemically complex and variable that its effects remain to be fully characterized.

In Europe, the lakes at peril are primarily those of the Scandinavian countries. In North America, the southeastern part of Canada and the northeastern part of the United States contain the most susceptible lakes. The full spectrum of the effects of high acidity has been the subject of much debate and research, beginning in Sweden in the 1970s and in North America a decade later, culminated by the National Acid Precipitation Assessment Program (NAPAP). A number of the dire predictions made prior to the start of these efforts have not been confirmed, but some have. Among them is the decreasing diversity of aquatic life capable of living and reproducing in highly acidic waters. A related effect is the ability of acid waters to dissolve aluminum from the soil, where it commonly exists as nontoxic Al_2O_3. When Al_2O_3 dissolves, however, the liberated Al^{3+} ion is harmful to many aquatic species. Because Al_2O_3 is more soluble in highly acidic solutions, the deposition of acid to soil carries a double penalty for living organisms in the vicinity.

The chemical concentrations measured in fresh waters are not merely reflections of the precipitation that fell from the atmosphere or the soil or rock through which the water moved; they can indicate active chemical processes as well. A nice example of such a situation is shown in Figure 9.3, which follows the concentration of dissolved

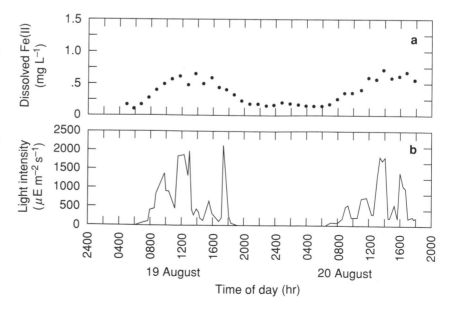

Figure 9.3 Changes in (a) dissolved Fe(II) and (b) light intensity [in microEinsteins (μE) per square meter per second] in an acidic mountain stream over a 2-day period. (Adapted with permission from D. M. McKnight, B. A. Kimball, and K. E. Bencala, Iron photoreduction and oxidation in an acidic mountain stream, *Science, 240*, 637–640. Copyright 1988 by the AAAS.)

Fe(II) in a mountain stream while simultaneously measuring the solar radiation. The researchers observed that the production of Fe(II) was as much as four times higher during daytime than during nighttime. This pattern is explained by the photoreduction of Fe(III) complexes, the same process discussed for fog and cloud droplets in Equations (8.51) and (8.52).

GROUNDWATER CHEMISTRY

Most of the water falling on the continents does not drain to lakes and rivers, at least not directly, but penetrates the soil and is eventually retained or transported in groundwater reservoirs. Groundwater first passes through the *vadose zone*, where the pores contain air and water. Further down is a region of saturation where the pores are filled only with water. Below about 1000 m, the soil is packed so tightly the water is essentially absent. At any of these levels, rock may intervene to constrain further percolation.

Groundwater generally reaches or approaches equilibrium with the soil and rock with which it comes in contact. Communities known for "hard" water are those whose water contains iron and calcium ions derived from the rock through which it flowed. Salts of these ions often

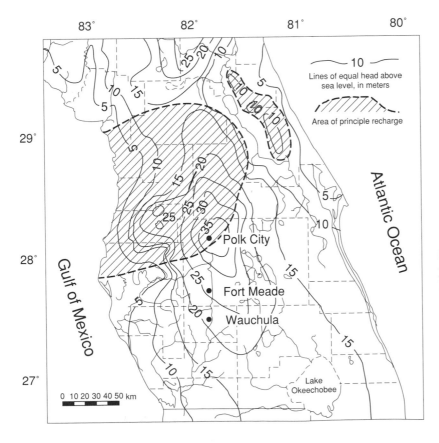

Figure 9.4 The locations of sampling wells (black circles) in central Florida used for a chemical study of Florida aquifer. The term "head" is a measure of water pressure. (Reproduced with permission from L. N. Plummer, D. L. Parkhurst, and D. C. Thorstenson, Development of reaction models for ground-water systems, *Geochimica et Cosmochimica Acta, 47,* 665–686. Copyright 1983 by Pergamon Press plc.)

precipitate in water systems, and hard water has a tendency to stain clothing. In contrast, "soft" water is rich in the alkali metals sodium and potassium; their salts are soluble and do not form precipitates.

Although biochemical processes are possible or probable in most aqueous Earth systems, percolating groundwater is a particularly obvious example of biogeochemical control. Many of the changes that occur are due to the action of bacteria on the percolating anions, changing NO_3^- to N_2O and SO_4^{2-} to H_2S, for example. The processes are sufficiently ubiquitous and vigorous that they may be important sources for several atmospheric trace gases.

An example of the chemical evolution of groundwater is provided by the study of Plummer and coworkers, in which computer model results are compared with water samples from several wells in a Florida aquifer (Figure 9.4). The field situation is one in which chemical data are available for three test wells: one (in Polk City) near the outer edge of the aquifer recharge area, a second (in Wauchula) 68 km away along a principal aquifer flow line, and a third (in Fort Meade) 47 km away at an intermediate point. From various evidence, it was determined that the water containing the cations Ca^{2+}, Mg^{2+}, Na^{2+}, and K^+ and the anions HCO_3^-, SO_4^{2-}, Cl^-, F^-, and NO_3^- was interacting with gypsum [$CaSO_4 \bullet 2H_2O$], calcite [$CaCO_3$], iron sulfides, and other soil and rock materials. A comparison of the data with several different assumptions for the concentration and dissolution of the solid phases is shown in Figure 9.5. None of the assumptions, nor others not shown here, fit the data perfectly, but the figure demonstrates the complexities and partial successes of groundwater chemical assessments.

Another topic of active research in groundwater chemistry is that of groundwater contamination, either through the direct injection of contaminants into the recharge system or the incorporation of surface contaminants into runoff that serves as a groundwater source. So long as the absorptive and retentive capacity of the soil is not exceeded, however, soil proves extremely effective as a purifier for groundwater. Thus, the hazard that results depends on the quantity and reactivity of the contaminants and on the characteristics of the aquifers into which they are discharged. Water in shallow aquifers has a relatively short lifetime, and these aquifers are thus highly susceptible to contamination and capable of moderately rapid recovery once the source of the contamination is removed. In contrast, deep aquifers have very long lifetimes and take decades to centuries to recover once contaminated.

CHEMISTRY OF THE OCEANS

The chemistry of the oceans, a subject in vigorous and broad development, has some strong and interesting areas of overlap with other applications of aqueous surface chemistry. Perhaps the most obvious of these are the relatively high ionic strength (about 0.7 for the oceans)

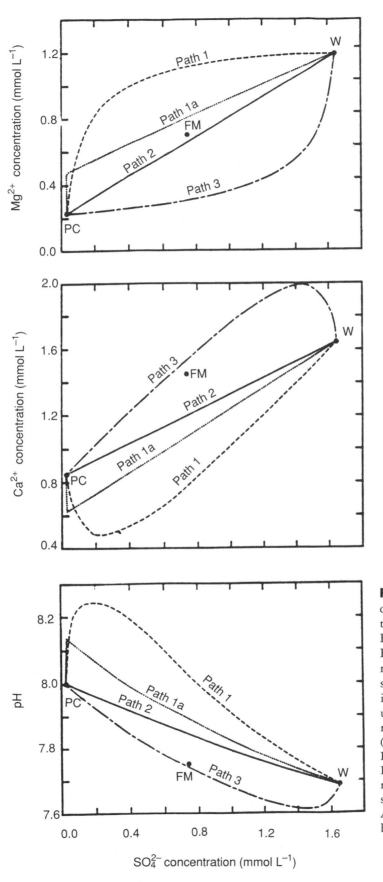

Figure 9.5 Changes in chemical constituents of groundwater during transport from Polk City (PC) past Ft. Meade (FM) to Wauchula (W), Florida. The filled circles show the measured values at the three sampling wells. The various curves indicate chemical calculations utilizing different groundwater and mineral dissolution assumptions. (Reproduced with permission from L. N. Plummer, D. L. Parkhurst, and D. C. Thorstenson, Development of reaction models for ground-water systems, *Geochimica et Cosmochimica Acta,* 47, 665–686. Copyright 1983 by Pergamon Press plc.)

TABLE 9.2. Concentrations of Major Salt Constituents in Seawater at 25 °C and Atmospheric Pressure[a]

CONSTITUENT	CONC. (mg L^{-1})
Na$^+$	0.48
Mg^{2+}	0.05
Ca^{2+}	0.01
K$^+$	0.01
Cl$^-$	0.56
SO$_4^{2-}$	0.03

a. These data are from M. Whitfield, Activity coefficients in natural waters, in *Activity Coefficients in Electrolyte Solutions*, R. M. Pytkowicz, ed., Vol. 2, pp. 153–299, 1979.

and commonality of ion content. Conditions typical of the world's oceans, which are of extremely uniform chemical makeup, are given in Table 9.2. Important related chemical parameters are the very high concentration of dissolved organic matter (an average of about 1 mg L^{-1}) in the surface waters of the ocean and a typical pH of 8.0 6 0.5 (i.e., slightly basic). An example of the distributions of constituents found in seawater as a function of position and depth is shown in Figure 9.6, which pictures dissolved organic carbon measured at a number of locations in the Philippine Sea. The concentrations fluctuate in an irregular manner, reflecting largely the biological complexity of the marine system.

The magnitudes of the fluxes of chemical constituents in deep ocean currents are truly awesome. Brewer and coworkers measured the concentrations of dissolved CO_2, O_2, and other chemical species in a section across the western Atlantic Ocean at 25 °N and combined these with estimates of oceanic mass transport. The result was a net CO_2 flux of 0.27 Pg C (10^{15} g C) per year moving southward. Even though the number is very large, it turns out not to be significant on the scale of the carbon budget. (The annual emission of CO_2 from fossil fuels is about 5.5 Pg C per year, for example.)

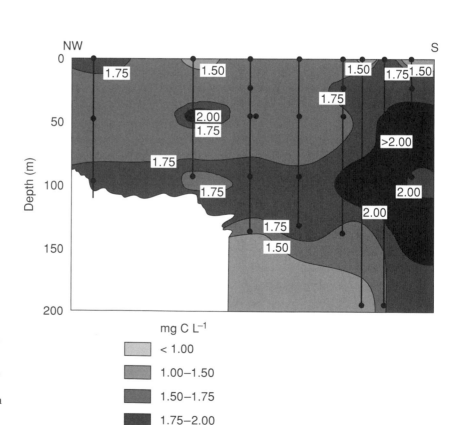

Figure 9.6 Concentrations of dissolved organic carbon in the Philippine Sea. The transect was along a NW→SE line from to 130 °E, 20 °N to 137 °E, 15 °N. (Reproduced with permission from E. A. Romankevich and S. V. Ljutsarev, Dissolved organic carbon in the ocean, *Marine Chemistry, 30,* 161–178. Copyright 1990 by Elsevier Science Publishers BV.)

Because of the high ionic strength of seawater, the distribution of dissolved metals among different chemical forms (the *speciation*) is substantially altered. Various techniques have been used to compute the speciation of the constituents; for the purposes of the present discussion, however, the important point is that some species show little tendency to complex to the ligands present, others exist predominantly in complexed forms, and still others are largely present as precipitates.

The chemistry of seawater is particularly interesting within the *photic zone*, the surface layer of the ocean within which penetrating solar radiation can enhance both chemical and biological activity. This zone is typically 20 m in depth, but much of the most energetic shorter wave radiation is absorbed in the upper meter or so, and it is there that marine photochemistry is most vigorous. The concentration of glyoxylic acid in the ocean, shown in Figure 9.7, illustrates this process. The glyoxylic acid is formed when seawater is exposed to solar radiation, presumably as a result of photochemical transformations of dissolved organic matter.

HYDROTHERMAL CHEMISTRY

Freshwater Hydrothermal Systems

Aqueous chemistry under high temperature conditions was first studied at geyser and caldera-forming systems on the continents. Freshwater hydrothermal systems form when magma or partially molten material is near enough to the surface to heat the groundwater with which it comes in contact. In many systems, the water several kilometers below the surface is at temperatures above 250 °C.

The water whose chemical composition has been altered by contact with deep, hot rock tends to come to the surface at fault lines, thereby creating boiling hot springs and geysers. The chemical composition of these solutions is controlled largely by equilibria between the groundwater and the rock and mineral structures through which the heated water migrates. These systems provide highly visible examples of evolving chemistry as a combination of surface and subsurface forcing.

A familiar example of a freshwater hydrothermal system is that of Yellowstone National Park, USA. Figure 9.8 (in the color insert) shows one of the hot springs in the park, together with the varied and abundant solid phases that have precipitated from the solutions as they reached the surface and cooled.

Marine Hydrothermal Vent Chemistry

In Chapter 2 we discussed the flow of magma up through the midocean ridges and the influence that this flow ultimately exerts on the motions of the continents and the generation of volcanoes and earthquakes. We

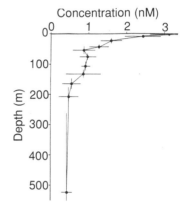

Figure 9.7 The concentration of glyoxylate anion as a function of depth for data collected during daytime at sites in the Gulf Stream off the coast of Florida, 1985. (Reproduced with permission from (D. J. Kieber and K. Mopper, Photochemical formation of glyoxylic and pyruvic acids in seawater, *Marine Chemistry, 21,* 135–149. Copyright 1990 by Elsevier Science Publishers BV.)

did not go into the chemical effects that must obviously result from the interaction of molten rock with cold seawater, partly because observing and sampling these systems at the bottom of the oceans was not an option until very recently. Within the last two decades, special research submarines have been diving to the ocean floor and examining the world beneath the sea.

One of the most dramatic discoveries made by the submersible vehicles has been the venting of hydrothermal solutions under the sea at the midocean ridge sites. Some of the fluid is little warmer than that of the surrounding seawater, but much is at temperatures exceeding 350 °C. As the solutions emerge from the vents, they may include substantial amounts of incorporated seawater. Indeed, estimates of the flux of seawater passing through the hydrothermal ridge crest system indicate that the entire volume of the oceans cycles through a hydrothermal stage about every 15–45 Myr; thus the vent chemistry has a significant role to play in historic and modern-day marine chemistry.

It is a difficult and dangerous task to sample hot vent fluids at depths of a thousand meters or more. Partly because of the difficulty and partly because the fluids can be captured only as they mix with seawater, some seawater is usually entrained in any samples that are taken. The degree to which the vent fluid has been diluted can be estimated by recognizing that magnesium is removed from seawater by rock–water interactions at temperatures above 150 °C. Thus, the amount of magnesium in a sample is an indication of the degree to which it has been diluted. Extrapolating all fluid compositions to $[Mg^{2+}] = 0$ permits comparison of what are presumably the undiluted vent liquid compositions.

As they emerge from the vents, the acidic, sulfide-rich solutions mix with alkaline bottom waters. The subsequent cooling and chemical transformations create precipitating sulfides and oxides, thus providing the name "black smokers" for the vents (see Figure 9.9 in the color insert). Iron, manganese, zinc, and other trace metals are highly enriched in the vent fluids relative to their seawater concentrations, and the gases CO_2, CH_4, and H_2S are highly enriched as well.

The warm temperatures and abundant supply of nutrients near the vents have resulted in vigorous and diverse communities of many different life forms, some familiar, some quite unfamiliar. It has been proposed, but not thus far determined, that the ocean floor near the vents may have been the origin of the precursor species to all Earth's living things.

SIMILARITIES AND DIFFERENCES IN THE CHEMISTRY OF EARTH'S WATERS

Having reviewed condensed water chemistry in a number of different regimes, it is instructive to assemble that information in various ways

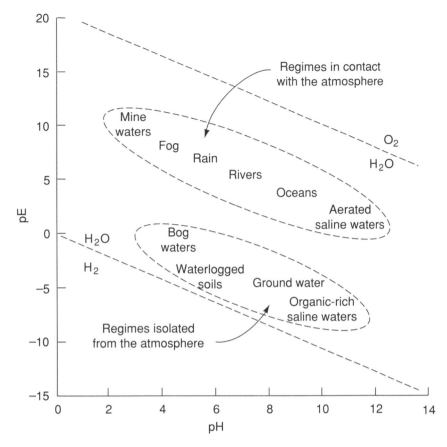

Figure 9.10 Approximate positions of some natural environments as characterized by pE and pH. The diagonal dashed lines represent the limits within which liquid water can occur. (Based on a diagram originally devised by R. M. Garrels, *Mineral Equilibria*, Harper, New York, p. 201, 1960.)

for comparative purposes. One of the first attempts to do that was Robert Garrels's plot of the positions of natural environments on a pE–pH diagram similar to that of Figure 9.10. He demonstrated that the waters of Earth span the range of oxidation–reduction potentials from the breakdown of water into hydrogen to its equilibration with oxygen of the atmosphere. He then used this tool in a long series of studies in which he related mineral assemblages laid down in past epochs to the environments in which they were formed.

Garrels's plot shows that most environments exposed to the atmosphere are oxidizing in nature, some rather significantly so. Among the strong oxidizers commonly available are ozone (O_3) and hydrogen peroxide (H_2O_2). Because the common strong acids HNO_3 and H_2SO_4 are highly oxidized forms of nitrogen and sulfur, strong oxidizing environments in the natural world are often strongly acidic. In contrast, environments isolated from the atmosphere tend to be reducing and are usually near-neutral or basic in pH level. Some environments, such as percolating groundwater, fall between these extremes.

Two decades ago, Ronald Gibbs of Northwestern University pointed out that many of the differences in the chemistry of Earth's surface waters could be understood by considering a small number of physical mechanisms. A version of a diagram that he used to illustrate

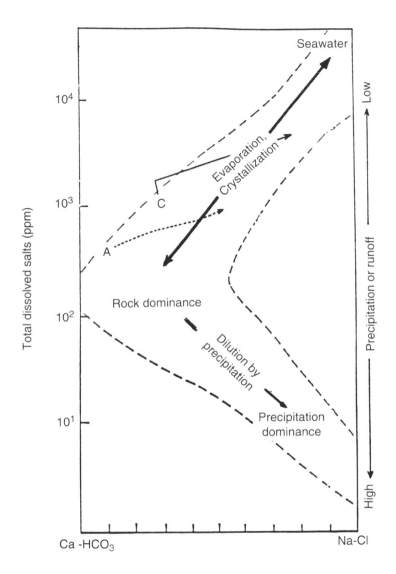

Figure 9.11 A diagrammatic representation of the mechanisms controlling the chemistry of Earth's surface waters. The lines labeled A and C represent the upstream-to-downstream traces of two rivers. (Adapted with permission from R. J. Gibbs, Mechanisms controlling world water chemistry, *Science, 170*, 1088–1090. Copyright 1970 by the AAAS.)

the concept is given in Figure 9.11, on which the left ordinate scale is the concentration of total dissolved salts and the abscissa reflects a high proportion of rock-derived calcium and bicarbonate ions at one extreme and a high proportion of sodium and chloride ions at the other. Nearly all surface waters fall within the butterfly pattern indicated by the dotted lines. Waters whose chemical composition locates them at the left center of the display are those whose chemistry is dominated by equilibration with local rock. If significant evaporation or crystallization occurs, the position of a water body on the diagram can move up or down the heavy line from left-center to upper-right. This shift tends to occur, for example, in rivers that are located in hot, arid regions and undergo evaporation and crystallization as they flow toward the ocean. Where substantial dilution by precipitation occurs, the position of a water body moves along the heavy line from left-center to lower-right, becoming more dilute and being much less strongly influenced by the rock through which the water travels. Many of the major chemical features of surface waters are thus the product of three mechanisms: atmospheric precipitation, equilibration with rock, and evaporation-crystallization from solution.

Figure 9.12 Chemical characteristics of some of Earth's natural waters. (a) temperatures; (b) pressures; (c) pH values; (d) ionic strengths.

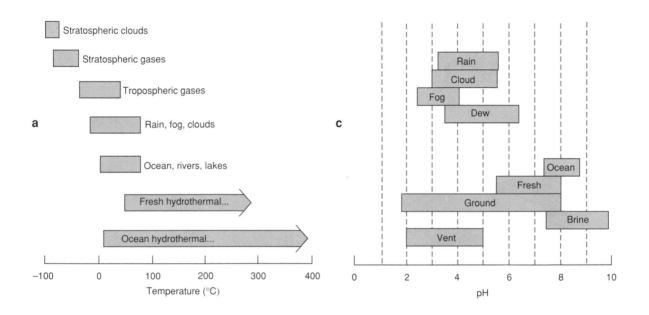

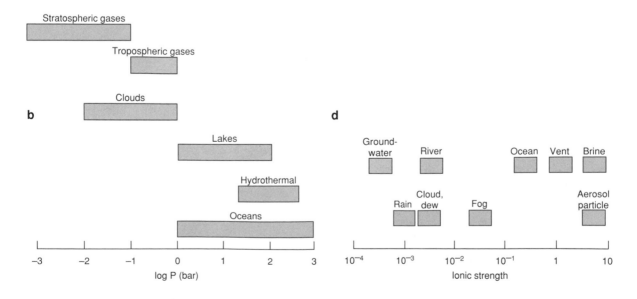

Figure 9.13 Concentrations of several chemical species in some of Earth's natural
waters: (a) chloride ion; (b) sulfate ion; (c) total iron; (d) H_2O_2.

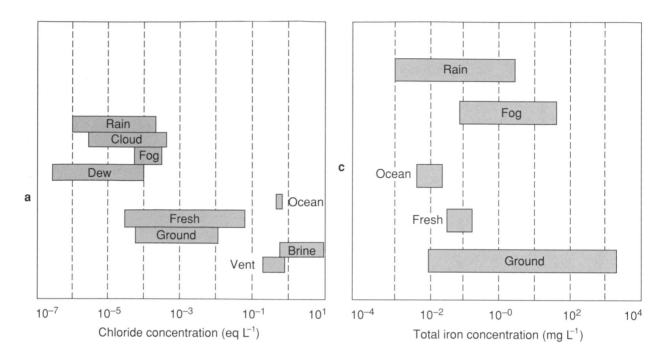

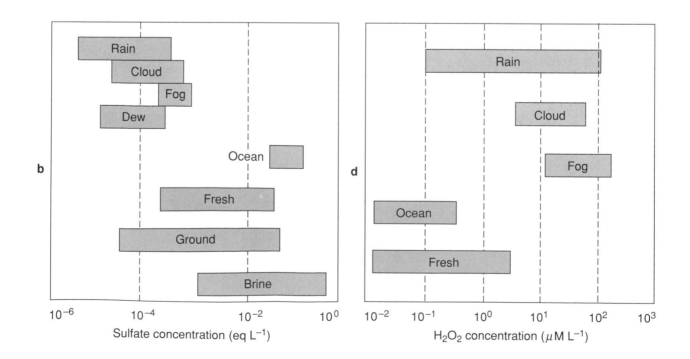

To carry further the concept of the unity of natural water chemistry, we show some physical properties of natural waters in Figure 9.12. Consider first the solution temperatures (Figure 9.12a). The values range from −90 °C for polar stratospheric clouds to about 400 °C for the hottest hydrothermal vent fluids. It is important to recognize that many of the regimes themselves have wide temperature ranges; deep ocean chemistry near hydrothermal orifices covers the approximate range −1 to 400 °C, for example.

In the majority of cases, the ambient pressure at which the chemistry occurs is 1 bar. High clouds, rain, and fog, however, exist at somewhat lower pressures, and polar stratospheric clouds at lower pressures still. Pressures substantially higher than 1 bar are encountered as one descends into deep freshwater lakes and into the deep ocean.

The acidity of geochemical water covers almost the entire pH range. Brines can have pH levels at least as high as 10, whereas that for aerosol particles can be well below 2. A consequence of this variability is that some regimes are conducive to rapid dissolution of minerals, some to rapid precipitation of minerals, and others to long-term stability. The degree of dilution of the solutions, as represented by their ionic strengths, is also very broad. In many cases in natural water chemistry, the ionic strength is high enough (greater than about $I = 10^{-3}$) to require correction for ion complexation. This exercise becomes increasingly difficult at ionic strengths above 3. Most atmospheric water is sufficiently dilute that corrections for complexation are unnecessary; most surface and subsurface waters, however, are concentrated enough to require corrections.

Among the many chemical parameters for which comparisons are of interest, Figure 9.13 illustrates four: chloride ion, sulfate ion, total iron, and hydrogen peroxide. In the case of chloride, the most abundant of Earth's common anions, the highest concentrations are encountered in evaporating surface waters, the lowest in raindrops. The range is almost seven orders of magnitude. The situation is slightly less diverse for sulfate, but the range is still some five orders of magnitude. The spread in iron concentrations is likewise very large, a fact that has important implications for the redox chemistry of solutions. The relative freedom with which iron oscillates between its two principal valence states, especially if exposed to solar radiation, establishes transition metal chemistry as an important field of study for geochemists. Hydrogen peroxide tends to be most abundant in atmospheric water, but is found to some degree wherever natural water is exposed to solar radiation.

Finally, it is appropriate to comment on the suitability of thermodynamics and kinetics for studies of Earth system aqueous solutions. The time scales that are relevant for this issue are those of fluid composition change due to external control (incorporation of reactive gases or solution evaporation, for example) and of composition change due to internal processes (chemical reaction or mineral formation, for example). If internal control dominates, the solution chemistry can be studied with thermodynamic techniques. Such an approach has been highly successful for brines and relatively successful for some applications in seawater and groundwater. Conversely, if external control dominates, the solution chemistry must be studied with kinetic techniques, as has been found necessary for nearly all atmospheric applications, marine and freshwater photic zones, acid mine drainage waters, and hydrothermal vent fluids.

No chapter could hope to treat the full diversity of Earth's water chemistry, nor has this chapter done so. What has been attempted, however, is to provide examples that demonstrate the great chemical overlap between regimes that differ very greatly in basic properties such as density, temperature, oxidizing capacity, and degree of dilution. It is not inaccurate to say that few aspects of physical inorganic chemistry are unexplored by the condensed water chemistry of Earth.

EXERCISES

9.1 The main constituents of runoff from mine tailings (discarded ore debris) into a mountain stream are given in the following table. Compute the ionic strength of the runoff.

Dissolved Constituents of Selected Fresh Waters		
	CONCENTRATION (mg L⁻¹)	
Species	Acidic tailings (pH 3.8)	Stream (pH 6.7)
Na^+	4.3	4.5
Mg^{2+}	13.0	3.9
Zn^{2+}	8.6	—
Ca^{2+}	2.7	25
SO_4^{2-}	91	90
Cl^-	4.0	12

9.2 Assume that the runoff from the previous exercise enters a flowing stream, where it constitutes 5% of the total flow. The stream's chemical characteristics are given in the table in Exercise 9.1. Compute the pH of the combined water, the concentrations of the principal ionic species therein, and the ionic strength.

9.3 Precipitation with a pH of 3.13 and ion concentrations of $[SO_4^{2-}]$ = 311 μeq L⁻¹, $[NO_3^-]$ = 212 μeq L⁻¹, $[Cl^-]$ = 23 μeq L⁻¹, $[NH_4^+]$ = 116 μeq L⁻¹ falls into a lake for 2 hr at a rate of 8 mm hr⁻¹. The lake has a surface area of 3.7×10^5 m², an average depth of 11 m, and an ion composition equal to that of the stream in the table in Exercise 9.1. Compute the pH and ion concentrations of the lake following the rainfall, neglecting both runoff into the lake and outflow from it.

9.4 Rivers are the main source of water to the oceans. The ionic strength of river water is typically about 0.03 and that for oceans is always very close to 0.7. Give qualitative explanations for how this occurs.

FURTHER READING

H. D. Holland, *The Chemistry of the Atmosphere and Oceans*, Wiley-Interscience, New York, 1979.

F. J. Millero, The physical chemistry of natural waters, *Pure and Applied Chemistry, 57*, 1015–1024, 1985.

R. A. Smith, R. B. Alexander, and M. G. Wolman, Water-quality trends in the nation's rivers, *Science, 235*, 1607–1615, 1987.

W. Stumm, and J. J. Morgan, *Aquatic Chemistry*, 2nd Ed., John Wiley, New York, 1981.

10

Ancient Earth: Climate Histories

It is hard to imagine a world in which climate, as characterized by available solar energy, temperature, and precipitation, is significantly different from that to which we have become accustomed. In part, this is because our experience and the written record with which we may be familiar describe temperature and precipitation conditions not unlike those typical of today. Climate histories are also recorded in rocks, fossils, and other records, however, and those histories do not support the notion of constancy; indeed, they depict a world in which climate change, rather than constancy, is the norm.

Recall from Chapter 1 that the age of Earth is approximately 4.6 Gyr. The oldest dated mineral grains (found in a sand deposited later) are about 4.2 Gyr old, the oldest igneous rocks about 3.9 Gyr old, the

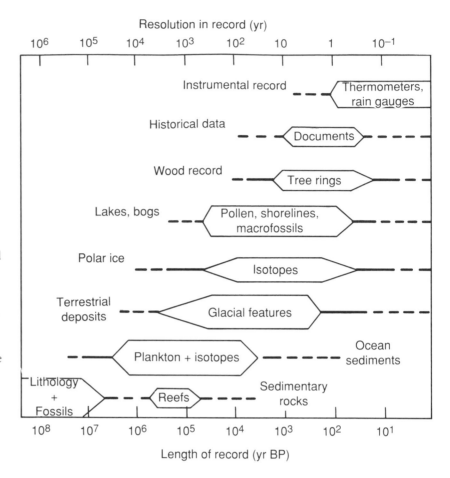

Figure 10.1 The time span of various climatic records from decades to hundreds of millions of years. The resolution in each record is also indicated, given the rule of thumb that it is about 1% of the length of the individual climatic record. The thickness of each polygon is an attempt to reflect the relative contribution for each contributing record at each time in the past. (Adapted with permission from T. Webb III, J. Kutzbach, and F. A. Street-Perrott, in *Global Change*, T. F. Malone and J. D. Roederer, eds., pp. 182-218. Copyright 1985 by Cambridge University Press, UK.)

oldest metamorphosed sediments about 3.8 Gyr old, and the oldest known fossils dated about 3.8 Gyr old. As will be seen, rock and fossil records provide us with a general idea of climate conditions since those very early times. The records are not at all detailed, however. Much later, at about 600 Myr BP, marine life with shells began to occur, evolve, and multiply rapidly, and fossil data provide a more complete, but still sketchy, climate record. The situation becomes a bit better after 170 Myr BP, when sediment was laid down in the Pacific Ocean. Some of this material has been recovered and examined as part of the Deep Sea Drilling Project and has provided much recent information. For the last 250 kyr, ice sheets have retained ancient samples that can be recovered today; this situation permits the first direct comparisons of detailed climate patterns with possible external causes to be made and begins to provide chemical signatures of atmospheric conditions as well. A final stage is the time span from about 20 kyr BP to the present, for which the evidence for climate and chemistry from a number of different sources permits us to know the environmental conditions in fair detail for a period substantially longer than that in recorded human history.

The time scale for the various climate records is shown in Figure 10.1. Logarithmic scales are used to construct the display; as the length of the record becomes longer, however, the time resolution and clarity in the record become poorer. For the relatively modern records, a number of different sources of information can be drawn upon. But for the older records, only one or two sources of data are available.

All these sources of past information provide a fascinating look at the history of our planet. They provide, as well, a perspective on where the planet is now and where it may be going. In this chapter, and several of those following, the different records will be used and discussed in some detail. Climate records of the distant past are the focus of the present chapter. The discussion is organized by geological period, and reference to Figure 1.9 will be helpful in putting the different periods into order and appreciating their differences in temporal scale.

PRECAMBRIAN CLIMATE

Temperatures on Early Earth

The oldest known metamorphosed sediments, that is, minerals laid down under water, are dated about 3.8 Gyr BP. This simple fact indicates that global average temperatures have exceeded 0 °C and that surface waters were common throughout almost the entire history of the planet. The presence of surface waters also implies the existence of at least a modest atmosphere, because liquid water would evaporate completely at atmospheric pressures below about 6 mbar. (Recall that today's atmospheric pressure at sea level is 1000 mbar = 1 bar.) Thus, except for the first 5% or so of the planet's life, for which we have no direct evidence, we can say with some assurance that the planet has always had substantial amounts of surface water and at least a modest atmosphere. Further, carbonate rocks are seen at all ages; they are precipitated from oceans containing CO_2 that are in equilibrium with an atmosphere containing CO_2. Thus, the atmosphere has contained carbon dioxide throughout most or all of its existence.

Sedimentary rocks vary as a consequence of the degree of turbidity (that is, the amount of suspended dirt, mud, and silt) in the overlying water. Research into this dependence has led to the ability of scientists to estimate the depth of the overlying water from the morphology of the rocks. Such studies suggest that some water depths during the Archean were of the order of a few kilometers. Depths of that magnitude indicate the presence of substantial oceans. Thus, very early Earth was much more likely to have been mostly covered by water than mostly covered by land.

The presence of water establishes a surface temperature above 0 °C and below the boiling point of water, but, because the boiling point is a function of atmospheric pressure (recall that water boils at lower temperatures on a mountaintop than at sea level), the planet's upper limit for surface temperature is not well defined merely by water's presence. It is possible, however, to derive limits to the temperature from the presence of a flourishing biological community at least as early as 3.5 Gyr BP. Also of use are temperatures derived from deposits of ancient glaciers, the *tillites*, which occur only under rather rigid constraints of temperature. Taken together (Figure 10.2), these considerations suggest that our water-covered world would have had reasonably stable temperatures averaging about 7 °C, although excursions of a few degrees below the freezing point of water on parts of the planet are not ruled out.

The Rock Record of Climate

As Earth's distance and inclination to the Sun follow their periodic patterns, as the oceans ebb and rise, and as continents are formed, the *albedo* (the reflectivity of Earth's surface) undergoes change. Acting in concert with the solar radiation, land areas can influence the albedo to greater or lesser degrees because of ice cap formation and growth of vegetation. The records of periods in which the land was covered by snow and ice are retained in the sizes, types, conditions, and positions of the rocks. This evidence has been used to show that Earth has passed through a number of cycles of *glaciation* (i.e., the formation and subsequent removal of substantial ice cover).

　　One of the active areas of experimental geology is the study of the temperatures, pressures, and physical and chemical environments under which different minerals tend to form. Among the more easily identified features are rocks that reflect the grinding action and temperature and pressure conditions peculiar to glaciation. When rocks showing patterns related to glaciation are radioactively dated, the results provide a history of ancient climate, limited largely by the

Figure 10.2 Constraints on the history of temperature near Earth's surface. The strongest constraint is imposed by the continuous existence of liquid water, which requires temperatures exceeding 0 °C at least locally. The next strongest constraint is probably the occurrence of distinctive lithographic features (such as tillites at 2.2 Byr BP). The temperature history suggested by lithographic features is indicated by the squiggly lines. The biota provides an upper constraint in that large groups of organisms are known to be intolerant of high temperatures; those tolerances decrease with increasing evolutionary rank, and a widespread and diverse biota of a given rank implies a constraint on ambient near-surface temperature. (Adapted from J. G. Cogley and A. Henderson-Sellers, The origin and earliest state of the Earth's hydrosphere, *Reviews of Geophysics and Space Physics*, 22, 131–175, 1984.)

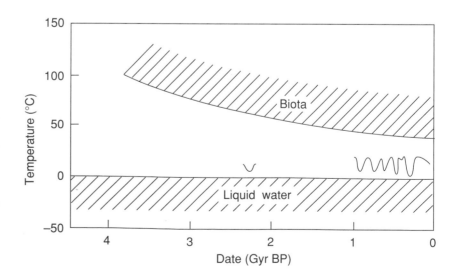

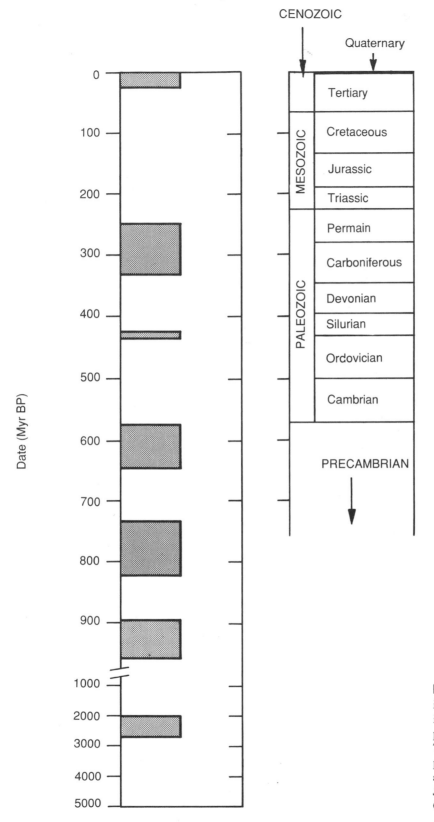

Figure 10.3 Major glacial epochs in Earth's history. Note the change in time scale at 1000 Myr BP. (Reproduced with permission from T. M. L. Wigley, Climate and paleoclimate: What can we learn about solar luminosity variations?, *Solar Physics, 74*, 435–471. Copyright 1981 by Kluwer Academic Publications.)

degree of completeness of the rock record. An example of such a history is shown in Figure 10.3, where the earliest verifiable glacial epoch (but not necessarily the earliest glacial epoch that occurred) is at about 2700–2300 Myr BP. The glaciation appears to have been extensive, although the supporting record is quite fragmentary. Its cause is uncertain; it may have been a consequence of rather low solar luminosity, of the presence of significant landmasses to reflect radiation, and of low concentrations of greenhouse gases, although these speculations are not supported by evidence.

Following the glaciation (at about the junction of the Archean and Proterozoic time periods), Earth was apparently warm and devoid of permanent snow or ice for 1000 Myr or so. The second known glaciation occurred at about 950 Myr BP, and two others followed at about 820–730 Myr BP and 640–580 Myr BP. The late Precambrian was a major period of mountain building on Earth, and the glaciations may have been related to tectonic motions and continental disruption.

The Weak Sun Paradox

An important aspect of early climate is the *Weak Sun Paradox*. The basis for this paradox is the knowledge that the Sun is a common type

Figure 10.4 The solar flux (energy per unit area and time) at the distance of Earth (relative to the present solar radiation) resulting from a standard Sun model and from three models all beginning with a mass 1.1 times that of the present Sun and allowing mass loss. Three different time scales of mass loss (τ_m) are shown because of a lack of constraints to choosing among them. (T. E. Graedel, I.-J. Sackmann, and A. I. Boothroyd, Early solar mass loss: A potential solution to the weak sun paradox, *Geophysical Research Letters, 18,* 1881–1884, 1991.)

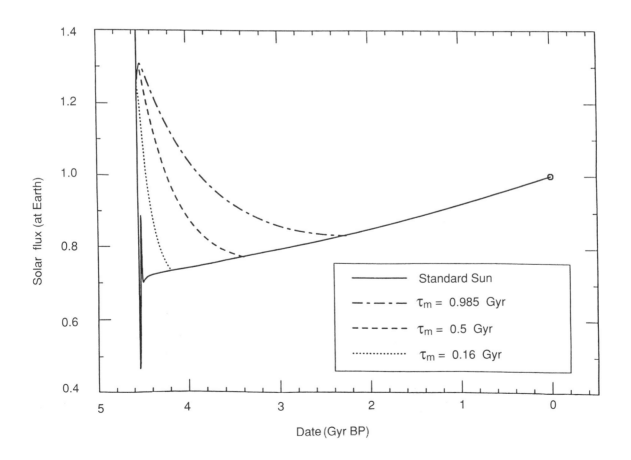

of star whose life cycle and properties are thought to be well known. Except for a brief period very early in the Sun's life, the "standard models" of solar physics have indicated that the luminosity of the early Sun was about 70–80% of the present value and that it has increased steadily since that time. (The luminosity is the rate of total energy output and has a present value estimated at 3.86×10^{33} erg s^{-1}.) As a result, the amount of solar radiation received by the young Earth would have been much less than is being received today, and the effective radiation temperature of the planet would be expected to be 10–15 K below today's value (see Chapter 3). Nonetheless, evidence from the rock record indicates that the surface temperature at both periods was quite similar.

How could significantly less radiation have been able to produce a warm Earth? Two answers have been posed. One relies on increased retention of infrared radiation, and the most plausible way in which that could happen is that a higher percentage concentration of carbon dioxide was present at that time, perhaps because the embryonic biosphere of that period was unable to sequester any significant amount of CO_2. Quite large CO_2 input during initial atmospheric formation would be required, however, because the presence of limestone ($CaCO_3$) by 3.8 Gyr BP indicates that removal of atmospheric CO_2 by limestone formation was substantial, and the rock record does not provide much in the way of constraints on what the CO_2 concentration might have been.

A second suggestion has to do with speculative new models of solar physics that suggest that the early Sun lost mass as it evolved and that this resulted in an early solar luminosity much higher than had previously been thought. The resulting solar fluxes at Earth predicted by the models are compared in Figure 10.4. The higher fluxes of the mass-loss model are readily able to explain early planetary temperatures in the liquid water range.

The resolution of the weak Sun paradox will not occur soon, and we may never know for certain the early workings of the planet. What does appear certain, however, is that despite our difficulties with the paradox, Earth found an effective way to provide water over most or all of the planet very early in its existence.

PALEOZOIC AND MESOZOIC CLIMATE

The Fossil Record of Climate

The beginning of the Paleozoic is identified by the occurrence and spread of marine life with shells. The preservation of the shells provides a much more detailed climatic record, which has been used to augment the rock information. A principal climate event of the early Paleozoic was a brief glaciation that occurred about 430 Myr BP. This glaciation event immediately preceded an interesting occurrence

Figure 10.5 Eustatic curves for the Phanerozoic. Great marine biological extinction events are noted by the asterisks. (Adapted with permission from A. Hallam, Pre-Quaternary sea-level changes, *Annual Reviews of Earth and Planetary Science, 12*, 205–243. Copyright 1984 by Annual Reviews, Inc.)

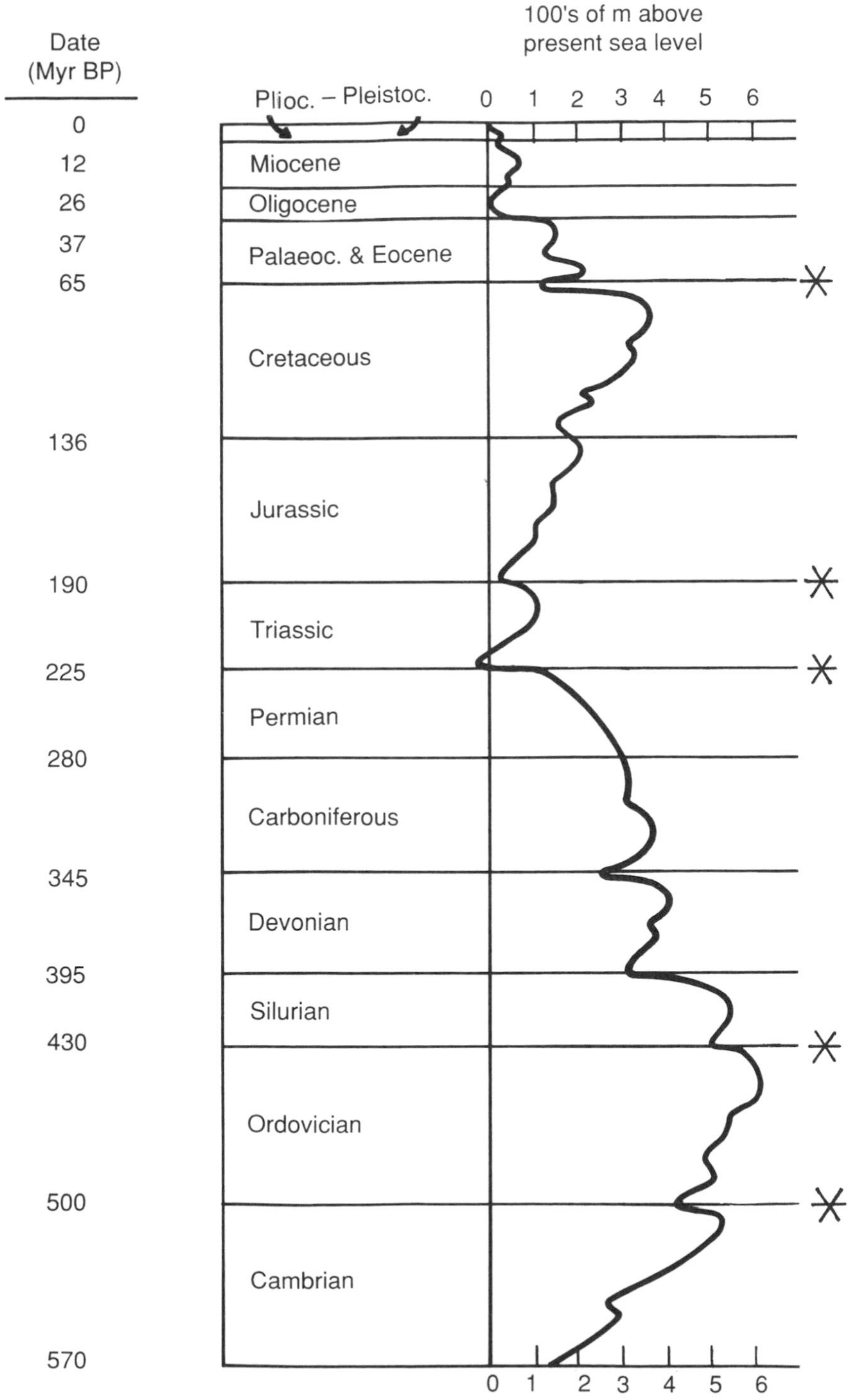

of the mid-Paleozoic, the evolution of land plants at about 400 Myr BP. (A moss and lichen cover might have been present much earlier.) The significance of the plants is that a fully developed vegetative land cover can decrease the albedo by as much as 10–15%, although in the earliest evolutionary stages the change would have been much smaller. The Paleozoic concluded with a long and somewhat mild glacial period, centered at about 300 Myr BP. This glaciation was probably related to continental drift, which cycled some of the largest continents through locations over the poles during this period.

The Mesozoic, spanning the period 225 to 65 Myr BP, appears to have been an extended period of widespread warmth, especially in the higher latitudes. Throughout much of this period, the continents were joined in a single supercontinent (Pangaea; see Chapter 2), which extended from high southern to high northern latitudes. This geographical situation may have facilitated the meridional transport of heat by ocean currents and thus resulted in a more even global temperature distribution than would be the case with dispersed landmasses.

The Eustatic Record

Because Earth is in a constant state of evolution, total global ocean cover, if it ever existed at all, did not exist in perpetuity. As tectonic processes gathered force and large landmasses began to grow, large altitude variations were created, the prospect of ice formation existed, and the water coverage of the planet decreased. Little can be said of the extent of water coverage until the formation of the present continental structures, for it is on their margins that the evidence for sea level changes is contained. The approach recognizes that sequences of sedimentary rock formations are different on land and at the ocean margins, and that by determining the ages and locations of such sequences, one can tell whether a given location near a continental margin was or was not under water at a particular time. If similar results are obtained from samples obtained in different locations within a region, then regional effects can be separated from purely local effects. If similar results are obtained for a number of regions over the planet, then the underlying cause can be considered to be *eustatic* forces, i.e., forces associated with global changes in the sea level.

Eustatic curves constructed by the methods outlined above trace the history of sea level from Cambrian times to that of the present. Such a curve (Figure 10.5) shows that throughout much of recorded Earth history the ocean depths have been far greater than they are at present. In addition to providing widespread flooding of the edges of continents, such conditions would have produced considerable flooding of continental interiors, creating the vast inland lakes whose existence has been well documented.

What is the cause of eustatic changes? Two reasonable possibilities have been suggested. One is the melting and freezing of polar caps;

the other is changes in the volume of ocean basins. Records of the polar ice caps are contained in the rock and debris that indicate the extent of glaciers; those records show that glacial expansion and contraction is insufficient to explain the eustatic changes shown in Figure 10.5. Changes in ocean basin volume, therefore, turn out to be a likely cause. As outlined in Chapter 2, tectonic activity results in enhanced seafloor spreading, the formation of substantial midocean ridges, and hence the interaction of plate tectonics and eustacy. The tectonoeustatic processes responsible for the sea level history of Figure 10.5 result from variations in heat flow from the mantle to Earth's surface, and the variations in eustasy are explainable in at least approximate terms by the interaction of internal heat with stress and strain relief in the overlying rock structures.

The relationship between eustasy and biology is shown in Figure 10.5, which indicates times of great extinctions of marine organisms. All of these events coincide with periods during which global water depths had suddenly decreased. Such changes, particularly if relatively rapid, would make adaptation very difficult. Other possible causes of extinctions, some of them discussed later in this chapter, may be involved as well.

Bolide Impacts

The time scale of change of many climate characteristics is slow with respect to geological time. There is increasing evidence, however, for the occasional occurrence of abrupt changes. Often the existing evidence for such a shift is biological rather than geological, such as a sudden extinction of large numbers of species of plants and animals.

Among the possible causes of abrupt climate events is the collision of a comet or asteroid (a *bolide*) with Earth. It has been calculated that an impacting object of 10 km or more in diameter would have a dramatic effect on the planetary atmosphere. Typical impact velocities of such a bolide would be 20 km s^{-1} or more, releasing 60 million megatons of energy and creating a crater 150 km in diameter. The high rate of shock heating associated with the impact would warm a substantial fraction of the atmosphere to 2000–3000 K, dissociating molecular oxygen and inaugurating the following series of reactions:

$$O_2 \rightarrow O + O \qquad (10.1)$$

$$O + N_2 \rightarrow NO + N \qquad (10.2)$$

$$N + O_2 \rightarrow NO + O \qquad (10.3)$$

$$NO + NO + O_2 \rightarrow NO_2 + NO_2 \qquad (10.4)$$

$$NO_2 + NO_2 + H_2O \rightarrow HNO_2 + HNO_3 . \qquad (10.5)$$

One immediate consequence of these processes is that the high concentrations of NO and NO_2 would rapidly remove ozone from the atmosphere, thus allowing the penetration of high-energy ultraviolet radiation to the surface. A second is that the nitric and nitrous acids will be incorporated into cloud and rain droplets, and will be deposited on the land and sea as precipitation. The acidity of the precipitation depends on the amount of acid that was created and on how fast it was taken up by the droplets; estimates of these factors suggest that very high acidities developed, dissolving the calcareous shells of animals, damaging vegetation, and creating difficult environments for aquatic life.

Changes in atmospheric chemistry would not be the only result of a bolide impact. Such an impact would also create a global dust cloud sufficiently dense to block out the Sun's radiation for several months. The decrease in solar radiation would inhibit photosynthesis, resulting in a rapid die-off of plants and thus a loss of food for herbivores. The heat would ignite extensive wildfires, which would consume a large fraction of Earth's biomass. (In apparent support of this idea, large amounts of soot, suggesting worldwide forest and grass fires, have been detected in the sediments at several interepoch boundaries.) High levels of soot particles in the troposphere would have led to a strong cooling of Earth. After the soot settled, high concentrations of CO_2, N_2O, CH_4, and other species from the fires and the rotting vegetation might have remained in the atmosphere for centuries to millennia, causing a substantial climate warming.

Although there have been a number of instances of abrupt biological extinctions noted in the fossil records, the one that seems most likely to have been due to a huge impact is the sudden extinction of about 70% of Earth's species some 65 Myr ago, at the Cretaceous–Tertiary (K–T) boundary. Besides the biological evidence of a major change, analyses of rocks deposited at that time reveal anomalously high levels of iridium, a rare earth element. The anomalous iridium concentration is global in extent and is attributed to the collision of an iridium-rich meteorite with Earth. In addition to the iridium evidence, the presence in the same strata of mineral grains thought to form only under tremendous shock impact and soot reflecting widespread fires indicate in completely different ways that a major bolide impact occurred.

Mass extinctions of much of the life of Earth have occurred a number of times during the Paleozoic and Mesozoic, and subsequently. Can each of these events be related to bolide impacts? Much research is underway to attempt to answer those questions, and no firm answers are available. It does appear, however, that bolides have struck Earth a number of times during its history. Arguments have been made for the probability of one such event about every 100 Myr. Some data, like those of Figure 10.6, suggest an impact cycle with a periodicity of about 26 Myr. Evolutionary biologists are thus led to

Figure 10.6 Species extinctions during the past 250 Myr. The arrows are spaced 26 Myr apart. The highest peaks represent loss of 50% or more of the observed families, but differences in species diversity in the fossil record make quantitative peak-to-peak comparisons problematic. These data are the basis of the theory of periodic extinctions in which a postulated solar companion star with a 26-Myr period, when near perihelion, perturbs the Oort Cloud and sends comets into the inner solar system. (Adapted with permission from L. W. Alvarez, Mass extinctions caused by large bolide impacts, *Physics Today, 40*(7), 24-33. Copyright 1987 by the American Institute of Physics.)

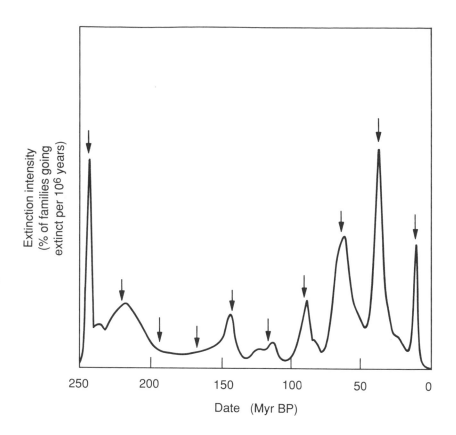

consider the possibility that the evolution of life on Earth has not been a continuous process but more like a series of restarts, each from a different starting point. As with many data concerning Earth of the past, the record is suggestive but not yet convincing.

CENOZOIC CLIMATE

The period from the late Cretaceous through much of the Tertiary is marked by reorganizations of Earth's landmasses, as the supercontinent fragmented and plate tectonic forces slowly drove the separated continents into the positions in which we find them at present. A number of climate events of various lengths and types occurred during this period, influenced at least in part by tectonic alterations of the solar heating cycle.

Much higher time resolution in climate histories is available during the Quaternary, a period that began about the time when members of the genus *Homo* first appeared on the planet. The dominant temporal climate pattern during much of that period was one of large-amplitude glacial–interglacial cycles with periodicities of about 100,000, 41,000, and 23,000 yr. These periodicities all turn out to be consequences of variations in the motion of Earth through space.

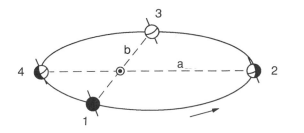

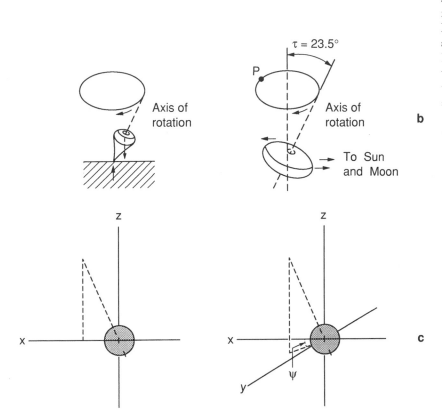

Figure 10.7 (a) The orientation of the axis of Earth in its orbit. In the course of one year, the orientation of the axis in space remains almost the same. A planet has seasons when its equator is inclined relative to the plane of its orbit. For Earth, the obliquity (tilt) is 23.5° and the seasons are pronounced. Points (1) and (3) are those of the vernal and autumnal equinoxes, when, as Earth rotates about its axis, all latitudes receive twelve hours of sunlight. Three months after the vernal equinox, at the summer solstice of point (2), the northern hemisphere receives sunlight for more than half the day, the southern hemisphere less. The situation is reversed at the winter solstice, point (4), when the southern hemisphere receives sunlight for a greater portion of the day than does the northern. (b) Precession of a spinning top (left) and of the spinning Earth (right). The attraction of Earth upon the center of gravity of the inclined top would allow it to fall. This is prevented by the rotation of the top. The result is a conical motion of the top's axis of rotation. In the case of Earth, axial rotation prevents the equatorial bulge from assuming the orientation of the ecliptic. The result is a conical motion of the axis of rotation of Earth. (c) A diagram illustrating precession in Earth's axis of rotation. In the diagram at the left, the axis of rotation falls within the plane created by the orbital path of Earth about the Sun (x) and the orthogonal direction to that path (z). (xy defines the plane of Earth's orbit.) The diagram at the right shows the movement of the axis of rotation out of that plane as a consequence of the gravitational attractive forces of Sun, Moon, and planets.

A schematic diagram of Earth's elliptical orbit about the Sun is shown in Figure 10.7a. The degree of ellipticity (the *eccentricity* = b/a) of the orbit is one of the three parameters describing orbital variations. The second is the angle of tilt between the axis of rotation of Earth relative to the Earth–Sun orbit (Figure 10.7b). Just as a spinning top wobbles around its angle of tilt, Earth wobbles as it *precesses* around its axis of rotation. The third orbital parameter is the orientation of the axis of rotation with respect to the orbital path (Figure 10.7c). These three parameters, which may be better appreciated by completing Exercise 10.4, obviously influence the amount of solar radiation received by particular regions of the planet. All of these factors undergo gradual and predictable change as a result of gravitational

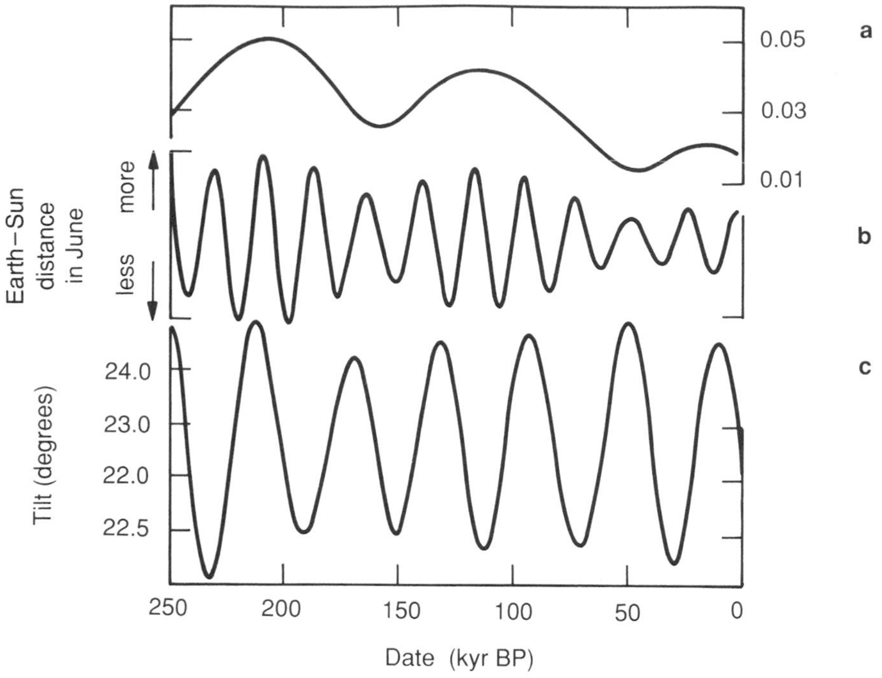

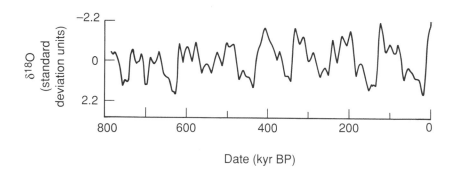

Figure 10.8 Long-term variations in three of Earth's orbital parameters that influence climate: (a) eccentricity; (b) precession; and (c) tilt. The time scale is from 250,000 yr BP to the present. (A. Berger, Milankovich theory and climate, *Reviews of Geophysics, 26,* 624–657, 1988.)

Figure 10.9 Composite oxygen isotope record from deep sea cores. The data are plotted so that interglacial conditions appear near the top of the graph and glacial conditions appear near the bottom. (Adapted with permission from J. Imbrie and J. Z. Imbrie, Modeling the climatic response to orbital variations, *Science, 207,* 943–952. Copyright 1980 by the AAAS.)

attractions for Earth by the Sun, the Moon, and the planets. As a consequence, these influences on climate have been computed for long periods into the past. The periods and timing of these variations are shown in Figure 10.8.

The influence of orbital variations on the solar radiation received by Earth is reflected in Earth's climate. A typical example is that of about 10,000 yr ago, when the inclination of Earth's rotational axis relative to the ecliptic plane was about 24.5° and the closest approach of the orbiting Earth to the Sun (perihelion) occurred in July. Compared with the present situation of a 23.5° inclination angle and a January perihelion, this past condition resulted in about 8% more solar radiation reaching the northern hemisphere in the summer and about 8% less in winter. As discussed in Chapter 3, the distribution and intensity of received radiation define the flow of the winds, the precipitation rates, and other climatic conditions, so it can readily be seen that solar radiation changes of that magnitude would have had direct and noticeable impacts on climate.

We will discuss in Chapter 11 how the proportion of the different oxygen isotopes in carbonate marine skeletons provides a rough measure of global climate. In Figure 10.9, we reproduce a composite plot of oxygen isotope variations for all the major ocean basins for the last billion years. Many fluctuations can be seen in these rather "noisy" data, but detailed mathematical analyses indicate a 100,000-yr long wave periodicity in the climate, with interspersed glacial and warm periods each lasting an average of 10,000 yr.

HOLOCENE CLIMATE

By far the most extensive record of past climates that we have is for the Holocene (the most recent 10,000 yr) and the period immediately preceding it. A number of different techniques have been used to derive climate conditions for various locales and to integrate such information into regional and global assessments of climate. In this section, we present a selection of Holocene climate records, discuss an overall climate summary prepared by a large, interdisciplinary working group, and describe some of the climatic effects that were produced in the Holocene by short-term events.

Climate Histories Preserved in Sediments

The sediments of lakes and wetlands are potentially rich sources of information about past environments, not only of the lakes or wetlands themselves, but also of the surrounding terrain. The sediments arise from two sources: internal processes within the lake itself, and deposition by wind transport or runoff to the water. Sediments that may be suitable for scientific studies must be carefully dated (a variety of methods are available), and some physical or chemical parameter of the sediment must be selected and evaluated at each stratigraphic horizon present in the sediment core. The property chosen must be one that does not undergo change after it is fixed into position; suitable properties include volcanic minerals, pollen grains, and microfossils.

An example of the information available from sediment cores is shown in Figure 10.10, which is a pollen diagram from a bog in northern Minnesota. The record traces the local pollen history for about 11,000 yr for 14 plant taxa. The record begins with spruce pollen being dominant, an observation implying that the climate was quite cool. Soon after, pine became dominant, a change indicating a significant warming. The pine dominance gave way to oak at about 8500 yr, a pattern typical of drier climates. Birch coexisted with oak for a period

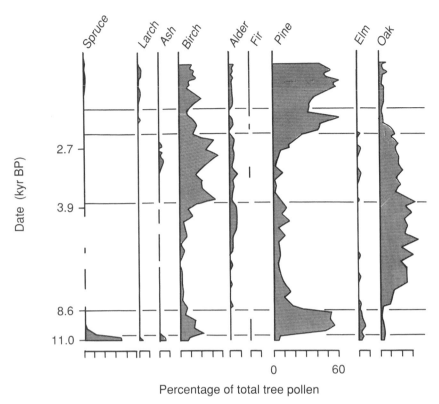

Figure 10.10 Pollen diagram for a bog in northwestern Minnesota over the period 11,000 yr BP to the present. (Adapted with permission from J. H. McAndrews, in *Quaternary Paleoecology*, E. J. Cushing and H. E. Wright, Jr., eds., pp. 218–236. Copyright 1967 by University of Minnesota Press.)

of about 2000 yr; this dual dominance gave way to the current overwhelming abundance of pine. Similar information is also available from sediment cores on grasses and other vegetative species.

Climate Histories Preserved in Trees

Trees have special characteristics as recorders of atmospheric information, because they interact with the atmosphere (and retain signatures of that interaction), because they reflect some of the properties of the climate in which they grow, because some trees are very old (thus providing long, continuous records), and because the records can be very accurately dated, generally to within a specific year. These properties are encapsulated in the rings produced by trees as they undergo their annual cycle of growth. Tree ring records may be obtained from dead but well-preserved trees and from tree stumps, and also from cores taken from living trees, the latter establishing an extensive and continuing climatological record.

The most extensive use of tree rings has been year-by-year determinations of their content of the radioactive isotope ^{14}C. The relatively short half-life of ^{14}C—5730 yr—in combination with the accurate dating allowed by the tree rings then allows analyses to be made of the amount of ^{14}C present in the atmosphere at the time of carbon uptake into the trees, thus providing a marker of solar activity.

An example of the use of tree ring carbon data is shown in Figure 10.11, which illustrates ^{14}C changes in tree rings from the west coast of the United States for the last millennium. The shorter term variability, with about a 200-yr period, is quite evident; this variation is tentatively ascribed to periodic variations in the pulsation of the core of the Sun, and thus of the solar wind that acts to inhibit or allow passage of galactic cosmic rays into Earth's atmosphere. A second feature worth noting in Figure 10.11 is the rapid decline in the ^{14}C signal after about 1850. This pattern is attributed to increased release of carbon dioxide from fossil fuel combustion, the carbon from that source being depleted in ^{14}C. The tree ring records are, therefore, also used as markers for the net release of carbon to the atmosphere due to fossil fuel burning, soil organic matter decay, and deforestation.

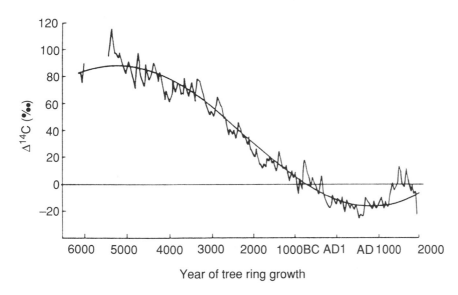

Figure 10.11 Changes in the relative abundance of the carbon-14 isotope in tree rings over the past 8000 yr. The sharp decline in the values after 1850 represents fossil fuel release of carbon dioxide, a process that alters the isotope ratios. (Adapted with permission from M. Stuiver and P. D. Quay, Changes in atmospheric carbon-14 attributed to a variable Sun, *Science, 207*, 11–19. Copyright 1980 by the AAAS.)

An Integrated Holocene Assessment

The most comprehensive look into paleoclimates has been that performed by COHMAP (Cooperative Holocene Mapping Project). This international research group has as its focus the understanding of the climates of the past 20 kyr. Its main goals are to assemble global sets of paleoclimatic data, to calibrate the data in climatic terms, and to use the data to test the ability of global circulation models to simulate past climates.

COHMAP performed its assessment primarily with data on pollen, lake levels, and marine plankton, although complementary geomorphic, archeological, and other evidence were used. We have discussed some, but not all, of these types of data earlier in this chapter and in Chapter 2. The principal finding is that the driving force behind much of the long-term climatic change is the varying insolation produced by changes in the orientation of Earth's axis. The orientation 18 kyr BP was similar to that of today. Between 15 kyr and 9 kyr BP, the Sun–Earth distance decreased in northern summer and the axial tilt increased. As a result, seasonality was more enhanced in the northern hemisphere and less in the southern hemisphere than is the case at present. By 9 kyr BP, the average solar radiation over the northern hemisphere was 8% higher in July and 8% lower in January than is the case today. Since 9 kyr BP, these levels have gradually approached modern values.

The major effect of these changes in insolation was to enhance the thermal contrast between oceans and land, causing stronger evaporation from the oceans and enhancing convection over land areas. This pattern produced strong summer monsoon precipitation in the northern hemisphere tropics and subtropics and hence warm and dry summers in the continental interiors of the northern midlatitudes. These features have been confirmed by the pollen, lake level, and fossil records and have been reproduced by meteorological models of atmospheric circulation and precipitation.

Some of the regional details of the COHMAP analysis are shown in Figure 10.12 (in the color insert). It is interesting to compare the present-day situation with those of the past, realizing that the changes that are noted are typical of the kinds of fluctuations that occur in the natural climate system.

In today's climate in North America and Eurasia, permanent land ice occurs mainly in Greenland, whereas sea ice covers most of the Arctic Ocean. Spruce forests cover much of the high latitude regions below the land ice latitudes, whereas oak is favored in the temperate latitudes. Subtropical conditions occur in the latitudes of Florida and northern Africa.

In contrast, the climate of 18 kyr ago in these regions was characterized by large ice sheets and substantial North Atlantic sea ice. Spruce and oak forests were absent or sparse in Europe, and the Mediterranean lowlands were without trees. Spruce dominated the forests of the North American Midwest. In the Southwest, lake levels were high and woodlands were abundant. As time progressed, the increasingly warm climate nudged many of the forests to the north, reaching their limit by about 6 kyr BP. Since that time, as summer temperatures have decreased somewhat, the southern limits of several of the tree species have begun once more to shift toward the south.

Africa, Asia, and Australia appear on the right side of Figure 10.12. The present-day climate in these regions is heavily influenced by the summer monsoons, which were discussed in Chapter 4. Great disparities in precipitation exist over these continents, with equatorial rain forest in some areas and dry desert regions in others. The data for 18 kyr ago show much smaller areas of forest, a pattern indicating that the Mediterranean and environs were cooler and perhaps dryer than today. In the period 12–6 kyr BP, greater precipitation resulted in increased vegetation, probably caused by the gradual formation of the strong monsoon pattern. Somewhat dryier conditions have since followed, as solar radiation gradually decreased to present values with the progression of the equinoctial precession.

The central message of the COHMAP study and of other similar efforts is that many aspects of climate change can be understood as arising naturally out of orbitally induced variations in the magnitude and seasonality of solar radiation. These periodic temperature changes modify not only the glacial ice sheets but also the sea-to-land circulation patterns, precipitation intensity and seasonality, and patterns of vegetative response (and hence that of the animal kingdom). Not only is climate inherently variable, but the broad aspects of that variability can be understood as consequences of the most basic aspects of Earth's astronomical and surface properties.

Volcanoes and Climate

The best-understood process of abrupt climate change, for which we have both historical and relatively modern evidence, is sulfur gas (SO_2) and ash injection into the atmosphere during a volcanic eruption. The ash particles tend to be a millimeter or so in diameter, and fall out of the atmosphere within a few days or weeks. The SO_2, however, is converted into very small (a few tenths of a micrometer) particles of sulfuric acid. Because of their very small settling velocity, these particles have residence times of months to years and thus can have long-term effects on climate.

A well-documented and dramatic example of such an eruption was that of the Tambora volcano on Sumbawa Island in Indonesia in April of 1815. The spectacularly large amount of ash released (perhaps 150 km³ in all) and the height that ash and gases attained (perhaps 50 km) combined to partially shield portions of Earth from the Sun for more than 2 yr. In this way the year 1816 became famous as "the year without a summer," when throughout parts of the northeastern United States and western Europe periods of freezing temperatures during the summer resembled those of midwinter. In different latitudes and regions, the average depression of temperature is estimated to have been between 0.4 and 1.0 °C.

The signatures of volcanoes too ancient or too remote to have been recorded historically can sometimes be captured by tree ring analysis. Two different techniques have been used. The first stems from the observation that as little as two successive nights with temperatures reaching −5 °C and an intervening day with temperatures remaining at or below 0 °C are sufficient to cause frost damage if it occurs during the growing season when the new cells are not yet fully matured. Freezing events of this type are readily caused by volcanic eruptions, if the concentration of volcanic particles is sufficient to screen much of Earth's surface from solar radiation. Frost rings are easily located in tree ring cores. If such rings are found in trees from

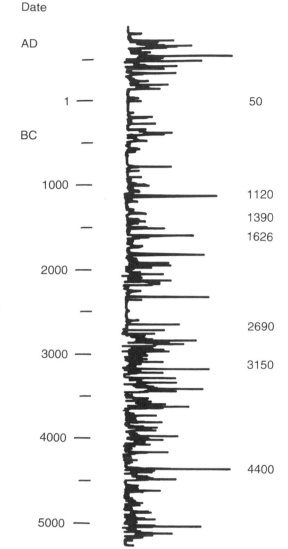

Figure 10.13 Tree ring width indications of volcanic eruptions sufficient to affect local climate. The data are "narrowness indices" for bog oaks from Northern Ireland, a high value indicating the presence of narrow growth rings for that year in a number of trees at several sites. The dates on the right of the figure correspond to acidity peaks identified in ice cores from Greenland, except 1626 BC, which is the date of a major frost event in trees from California. Both the frost data and the narrowness indices suggest a major volcanic event at this time, probably that of Santorini in the Aegean Sea. (Adapted with permission from M. G. L. Baillie and M. A. R. Munro, Irish tree rings, Santorini and volcanic dust veils, *Nature,* *332,* 344–346. Copyright 1988 by Macmillan Magazines Limited.)

different geographical locations but for the same time period, they can be ascribed to a regional or global event rather than local factors such as drought or a shortage of nutrients. Frost ring analyses have been used to confirm the dates of several relatively recent volcanic eruptions, as well as to propose dates for several for which historical records do not exist.

A technique complementary to that of frost rings is the measurement of the thickness or narrowness of the rings from one or several cores. Periods during which the rings are very narrow are periods in which tree growth was constrained for some reason. The constraint may involve temperature, water supply, air pollution, or other environmental factors, but most of these climatically related limits involve gradual changes in the tree ring thicknesses, unlike the abrupt changes produced by major volcanic eruptions. Abrupt changes are revealed by searching the data for intensely narrow short-term peaks of tree ring growth. Examples of such behavior are pictured in Figure 10.13, where eruption dates for a number of large volcanoes are confirmed or cited for the first time.

It should be noted that the past century has been characterized by relatively low volcanic activity. Two recent eruptions—El Chichón, Mexico (1982), and Pinatubo, Philippines (1991)—may, however, signal a period of increased volcanic effects on climate. The sulfur emissions from Pinatubo, the more climatically significant of the two, are shown in Figure 10.14 (in the color insert) in a satellite image taken only a few days after eruption. It is clear that large volcanic eruptions can have major climatic effects; it is also clear that volcanoes are the most unpredictable of the important natural climatic forcing functions.

A SYNOPSIS OF CLIMATE HISTORIES

It is truly startling that we possess a general idea of the climate of Earth almost from the time of its beginning more than 4 Gyr BP. During the interval 600 to 0.2 Myr, somewhat more information is available; and for the past 250,000 yr the climate record is moderately well detailed. For the past 20,000 yr, the records are so complete that we know not only the typical temperature and precipitation patterns but also features as detailed as the geographical ranges of specific trees.

Climate histories on several different time scales (inferred from global temperature reconstructions) are shown in Figure 10.15 for the past 150,000 yr. The data for the period with the longest time span and the least resolution appear in Figure 10.15d. The figure shows a rapid warming from about 150 to about 120 kyr BP, followed by irregular cooling over the following 100 kyr.

As shown in Figure 10.15c, the climate changed from a period of extreme glaciation to a warm interglacial in the past 20,000 yr. Computer simulations have shown that midway through this period the atmospheric circulation increased monsoon precipitation and warmed continental interiors. The result was that favorable climatic conditions existed in the northern latitudes, which doubtless contributed to the expansion of human activity in those regions.

The past thousand years have seen significant changes in global climate as well, as shown in Figure 10.15b. (Note, however, that the vertical scale of this figure is only about a fourth of that in panels (c) and (d) of Figure 10.15.) The most striking features of the period are the sharp coolings in about AD 1400 and AD 1650, during the so-called Little Ice Age when the Baltic Sea and rivers such as the Thames

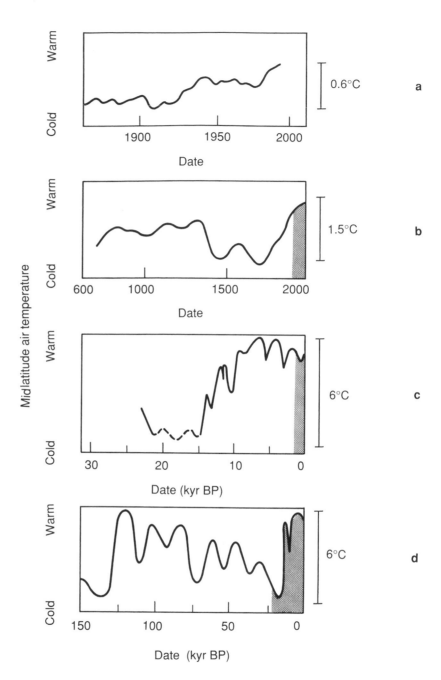

Figure 10.15 General trends in global temperature on various time scales ranging from decades to hundreds of thousands of years. On the bottom three displays, the shaded area indicates the time segment that is expanded in the display immediately above. The data reconstructions are based primarily on the following records: (a) instrumental data, (b) historical information, (c) pollen data and alpine glaciers, and (d) marine plankton and sea level terraces. (Adapted with permission from T. Webb III, J. Kutzbach, and F. A. Street-Perrott, in *Global Change*, T. Malone and J. Roederer, eds., pp. 182–218. Copyright 1985 by Cambridge University Press, UK.)

in England and the Tagus in Spain were regularly covered with ice. It has been proposed that this period was related to an absence of magnetic activity at the surface of the Sun, because sunspots were virtually absent during that period. No strong case has yet been made, however, for a process to link solar activity to Earth climate.

Air temperature as an indicator of global climate during the past century is shown in Figure 10.15a. The pattern, which was discussed in some detail in Chapter 1, shows a definite increase in temperature over the past 20 or 30 yr relative to that earlier in the century.

Figure 5.12 Views of Denver, Colorado, USA. (top) A computer-produced view of what Denver visibility would be like without the presence of anthropogenic aerosol particles, and (bottom) a photograph of current smoggy conditions. (Courtesy of L.L. Mauch, US National Park Service.)

Figure 6.9 Examples of stratospheric and mesospheric clouds. (top) Nacreous clouds; (bottom) Noctiluscent clouds over Stockholm. (Courtesy of Jacek Stegman and Georg Witt, University of Stockholm.)

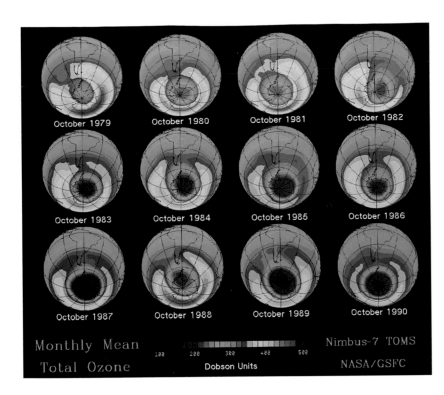

Figure 8.2 Total column ozone concentrations (in Dobson units) over Antarctica as determined by the TOMS (Total Ozone Monitoring Satellite) during October from 1979 to 1990. (Courtesy of Arlin Krueger, National Aeronautics and Space Administration.)

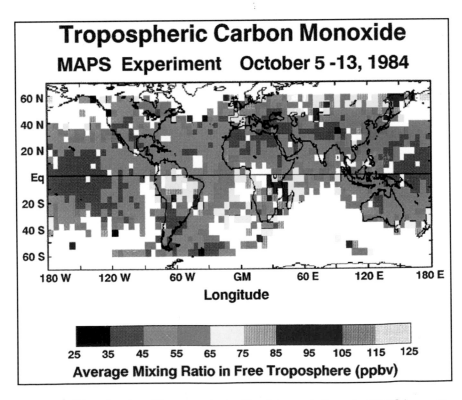

Figure 8.12 Inferred middle tropospheric CO mixing ratio from the MAPS (measurement of air pollution from satellites) experiment on the flight of the NASA space shuttle Challenger, 5–13 October 1984. (Courtesy of H.G. Reichle, Jr., National Aeronautics and Space Administration.)

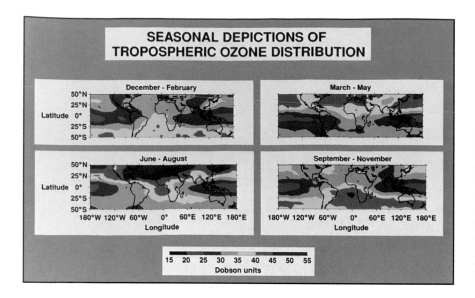

Figure 8.13 Tropospheric column ozone concentrations (in Dobson units; see Figure 4.9) derived from orbiting satellite observations, averaged over 3-month periods for the tropics and midlatitudes over the period 1979–1990. (Courtesy of J. Fishman, National Aeronautics and Space Administration.)

Figure 9.2 The acidity of lakes in Sweden in 1985. (Swedish Environmental Protection Board, *Sura och forsurade vatten. Monitor 1986*, Solna, Sweden, 1986.)

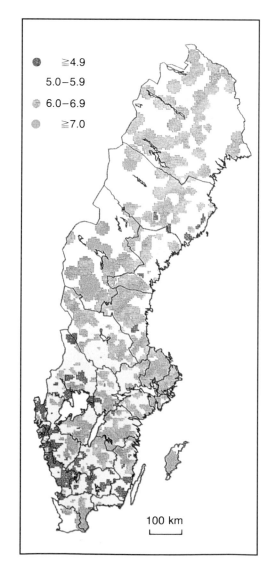

Figure 9.8 Mammoth Hot Springs, a product of freshwater hydrothermal activity in Yellowstone National Park, USA. (Photograph by Susannah K. Graedel.)

Figure 9.9 A "black smoker" hydrothermal vent at the "Hanging Garden" site on the East Pacific Rise at 21 °N. Appendages of the research submarine appear at the bottom of this picture, crabs living in the warm vent environment can be seen at the left center, and the vent fluid is emerging near the top left. (Courtesy of Kyung-Ryul Kim, University of California, San Diego.)

Figure 10.12 Changes in the atmosphere, geosphere, and biosphere that accompanied the transition from glacial to interglacial conditions during the past 18,000 yr, as illustrated by geological and paleoecological evidence for eastern North America and Europe (four left panels plus top right panel) and Africa, Asia, and Australia (three lower right panels). The individual panels show the extent of ice sheets and of year-round and winter-only sea ice; from 18 to 9 kyr BP they show the broadened land areas resulting from lowered sea level. The distributions of oak and spruce as inferred from pollen data are shown for eastern North America and Europe. Moisture status relative to the present is displayed for all areas from 18 to 6 kyr BP, as is the present region where annual precipitation is less than 300 mm. (Adapted with permission from COHMAP Members, Climatic changes of the last 18,000 years: Observations and model simulations, *Science, 241*, 1043–1052. Copyright 1988 by the AAAS.)

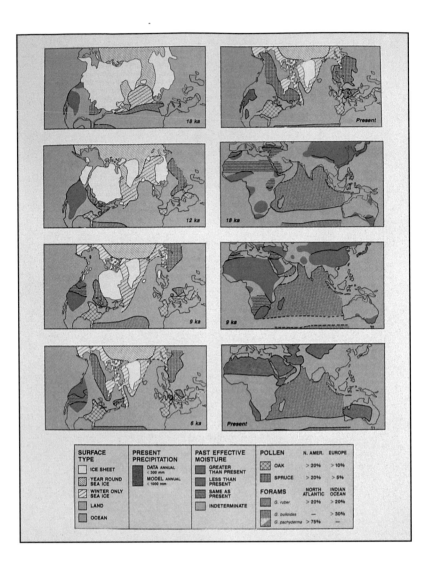

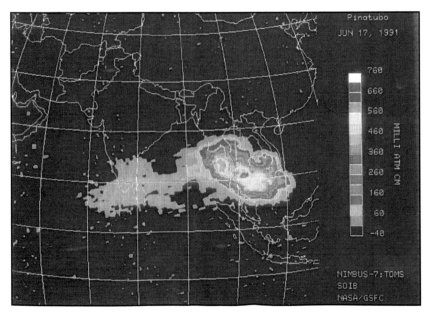

Figure 10.14 Map of total column concentrations of sulfur dioxide from the eruption of Mt. Pinatubo, Philippines, on June 17, 1991. This figure was compiled by using data from the Total Ozone Mapping Spectrometer (TOMS) instrument on the *Nimbus 7* satellite. (Courtesy of A.J. Krueger, National Aeronautics and Space Administration.)

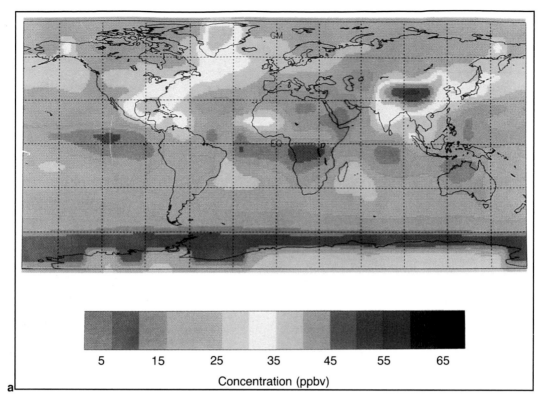

a

Concentration (ppbv)

5   15   25   35   45   55   65

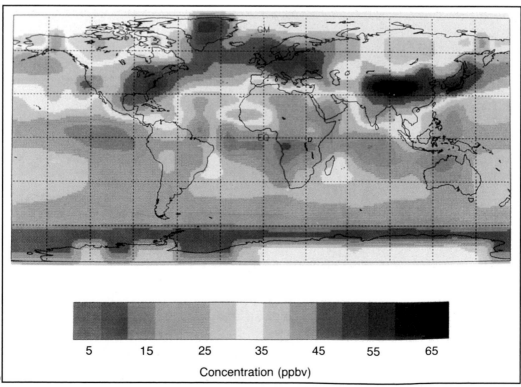

b

Concentration (ppbv)

5   15   25   35   45   55   65

Figure 15.14 July surface ozone mixing ratios for the preindustrial era (top) and for the present (bottom), as computed by the 3D global transport and chemistry model of Crutzen and colleagues.

Figure 15.15 Top panel: Global ocean temperatures at 160 m depth as computed by a 3D ocean model. Bottom panel: 3D ocean model computations of the volume transport stream function in the world oceans. (A. J. Semptner, Jr. and R. M. Chervin, A simulation of the global ocean circulation with resolved eddies, *Journal of Geophysical Research, 93,* 15502–15522, 1988.)

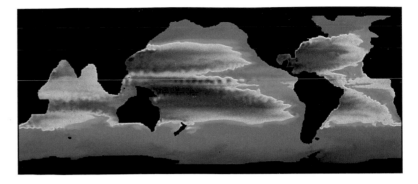

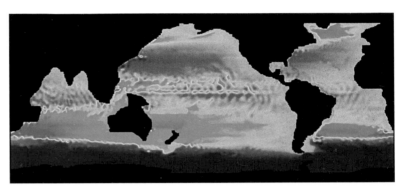

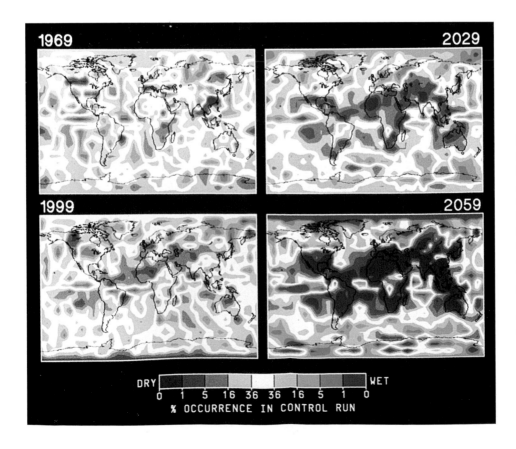

Figure 17.7 The occurrence of drought for the period June–August in a global climate model simulation. The color scale is set by the frequency of occurrence of drought in a 100-yr control run that had an 1958 atmospheric composition. (D. Rind, R. Goldberg, J. Hansen, C. Rosenzweig, and R. Ruedy, Potential evapotranspiration and the likelihood of future drought, *Journal of Geophysical Research, 95,* 9983–10004, 1990.)

One way in which the history of climate can be summarized is to assume that a record of global mean temperature is available throughout Earth's history and then to plot its variance spectrum. Many aspects of such a project cannot be specified with certainty, but a current effort is shown in Figure 10.16. The spectrum shows a background level of variability, probably largely stochastic and perhaps increasing in amplitude toward the longer time scales. Astronomically related periodic components with diurnal, seasonal, and annual variations are seen, as is a 3–7 day peak reflecting typical weather front duration. The peak near 2500 yr is not understood; it may be related to a characteristic mixing time for the deep ocean. The next three peaks are those related to variations in Earth's orbital parameters. The final two, at about 45 and 300 Myr, are probably related to orographic and tectonic effects and to the drift of the continents.

Many aspects of Figure 10.16 are well understood, an example being the relationship of climate to variations in Earth's orbit. Some climate factors are less well comprehended, such as the interactions between the ocean–atmosphere heat budgets. The overall message is a clear one, however: that a history of climate is a history of change and not a history of stability. At present, Earth has an unusually high average temperature, only comparable with the climatic optimum some 7 kyr ago. From this warm base point, humankind's activities will probably warm the planet further by several degrees on a time scale of only a century or so. There is no historical precedent for the reaction of the planet to a perturbation of this intensity and direction.

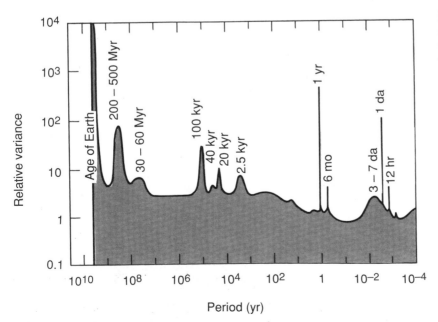

Figure 10.16 An estimate of the relative variance of global mean temperature over all periods of variation. The probable causes of the individual features are discussed in the text. (A. Berger, Milankovich theory and climate, *Reviews of Geophysics, 26,* 624–657, 1988.)

EXERCISES

10.1 The volume of the oceans is 1.35×10^9 km³. If new magma is injected along midocean ridges in volume such that it exceeds by 40 km³ the amount lost to subduction, thus raising the seafloor, what will be the average rise in sea level around the Earth?

10.2 The $\delta^{13}C$ of methane and other carbon-containing gases is defined in the same manner as is $\delta^{14}C$ (see Chapter 2). If $(^{13}C/^{12}C_{std} = 0.0112372$ and $(^{13}C/^{12}C)$ from carbon stored in vegetation is 0.0109365, find $\delta^{13}C$ for the vegetation sample. Also, given $(^{13}C/^{12}C) = 0.0106416$ in methane from landfills, find $\delta^{13}C$ for that sample. (Atmospheric methane currently has $\delta^{13}C$ of about $-49.6^o/_{oo}$, and is regarded as being composed of part vegetative methane and part landfill methane, in addition to other sources.)

10.3 A most dramatic extinction of life on Earth happened about 65 million years ago, at what is termed the Cretaceous–Tertiary boundary. This extinction is thought most likely to have occurred as a consequence of a bolide impact (L. W. Alvarez, Mass extinctions caused by large bolide impacts, *Physics Today, 40*(7), 24–33, 1987; M. Zhao and J. L. Bada, Extraterrestrial amino acids in Cretaceous/Tertiary boundary sediments at Stevns Klint, Denmark, *Nature, 339,* 463–465, [and 423–424], 1989) but has alternatively been attributed to extensive volcanism (C. B. Officer and C. L. Drake, Terminal Cretaceous environmental events, *Science, 227,* 1161–1167, 1985) and to terrestrial instabilities resulting from continental drift (T. J. Crowley and G. R. North, Abrupt climate change and extinction events in Earth history, *Science, 240,* 996–1002, 1988). Using the cited papers, their references, and any other sources of your choice, present the arguments in favor of each of these mechanisms. May some mechanisms have been more effective at some epochs of history than at other times?

10.4 (Group Exercise). The conceptual picture of Earth's orbital variations is one of those topics difficult to describe in words and two-dimensional diagrams but fairly easy to see as a three-dimensional demonstration. To do so, get an orange, draw a black line around it to indicate the equator, and pierce it through the poles with a metal rod about 15 cm long to indicate the axis of rotation. Place a table in the center of the room to serve as the plane of the Earth–Sun orbit and mark it to indicate the direction of perihelion, the point of closest Earth–Sun distance. (The location of this mark is arbitrary.) Move the artificial Earth in the orbital plane (the tilt of the orbital axis does not matter), bringing it closest to the Sun when it passes the perihelion position. Now repeat the process, but increase the perihelion distance slightly, leaving all other distances the same. This is a change *in ellipticity*. Next, freeze the tilt of the orbital axis and move Earth around the Sun (see Figure 10.7). Note the different orientations of the northern and southern hemispheres, defining the seasons. Repeat the exercise, keeping the orbital axis in exactly the same orientation with respect to the orbital plane but decreasing the angle of inclination to that plane slightly. This is a change in *tilt*. Study how that change would influence the received solar radiation at different latitudes and at different points in the orbit. Finally, position Earth at one point of the orbit, maintaining its previous orientation. Without changing the tilt, rotate Earth a quarter turn left or right. This is a change in *precession*. Study its effect on the amount of solar radiation received by the different hemispheres at different seasons. Note that during changes in tilt and precession, the location of the orbital plane and the direction of perihelion can remain the same.

FURTHER READING

W. Alvarez, E. G. Kauffman, F. Surlyk, L. W. Alvarez, F. Asaro, and H. V. Michel, Impact theory of mass extinctions and the invertebrate fossil record, *Science, 223*, 1135–1141, 1984.

A. Berger, Milankovich theory and climate, *Reviews of Geophysics, 26*, 624–657, 1988.

T. J. Crowley, The geologic record of climatic change, *Reviews of Geophysics and Space Physics, 21*, 828–877, 1983.

COHMAP Members, Climatic changes of the last 18,000 years: Observations and model simulations, *Science, 241*, 1043–1052, 1988.

T. E. Graedel, I.-J. Sackmann, and A. I. Boothroyd, Early solar mass loss: A potential solution to the weak sun paradox, *Geophysical Research Letters, 18*, 1881–1884, 1991.

H. D. Holland, B. Lazar, and M. McCaffrey, Evolution of the atmosphere and oceans, *Nature, 320*, 27–33, 1986.

W. R. Kuhn, J. C. G. Walker, and H. G. Marshall, The effect on Earth's surface temperature from variations in rotation rate, continent formation, solar luminosity, and carbon dioxide, *Journal of Geophysical Research, 94*, 11129–11136, 1989.

R. G. Prinn and B. Fegley, Jr., Bolide impacts, acid rain, and biospheric traumas at the Cretaceous–Tertiary boundary, *Earth and Planetary Science Letters, 83*, 1–15, 1987.

CHAPTER 11

Ancient Earth: Chemical Histories

At first thought, it seems almost inconceivable that we could say very much about the chemistry of the atmosphere and surface further back in time than the 30 yr or so during which modern analytical instruments have been developed and used. It turns out, however, that there exist at least five techniques that have the potential for determining historical chemistry:

- Interactions occur between atmospheric constituents and the surfaces with which they come into contact. If these interactions are understood and if the interacting surfaces preserve the signature of the interaction, then the surfaces can be useful for establishing past atmospheric characteristics. The most fruitful data of this type are from the rock record, because it is known that certain dated rocks could be formed only under specific atmospheric conditions. Even though this technique cannot be used to derive trace species concentrations, it is useful for major constituents, particularly molecular oxygen. Variations of the technique have been proposed to study more recent atmospheric compositions, and several potentially useful reactant–surface interaction pairs have been suggested, but their interaction relationships are not understood sufficiently for the surfaces to function as surrogates for determining atmospheric compositions.

- If, in some way, samples of ancient atmospheric constituents have been preserved, then they can be analyzed by modern analytical techniques. The best examples of such preserved samples are those from polar or glacial ice cores. The source of the ice is snow that falls each year on the glacier and then consolidates. (Snow that has been partly consolidated into glacial ice is termed *firn*.) Particles and soluble gases deposited together with the snowflakes or at the snow surface remain in the layer corresponding to the period of deposition. The compaction and annual deposition processes mark each year's boundaries (see Figure 1.3). In this way, the age of the ice core at various depths can be estimated, much as a forester tells tree age by the rings. Reference dates are also established by deposition events dated by historical markers, such as radioactive fallout from nuclear weapons tests or ash deposition from large volcanic eruptions.

In addition to establishing a dating chronology for ice cores, it has proved possible to determine the average Earth surface temperature at the time each year's snow fell. The basis for this determination is the measurement of the abundance of the isotopes deuterium and oxygen-18. Water molecules containing either of these isotopes are slightly less volatile than those made up only of hydrogen and ^{16}O, and snow formed from atmospheric water vapor over cold regions will hence be deficient in D and ^{18}O relative to snow from warmer regions. It turns out that a linear relationship exists between the mean annual surface temperature at a given location and the mean $d^{18}O$ or $d\,D$ of deposited snow. These chemical histories buried in the ice cores thus provide a record of average climates of the past.

Bubbles of air are also trapped in glacial ice. This enclosed air is younger than the ice that surrounds it, because the closure of the ice matrix occurs tens of meters deep in the ice; at shallower depths atmospheric air still circulates through open pores. Careful analytical work is thus required to determine the age difference between ice and air at the enclosure time. A typical number for the age difference is 20–80 yr, but the age difference may be as much as 2500 yr in regions of very low snow accumulation.

Measurements of perhaps a dozen chemical species in the ice have been made in various glacial locations. Other ancient air repositories have been suggested, such as air in weld-sealed compartments in sunken ships or air in un-opened Egyptian royal tombs. However, microbial contamination and other factors are concerns, and no satisfactory analyses from such sources have yet been accomplished.

- Lake sediments can provide data on the deposition of soot particles and on airborne trace metals such as mercury and lead, thus supplying a record of natural and human-ignited forest fires, agriculture, and industry. Indirectly, the sediment record also reflects pH, salinity, and dissolved gas content through the relative abundance and mineralization of different diatom species.

- Analytical chemistry as a discipline is well over 100 yr old, and in some cases data exist from detailed atmospheric measurements programs of a century ago. If the data are to be used today, the measurement technique must be reproduced and the response certified. For ozone and carbon dioxide, such certification has proved to be possible. With time, the quality of the data improves, so that for the last 50 or 60 yr some rather reliable data are occasionally available.

- For all but the most reactive species, the atmospheric concentrations that are measured can be related to the emission fluxes from various sources. Modern source–concentration relationships can (perhaps with the help of advanced computer models) provide information on trace constituents in past atmospheres provided historical data related to emission fluxes are known or can be estimated.

In this chapter and the two following, we report on the results of studies into the compositions of the environments of the past, using the techniques outlined above. The time periods involved are far from uniform in length or in available information. We begin with ancient times, about which we know something, but very little. We then proceed to the past several centuries, about which we know a moderate amount. We complete the exposition by discussing the environments of the past few decades. Because this latter period is coincident with the development of much of modern analytical chemistry, substantial amounts of data are extant.

Nowhere is the concept of an Earth system, intimately connecting biology, geology, hydrology, and atmospheric science, better illustrated than in the story of the formation and evolution of Earth's atmosphere. The sequence began with the evolution of the Sun from a juvenile star to one of maturity, a process that included the gravitational accretion of the grains, dust, and gas in the solar nebula to form the planets. Various-sized smaller solid bodies were formed as well; they generally had highly perturbable orbits. Theoretical analyses suggest that the collapse of the solar nebula and accretion to form Earth and the other planets occurred within the geologically short time scale of perhaps 50–100 Myr.

Spaceborne and ground-based observations of the last two decades provide us with much information concerning the present atmospheres of the planets of the solar system. Table 3.1 compared some of the characteristics of the terrestrial planets Earth, Venus, and Mars. Those data indicated that the three planets have, at present, very different chemistries and climates. Compared with Earth's atmosphere, that of Venus has a smaller amount of water, much less N_2 and O_2, a very high concentration of carbon dioxide, and a very high relative concentration of sulfur dioxide. The Martian atmosphere is also dominated by carbon dioxide, has little N_2 and O_2, very little water, and no detectable sulfur. The surface temperatures are dramatically different, Venus being hundreds of degrees warmer than Earth, and Mars, much colder.

Are these differences a result solely of the relative positions of the planets with respect to the Sun? This question has not been answered with much certainty, but it appears most likely that the differences relate in part to the retention of volatile species (H_2O, C, Cl, S, and N) by the solid particle grains that accreted to form the planets. Planets forming near the Sun are less likely because of the higher temperatures to assemble hydrated phases such as talc [$Mg_3Si_4O_{10}(OH)_2$] or serpentine [$Mg_3Si_2O_5(OH)_4$]. Thus, Venus probably formed with little water, Earth with more, and Mars with more still. Temperature and mineral considerations suggest that Venus and Earth originally had similar amounts of CO_2, Cl, and S, whereas Mars had more S and Cl than Earth but less CO_2.

Once the planets were formed, volatilization of gases could proceed. That process is enhanced at high temperatures, and Venus probably outgassed (i.e., expelled gases from its interior) to a greater extent than Earth. Having little water, much of the outgassing was CO_2 and SO_2, and the large amounts of CO_2 immediately became efficient absorbers and set up a "runaway" planetary greenhouse. Without water to dissolve the CO_2 or plants and animals to recycle it, the greenhouse on Venus stabilized at an extremely high value, much too warm for life as we know it.

The opposite extreme is exemplified by Mars. Although the low pressures resulted in some outgassing, CO_2 on Mars was apparently never abundant enough to produce a substantial greenhouse effect. The resulting cold planet was unable to volatilize much of its H_2O, and without liquid water many of the dissolution and precipitation processes that make Earth's climate so hospitable to life failed to develop.

Thus, Earth was especially favored at its birth. Its early atmosphere was likely influenced in a major way by the preponderance of hydrogen atoms in the original solar nebula. Because of high hydrogen abundances, in fact, virtually all members of the solar system feature reducing conditions rather than the oxidizing conditions that are now dominant on Earth. How did our oxidizing atmosphere come about?

Provided with reasonable supplies of water locked in its forming particles, Earth was located at an appropriate distance from the Sun for that water to be in liquid form. Even given its supply of water, a delicate balance of some amount of infrared-absorbing gases (but not too much!) was apparently necessary for establishing and maintaining the principal features of Earth's climate. Carbon dioxide was available to assist in maintaining surface temperatures within the liquid water range. The other major constituents were nitrogen, hydrogen, and water, with perhaps smaller amounts of hydrogen sulfide, ammonia, and methane. There was originally no free oxygen. Because iron was readily available, it is hypothesized that additional water formed by a process that can be represented stoichiometrically as

$$FeO + H_2 \rightarrow Fe + H_2O \quad (11.1)$$

with the water remaining at the surface or in the atmosphere and the free iron gradually collecting in the hot core and mantle.

Early in the planet's life cycle, and within an atmosphere like that described above, the first forms of life appeared. The life forms were bacteria that lived by breaking down molecules in oxygen-free conditions. Eventually, about 3.5–3.8 Byr BP, some of the bacteria began to develop a photosynthesis capability that used water instead of non–oxygen-containing molecules. The hydrogen in the water was used to form molecules useful to the bacteria themselves, such as formaldehyde; the oxygen was a waste product, ultimately escaping to the atmosphere. The process may be represented as

$$CO_2 + H_2O \rightarrow HCHO + O_2 . \quad (11.2)$$

Oxygen was also being slowly produced by the photodissociation of water in the atmosphere, followed by the thermal loss of the light hydrogen atoms to space at altitudes near 500 km. For a long period, the oxygen released near the ground continued to be immediately consumed in oxidizing the abundant iron ions dissolved in the water, a process reflected in the ancient "red bed" deposits of oxidized iron. Eventually the rate of oxygen production exceeded the rate at which iron deposits were brought to Earth's surface by geological processes. At that point, oxygen began to accumulate in the atmosphere. This change brought about a biological revolution; even though the oxygen-rich atmosphere was lethal to most existing species, it stimulated the development of species able to utilize oxygen and thus solar energy much more efficiently. The evolutionary history of molecular oxygen can be derived from the chemical compositions of age-dated rocks and is shown in Figure 11.1. Some of the more significant milestones are as follows:

- Atmospheric O_2 reaches 1% of present value: 2.0 Gyr BP

- Atmospheric O_2 reaches 10% of present value. As a consequence, sufficient O_3 was generated to begin successfully shielding the surface from solar ultraviolet radiation: 700 Myr BP

- Atmospheric O_2 reaches 100% of present value: 350 Myr BP

The accumulation of O_2 gradually produced the radiatively important gas ozone (O_3), by the absorption of solar radiation:

$$O_2 + h\nu \ (\lambda < 242 \ nm) \rightarrow O + O, \qquad (11.3)$$

Figure 11.1 A reconstruction of the evolutionary development of oxygen and ozone in Earth's atmosphere. The oxygen curve is based on evidence from rocks and fossils, ozone from a photochemical model. Land-based life could not have become established until there was sufficient ozone to provide protection against ultraviolet radiation. (Reproduced with permission from R. P. Wayne, *Chemistry of Atmospheres*, 2nd Ed. Copyright 1991 by Clarendon Press, Oxford, UK.)

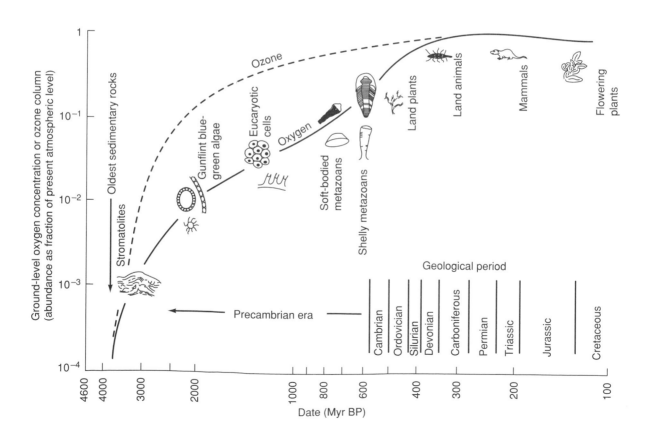

followed by

$$O_2 + O + M \rightarrow O_3 + M. \quad (11.4)$$

Increasing levels of atmospheric O_2 and O_3 began to shield Earth's surface from the lethal solar ultraviolet radiation. This shielding eventually permitted enhanced biological evolution on land. With the existence of the water-rich, oxygenated atmosphere, climate and chemistry were capable of playing important roles in Earth's physical, chemical, and biological development and of being themselves affected as that development assumed scales comparable with those of nature.

CO$_2$ FROM THE MID-CRETACEOUS SUPERPLUME

We have no direct information on the chemistry of the hydrosphere and atmosphere for the first 99.98% of Earth's existence and must therefore use the rock record as inferential evidence (as we did earlier in deriving the evolution of atmospheric oxygen). One of the most interesting attempts to do so in recent years is the "superplume" proposal of Roger Larson of the University of Rhode Island. Larson has studied massive undersea lava plateaus, formed by lava eruptions and venting activity beneath the sea (see Chapters 2 and 9). The evidence indicates that a major event of this type occurred in the Pacific Ocean basin about 120 Myr ago. The large amounts of magma rising from Earth's interior would have released great amounts of carbon dioxide, much more than the oceans could hold. The excess CO_2 would have entered the atmosphere and created an enhanced greenhouse effect, thereby causing a warm planet for an extended period of time.

Indirect chemical information from this period of Earth history is consistent with Larson's proposal. For example, sediment degradation linked to carbonic acid has been used to infer that ocean waters during the mid-Cretaceous were indeed rich in dissolved carbon dioxide. A similar conclusion is reached by studies of the carbon isotopic composition of marine sediments. It thus appears that submarine volcanism can be linked to atmospheric chemistry and climate in Earth's past. If so, we cannot rule out the possibility of a similar event or events in the future.

NATURAL SOURCES OF ATMOSPHERIC COMPOUNDS

Natural vegetation and soils are sources of carbon dioxide, carbon monoxide, nitric oxide, and a very large number of organic compounds, as will be immediately recognized by recalling the diversity of floral scents. The vegetative emittants include many types of hydrocarbons and a number of oxygenated species, most of which are subject to reaction and transformation in the atmosphere. The emissions arise naturally as a consequence of microbiological and plant physiological processes. For example, CO_2 and CH_4 are the respective end products of aerobic and anaerobic respiration; N_2, N_2O, and NO are produced by biological oxidations and reductions of inorganic nitrogen in soils and waters; and plants produce and emit hydrocarbons as by-products of photosynthesis and as defensive agents. Although the ultimate product of the oxidation reactions is CO_2, many intermediate products play an important role in atmospheric photochemistry. These intermediates may appear in the gas or liquid phase.

The effects of emissions from vegetation and soils to the atmosphere will be felt over substantial ranges of both distance and time. Reactive hydrocarbons and NO_x will involve themselves in local and regional atmospheric processes (as we saw in Chapter 8). N_2O is so long lived that it is even important to stratospheric ozone chemistry, where its effects are manifested on global distance and centurial time scales. Estimates of CO_2 release from soils and forest cutting and from uptake in the oceans are important in connection with the interpretation of the observed CO_2 increase in the atmosphere over the past century and its anticipated increases due to increased fossil fuel burning.

Aquatic ecosystems can be important sources of species involved in atmospheric chemical impacts. In most cases, emissions arise as a consequence of biological processes. A case in point is that of nitrous oxide, which is a biological nitrification–denitrification product. Nitrous oxide is produced as a volatile intermediate waste product during the microbiological oxidation of ammonium to nitrate (*nitrification*) or of nitrate to N_2 (*denitrification*). The latter occurs in anoxic (anaerobic) environments. Substantial N_2O fluxes are seen from coastal waters and estuaries or from oceanic regimes of high productivity, which are generally found in upwelling regions. Reduced sulfur gases are also emitted from aquatic ecosystems as by-products of biological activity. The most geographically widespread of the emissions from aquatic ecosystems is that of dimethyl sulfide from the oceans.

For the atmosphere, the most important process in wetlands (marshes, swamps, bogs, shallow lakes) and anaerobic waterlogged soils is the bacterial production of methane. The methane budget is not well understood, but freshwater wetlands are thought to be among the sources with the highest methane flux. The emission rate varies with wetland type and with season, and shows a positive correlation with temperature. As a result, the slow warming of the atmosphere expected from the greenhouse gas increases may indirectly produce heightened CH_4 concentrations from increased wetlands emission. However, if the temperature increase is accompanied by lower precipitation amounts and more evaporation, the resulting drier conditions may lead to enhanced microbial oxidation and less methane release. Because it is thought that future climatic warming will be greater at high latitudes, methane emission and changes in it can be expected to be latitude dependent.

Large emissions of chlorine occur at coastal locations as a consequence of the generation of seasalt particles by breaking waves. Hydrogen chloride (HCl) is known to be volatilized as a consequence of uptake of sulfuric acid and nitric acid by seasalt aerosols. But, because it is a highly water-soluble gas, HCl will be easily removed from the atmosphere. The chemical influence of HCl and derived species will thus likely be restricted to the lowest 1–2 km of the troposphere near its sources.

Biological activity in coastal marshes and in the open ocean also plays a role in the production of various halogenated organic compounds. For example, the bioformation of methyl chloride (CH_3Cl) in these environments provides the main natural source of chlorine to the stratosphere, although its importance is at present only about one-fifth that of the anthropogenic halocarbon compounds.

Volcanoes are a sporadic and very important natural source of emissions to the atmosphere. We discussed their effects on climate in earlier chapters; their chemical effects are also important. The typical volcano emits significant amounts of ash, which is chemically unreactive, and a suite of sulfur gases, especially SO_2, which is significant. When volcanically injected into the stratosphere, sulfur dioxide is oxidized to sulfuric acid (see Chapter 8) and forms small, mostly liquid solution droplets of water and H_2SO_4. These small particles can remain in the stratosphere for several years, scattering solar radiation and providing surfaces for chemical reactions. Additionally, volcanoes are sources of HCl and so can enhance the catalytic halogen chemistry of the stratosphere.

ICE CORE CHEMISTRY

The use of ice cores to study the chemistry of atmospheres of the past is a development of the past two decades. It has been applied at several areas of Earth where temperatures seldom or never exceed the freezing point of water: Antarctica, Greenland, and some mountain glaciers. The technique is to locate an appropriate ice mass, drill a core in such a way that the sample is retrieved in a minimally contaminated state, and return the frozen core to the laboratory for study. The measurement of soluble species frozen into the ice is made by slicing off a section of the ice corresponding to the desired age period, removing the contaminated external surface, melting the remainder under ultraclean conditions, and analyzing the resulting solution. If gas concentrations are to be measured, the uncontaminated portion of the ice sample is placed in a vacuum chamber and cracked; the gases are then sucked out and analyzed by gas chromatographic or spectrometric methods. If ionic concentrations from the frozen snow are to be determined, the melted sample is analyzed directly by ion chromatography.

Ice core techniques are particularly appropriate for species that are relatively unreactive, because the possible complication of reactive loss during or after freezing is minimized. Species of this type are long lived and well mixed in the atmosphere, so the ice sample provides an approximate average of their global atmospheric concentrations. Aerosol particles are more reflective of local or regional than global conditions, although they may in some cases have been transported to the ice from quite distant locations.

Atmospheric records in ice do not extend back in time indefinitely, because the lowest layers of ice gradually become so compressed by the weight of the new ice formed atop them that they flow under pressure and their individuality and identity are lost. The oldest ice cores drilled and analyzed thus far have been dated to nearly 100 kyr in Greenland and about 160 kyr in Antarctica. Cores from both locations are about 2000 m long and about 10 cm in diameter. Dating is accomplished for most of the present interglacial period by counting annual ice layers. This analysis can be done by various physical and chemical methods, such as relying on the seasonal deposition pattern of various cations (e.g., H^+) and anions (e.g., SO_4^{2-}, NO_3^-, Cl^-). From that period back to about 30 kyr ago, ^{14}C dates provide a good chronology. The oldest ice cores can only be dated by theoretical studies of the flow of the ice sheet floor as it is distorted by the pressure of the overlying ice. In general, the further back one goes in time, the less accurate the dating becomes. The bottom of the deepest Antarctic core is older than is the bottom of the deepest Greenland core because the annual snowfall is less in Antarctica than is that in Greenland; a corollary, however, is that the temporal resolution of data from the Antarctic ice cores is lower than is that for the Greenland cores.

GAS HISTORIES PRESERVED IN ICE

Although the deep Antarctic ice cores reflect the characteristics of that continent, they also serve as an integrated record of the global atmosphere for gases with lifetimes of at least several years, because that is the approximate time scale for effective mixing between the air in the different hemispheres. In Figure 1.4 we reproduced the CO_2 concentration record from the Vostok Antarctic ice core, the oldest direct record we have of the chemical composition of the atmosphere. Also appearing in the figure is the atmospheric temperature profile deduced from the same ice core. The relationship between global temperature and atmospheric CO_2 content is striking. The CO_2 pattern tracks the warm interglacial peak at 130–140 kyr BP, the gradual decline to the glacial epoch at 20 kyr BP, and the rapid warming from that point to the present day. It should be noted that the highest CO_2 concentrations in history prior to very recent times are those of 130–140 kyr BP; they peak at over 290 ppmv and decrease over a period of 4 kyr to about 270 ppmv, the average rate being about 1 ppmv per 200 yr. (The present average CO_2 concentration in the remote atmosphere is about 350 ppmv.) It is an interesting perspective that the present rate of increase as a consequence of fossil fuel combustion is larger than 1 ppmv per *one year*! On a historical scale, therefore, the modern rate of change of a major atmospheric constituent of crucial importance to the biosphere and climate is more than two orders of magnitude greater than the highest average rate that had occurred prior to the present century and is occurring at a time when the absolute concentration of CO_2 is higher than it has been at any time during at least the past 160 kyr.

Because the concentrations of other trace gases in the atmosphere are so much lower than that of carbon dioxide, detection of their presence in small samples of ice is more difficult and the records are thus sparser, particularly for the very old, compressed ice core samples. Of the data that are available, perhaps the most interesting are those of methane concentrations in the Vostok ice core (Figure 1.4). Like CO_2, methane appears to have undergone changes in concentration that mimic those of temperature. This pattern could have occurred if the biosphere was more productive and the bacterial generation of methane from freshwater wetlands was a stronger source in warmer periods as a result of a wetter climate, a greater deglaciated area, and higher emission rates (the rates are functions of temperature).

ION CHEMICAL HISTORIES PRESERVED IN ICE

Ice contains a variety of trace constituents in addition to gases. Among those that are easiest to determine are the inorganic ions, which represent material incorporated into precipitation (frozen and otherwise) prior to its deposition, together with probably minor amounts from the deposition of gases on the frozen surface, solubilized ions from aerosol particles, and occasional major injections from particles produced by volcanic eruptions. A time record for Greenland is reproduced in Figure 11.2, where samples showing the influence of volcanic eruptions have been omitted so that the background levels of the ions can be deduced. Perhaps for chloride, but definitely for nitrate and sulfate, the figure shows a relatively stable mean value over nearly the entire time period, followed by a very sharp recent increase. In each case, the increase is attributed to emissions from increased fossil fuel burning, in combination with long-range meteorological airflows from industrial regions to the Arctic. This picture contrasts strongly with that of the Antarctic, which is so far from combustion sources that it receives little ionic sulfate or nitrate, except that attributable to volcanic eruptions and marine sources of biogenic sulfur.

Another measurement reflecting the chemistry of ancient precipitation is that of ice core acidity, a determination that largely relates to the concentrations of the complementary anions SO_4^{2-} and Cl^-.

Figure 11.2 Concentrations of inorganic ions in precipitation preserved in ice from the Greenland ice sheet. (R. C. Finkel, C. C. Langway, Jr., and H. B. Clausen, Changes in precipitation chemistry at Dye 3, Greenland, *Journal of Geophysical Research, 91,* 9849–9855, 1986.)

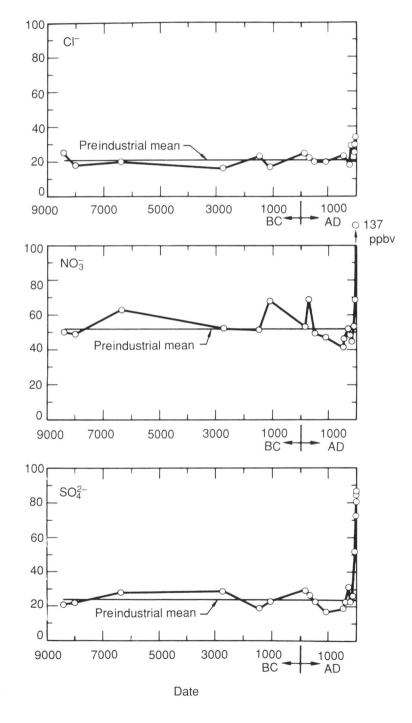

Figure 11.3 Mean acidity of annual layers from AD 1972 to AD 553 in an ice core from central Greenland. Acidities above the background, 1.2 ± 0.1 μeq H⁺ kg⁻¹ ice, are due to fallout of volcanic acids, mainly H_2SO_4, from eruptions north of 20 °S latitude. The ice core is dated with an uncertainty of ± 1 yr in the past 900 yr, increasing to ± 3 yr at AD 553. This precision makes possible the identification of several large volcanic eruptions known from historical sources, as well as the accurate dating of the Icelandic Eldgja eruption shortly after the Norwegian settlement of Iceland in AD 930. (Adapted with permission from C. U. Hammer, H. B. Clausen, and W. Dansgaard, Greenland ice sheet evidence of post-glacial volcanism and its climatic impact, *Nature*, 288, 230–235. Copyright 1980 by Macmillan Magazines Limited.)

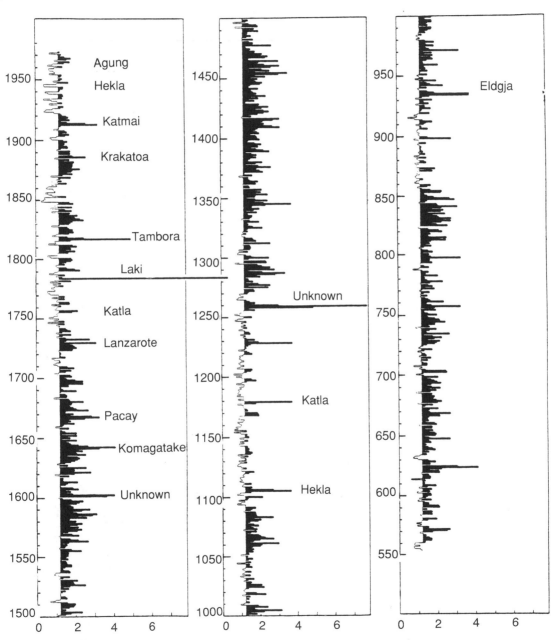

Mean acidity (μeq H⁺ kg⁻¹ ice)

Acidity measurements can be used in combination with the inorganic ion concentrations to derive considerable information about ambient chemical conditions. They have proved particularly useful as well in demonstrating signals related to volcanic eruptions. The information is useful in two distinct ways. If the eruption is already dated from historical records, as is the case for many events in the last few hundred years, the timing and intensity of the signal in the ice can be used to study atmospheric motions over long periods of time. If an appropriate volcanic eruption is not known at the time of an ice acidity event, the ice record helps expand the historical data available to volcanologists. A detailed record of acidity in Greenland ice is shown in Figure 11.3.

Finally, it is worth noting that even reactive chemical species have been measured in ice cores. For example, hydrogen peroxide has been clearly identified in ice as old as 22 kyr, and seasonal variations (higher in summer, lower in winter) have been observed. However, the data do not seem to follow any regular pattern, so they cannot be definitely linked to the hydrogen peroxide content of the atmosphere. It is thus far uncertain whether heavy amounts of dust can catalyze H_2O_2 destruction or whether other loss processes exist. If these uncertainties can be resolved, however, there is some potential for deriving substantial perspective on the chemical processing that may have taken place in ancient atmospheres.

Figure 11.4 Concentrations of aluminum (an indicator of land-generated atmospheric aerosol particles) in ice from an Antarctic ice core. (Reproduced with permission from M. De Angelis, N. I. Barkov, and V. N. Petrov, Aerosol concentrations over the last climatic cycle (160 kyr) from an Antarctic ice core, *Nature, 325,* 318–321. Copyright 1987 by Macmillan Magazines Limited.)

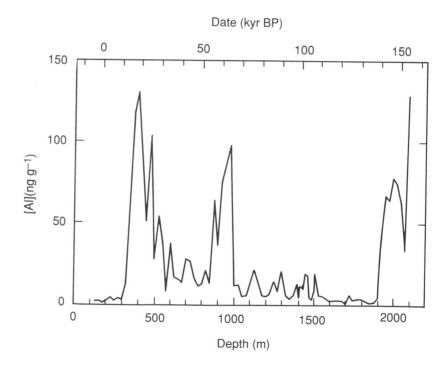

Insoluble atmospheric aerosol particles can readily be seen in ice cores, thus providing an additional tool for the study of historic atmospheric chemistry. A precise technique that has been developed involves selecting an element that is found in aerosol particles and not elsewhere and determining the composition of that element. For aerosol particles generated over continents (as windblown dust, for example), aluminum is particularly suitable, because it is a common ingredient of Earth's crust. Such an analysis is shown in Figure 11.4. Two features of the data are worth noting. One is that the aluminum concentration during the most recent period (the "last interglacial") is similar to the lower values seen in earlier times. A second is that the three major aluminum concentration peaks occur at minima in the oxygen isotope record, i.e., at periods of low global temperatures. It is thought that these periods may have been times of extended arid regions and intensification of atmospheric circulation, each of which would contribute to enhanced generation and long-distance transport of windblown aerosol particles.

SEDIMENT HISTORIES OF SURFACE WATER CHEMISTRY

Among the proxy data that can tell us something of the environmental chemistry of the past, the existence of particular species of plant and animal life in sediments has often been used to good purpose. An example is the ability of diatom analyses to provide information on properties of local hydrologic chemistry such as acidity and salinity.

Figure 11.5 contains selected results of a study of diatom populations in Devils Lake, North Dakota. For this study, a sediment core was extracted from the lake, dated, and analyzed for the occurrence and abundance of different diatom species. Some of the species studied are known to flourish in fresh water, others in salt water. The patterns of occurrence were used to infer the salinity of the lake and thus the aridity of the region. At least seven oscillations of the lake between fresh and saline conditions are indicated, and the data suggest that the lake was much more saline and the local climate much drier in the past than it is at present.

Figure 11.5 The salinity of Devils Lake, North Dakota, inferred from diatom populations in sediment cores. Several independent techniques were used to relate core depth to age. (Adapted with permission from S. C. Fritz, S. Juggins, R. W. Batterbee, and D. R. Engstrom, Reconstruction of past changes in salinity and climate using a diatom-based transfer function, *Nature, 352,* 706–708. Copyright 1991 by Macmillan Magazines Limited.)

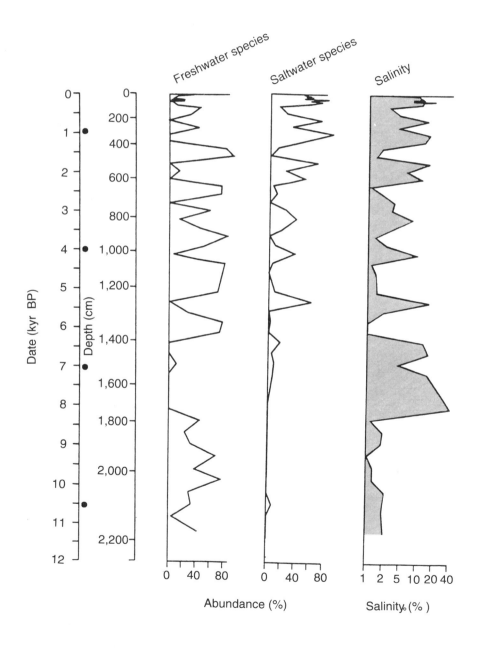

Although it is remarkable that we have any samples or signatures at all of the temperature and composition of ancient atmospheres, it is important to realize that those data are indeed few. No direct information on the environmental chemistry of any period in Earth history until the late Pleistocene is extant, nor is there any reasonable prospect of ever getting any. A very lengthy part of the history of Earth's atmosphere therefore can only be inferred from indirect information and scientific model studies, requiring an understanding of the interrelationships of Earth system processes.

For more recent periods (still ancient by human standards), the data situation becomes somewhat better. As shown in Figure 11.6, atmospheric samples from ice cores provide us with information on at least the more abundant trace atmospheric constituents of the past 100–200 kyr. The time span for data from relatively recent times is indicated in this figure as well; it will be discussed in the following chapters. Perhaps the most important aspect of these data is that they overlap time periods containing cycles of glacial and nonglacial periods on Earth, thus providing some potential for eventually understanding the forces that shift Earth from one regime of stability to another.

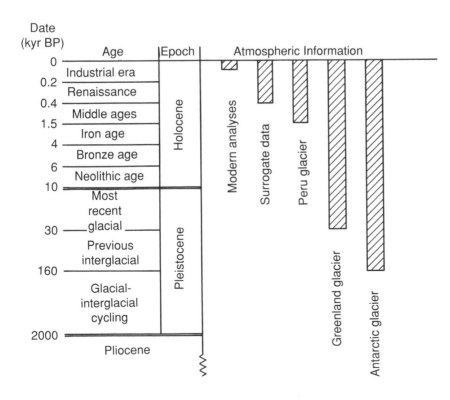

Figure 11.6 A time scale for the Holocene (the most recent 10 kyr), and the remainder of the Quaternary, indicating the periods covered by various types of atmospheric information. Numbers at the sides of the columns are ages in thousands of years before the present. The vertical scale is not linear and covers some four orders of magnitude in time. Information older than 30 kyr exists for Greenland, but precise dating remains to be accomplished.

EXERCISES

11.1 The manner in which the Earth's atmosphere was originally created and subsequently evolved is thought to be understood in broad outline, but the actual evidence is slight and the topic generates vigorous discussion about nearly every detail. Prepare a brief report on the areas of agreement and contention, using references of your choice. Among those appropriate are the following papers and the references contained within them: V. M. Canuto, J. S. Levine, T. R. Augustsson, C. L. Imhoff, and M. S. Giampapa, The young Sun and the atmosphere and photochemistry of the early Earth, *Nature, 305,* 281–286, 1983; J. F. Kasting, H. D. Holland, and J. P. Pinto, Oxidant abundances in rainwater and the evolution of atmospheric oxygen, *Journal of Geophysical Research, 90,* 10497–10510, 1985; H. D. Holland, B. Lazar, and M. McCaffrey, Evolution of the atmosphere and oceans, *Nature, 320,* 27–33, 1986.

11.2 The overlap in both time and locale between the pollen data of Figure 10.10 and the diatom data of Figure 11.5 is substantial. Compare the climate information for the north central United States that can be inferred from the data. Where is it similar and where does it differ? What are some reasons that might explain the differences?

11.3 Describe the chemical differences and similarities between today's atmosphere near a large city such as Moscow, Berlin, or Chicago and a similar region at ages of 1 Gyr BP, 1 Myr BP, and 1 kyr BP.

11.4 A central question of biochemistry and geochemistry is how life began on Earth. One suggestion is that amino acids and other "building block" molecules were formed as a result of chemical reactions between simple molecules in warm pools of water at Earth's surface in the presence of such catalysts as clays and metal ions while exposed to an energy source such as lightning, heat, or solar radiation. Another suggestion is that life formed beneath the ocean at hydrothermal vents. A third is that the key molecules were formed in interstellar space and Earth was seeded with the molecules as a result of cometary collisions. Using the following sources or others of your choice, present the arguments for and against each of these theories: C. Ponnamperuma, *Cosmochemistry and the Origins of Life*, Reidel Pubs., Dordrecht, 1983; Z. Borowska and D. Mauzerall, Photoreduction of carbon dioxide by aqueous ferrous ion: An alternative to the strongly reducing atmosphere for the chemical origin of life, *Proceedings of the National Academy of Sciences, USA, 85,* 6577–6580, 1988; C. F. Chyba, P. J. Thomas, L. Briikshaw, and C. Sagan, Cometary delivery of organic molecules to the early Earth, *Science, 249,* 366–373, 1990; S. L. Miller and J. L. Bada, Submarine hot springs and the origin of life, *Nature, 334,* 609-611, 1988; and issues of the journal *Origins of Life*.

FURTHER READING

J. M. Barnola, D. Raynaud, Y. S. Korotkevich, and C. Lorius, Vostok ice core provides 160,000-year record of atmospheric CO_2, *Nature, 329,* 408–414, 1987.

M. A. K. Khalil and R. A. Rasmussen, Atmospheric methane: Trends over the last 10,000 years, *Atmospheric Environment, 21,* 2445–2452, 1987.

C. Lorius, J. Jouzel, D. Raynaud, J. Hansen, and H. LeTrent, The ice-core record: Climate sensitivity and future greenhouse warming, *Nature, 347,* 139–145, 1990.

CHAPTER *12*

Global Change: The Last Few Centuries

Large-scale environmental change due to anthropogenic activities began with slash-and-burn agriculture and other exploitation of soils and forest resources in China, Roman North Africa, and elsewhere. While resulting in noticeable local pollution, these activities had little effect on the global environment, nor did the use of wood (and later coal) for residential heating have much global impact. This situation changed markedly following the early part of the eighteenth century during what has come to be called the Industrial Revolution.

The Industrial Revolution was based on coal and iron. Coal provided the power for the newly invented steam engines and the energy to produce iron. The abundant availability of iron permitted the manufacture of high-quality tools and machines, the building of improved bridges and ships, and, ultimately, the mechanization of agriculture and thus the impetus for transformation from agrarian life to urban life for a sizable portion of Earth's population. The Revolution was particularly successful in societies, first in Europe and then in North America, that were able to bring together within a small geographical region the necessary scientific and technical knowledge and substantial quantities of energy, work force, and capital.

As industrialization developed, metals other than iron were needed and techniques were developed to recover them from their ores. Copper was particularly important because of its high electrical conductivity and was produced in large quantity in the latter half of the nineteenth century. In the present century, the production of zinc, aluminum, and other metals from their ores has become common.

The advances of the Industrial Revolution did not come without price. Of the mass of materials extracted from the earth and converted into products, more than 90% was discarded as waste by-products. Concomitantly, the first large-scale anthropogenic emissions of gases and particles into the atmosphere occurred. These emissions continue today. It is interesting that the processes that began the Industrial Revolution more than two centuries ago—the generation of energy by the combustion of fossil fuels and the smelting of ores to recover metals—are still processes that have major impacts on the atmosphere of Earth, over long and short distance scales and over long and short time scales.

Coincident with these industrial emissions were changes produced by an evolving agriculture. Agriculture developed rapidly in many parts of the world during this period, driven by population growth and by immigration to areas only sparsely populated and developed. In these regions, huge tracts of land were, and are still being, deforested to provide farmland. This deforestation released CO_2 to the atmosphere from decaying tree biomass and oxidized soil organic matter, as well as a suite of compounds emitted from biomass burning.

Historical chemical records for the past few centuries are largely drawn from the same type of samples as are those representing the Cenozoic atmosphere—Earth's ice cores. The principal distinction between the data sets is that the record of the past few centuries is much more detailed than that of the past few millennia, because the recovery and analysis of the newer samples are easier; the older deposits are at greater depths and have suffered much greater compaction. In addition, the polar ice data are supplemented by data from midlatitude glaciers, these glaciers having had several hundred years of continuous existence. Accurately dated sediments provide an additional source of information.

Biomass Combustion

The combustion of *biomass* (fuel wood, brush, vegetation, grass, and other organic materials) is a rich source of atmospheric emissions and is a topic that is receiving much increased study. The species emitted comprise, besides CO_2, a whole suite of chemically reactive gases, including CO, NO_x, CH_4, CH_3Cl, various other hydrocarbons, and particulate matter. Different types of burning and different biomass materials produce variable amounts of trace substances.

Biomass burning may be natural (e.g., fires started by lightning) but is mostly anthropogenic. It is principally conducted in the tropics during the dry season to remove forest material to clear land for agriculture and grazing, to burn agricultural wastes and dry savanna grasses, and, of course, for heating and cooking. Agricultural burning was still very common as late as the past century in Europe and North America, where it was probably the main source of air pollution at that time.

The current rate of biomass burning is very uncertain but is thought to be in the range of 2–5×10^{15} g C yr^{-1}. Of far less quantitative importance are fires in temperate and boreal forests. In these regions, especially the United States, Canada, and Australia, extensive efforts have been made over the past century to control natural forest fires. For example, it is estimated that the area burned annually in the United States decreased by a factor of 10 between 1930 and 1980. More recently, however, foresters in the midlatitudes are again beginning to recognize the beneficial effects of natural fires on forests and have tended to control fires only near populated areas.

Crop Production

Crop production involves changes in land use and in soil chemistry. As such, it has the potential to affect the atmosphere. Perhaps the greatest potential effect has to do with carbon dioxide, which is released to the atmosphere when soil organic matter is oxidized by tilling.

Certain crops, especially rice, grow under anaerobic conditions, leading to the formation of reduced gases. Methane emission is of the greatest concern because the flux is relatively large. Not only has the total area in rice production increased over the past half-century, but multiple cropping has increased the total area under cultivation. As a consequence, rice production appears to have contributed significantly to the observed atmospheric increase in methane.

Tilling of the soil may also lead to substantial but not very well quantified emission of N_2O, as implied from recent studies that showed that 2.7% of the fixed nitrogen that was lost in drained, cultivated soils in south Florida appeared as N_2O. Fertilization of cultivated land is a common practice that results in the emission of

N_2O. So far, field measurements suggest that this source of N_2O is not very large, but measurements have been made at temperate latitudes only; tropical soils may behave differently. This potential source of atmospheric N_2O is one that may be expected to grow with the globally increasing use of fertilizer nitrogen. However, much field research remains to be done in order to quantify its importance.

As the land used to grow crops has increased and as agricultural practices have permitted land to be farmed more intensively, emissions to the atmosphere from these activities have increased apace. It is unfortunately the case that reliable data on world land use, from which areas devoted to crop production could be derived, are generally unavailable on a global scale before about 1950. It has been estimated, however, that during each of the periods 1860–1919 and 1920–1978 the land areas converted to regular cropping grew by about 400 million hectares, an area larger than the size of India. Another aspect of the situation pertinent to the emission of nitrous oxide and of ammonia is the amount of fertilizer used in the production of crops. Even though the land area per person devoted to the production of grain has decreased over the years, the use of fertilizer has increased sharply.

Domestic Animals

The animal kingdom is a source of several atmospheric trace gases. Methane is produced by fermentation in the guts of animals and by the digestion of wood by termites. Although difficult to quantify in detail, the fluxes from the undomesticated animal kingdom are small compared with other methane sources, whereas those from domestic ruminant animals are very significant, accounting for some 15% of the total methane source. In addition to methane, animals produce ammonia, hydrogen sulfide, and other volatile products of excrement. The tendency for domestic animals to be gathered together in feedlots and pastures in the developed countries is thought to cause substantial local impacts, especially corrosion of metals, although these effects are very poorly quantified.

The number and mass of nondomestic animals on Earth have almost certainly decreased in recent times because of the steady expansion of human influence. The effects of this decrease on atmospheric emissions of CH_4 and NH_3 are, however, overwhelmed by the growth in domestic animal populations and the concomitant conversion of natural lands, especially forests, to pasture and agricultural lands under highly productive cultivation. Since 1950, the number of domestic animals has increased dramatically, with proportional increases in the emissions of methane and ammonia. The approximate methane emission rate for range and dairy cattle is about 55 kg per animal per year. For ammonia, the corresponding figure is much less. Because the consumption of higher levels of meat is a central characteristic of the diet of more highly industrialized countries, the rapid increase in the number of domestic animals may be expected to continue as long as the supply of grain does not limit that increase.

Coal Production and Use

Coal use results in large and diverse emissions into the atmosphere. In addition, methane is emitted during the mining and processing of coal. During combustion, coal produces major amounts of CO_2, CO, hydrocarbons, NO and NO_2, SO_2, and soot. In addition, coal combustion produces hydrogen chloride (HCl), ammonia (NH_3), and several trace metals. In general, the combustion of coal produces less energy per unit of fuel weight than do petroleum and natural gas, and, because coal has a higher impurity content, its emission factors are larger.

The importance of coal combustion as a source of atmospheric trace gases will almost certainly grow with time, because the world's reserves of coal are very large in comparison with those of petroleum and natural gas. In addition, coal is generally the least expensive reliable source of energy. However, these desirable characteristics are counterbalanced by the fact that coal is the most detrimental energy source from the standpoint of atmospheric impacts. This unfortunate combination of properties means that coal combustion will be one of the most serious problems in planning for the long-term sustainable development of the biosphere. Emissions of CO_2, NO_x, and SO_2 are of especial concern. The result may be that future use of coal will be permitted only after a precleaning step to remove sulfur and other trace constituents or after gasification. Should either of those requirements be implemented, the cost advantage coal now possesses over other forms of fuel will diminish or disappear.

The use of coal has traditionally been very directly related to industrialization and home heating, and has risen rapidly in the last century as nations increasingly joined the Industrial Revolution. Figure 12.1 shows the history of coal use for about the last 120 years. The scale used on the vertical axis is the amount of carbon dioxide emitted by the combustion process, a normalizing factor that makes it easy to compare the atmospheric effects of different energy sources.

The emissions of other species from coal combustion do not necessarily follow that of CO_2. For example, the overall sulfur dioxide emissions from coal combustion have been reduced sharply in some parts of the developed world. Most of that reduction is due to the use of coal with lower sulfur content, not to increased emissions control.

Petroleum Production and Use

The combustion of petroleum in its various forms is a very substantial source of trace gases, many of which have negative atmospheric impacts. The primary gaseous emittant is carbon dioxide, but incom-

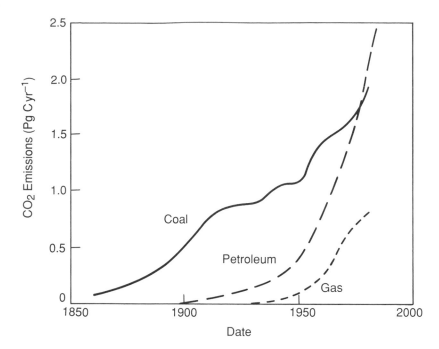

Figure 12.1 World use of coal, petroleum, and natural gas, 1860–1980, as measured by the emission of carbon dioxide to the atmosphere (expressed as Petagrams $C = 10^{15}$ g C per yr).

plete combustion produces carbon monoxide and a wide variety of hydrocarbon compounds as well. In addition to these carbon-based products, the high-temperature combustion of petroleum inevitably produces various oxides of nitrogen from the combustion of nitrogen impurities and from the oxygen and nitrogen in the air, which partially dissociate at the high combustion temperatures. NO and NO_2 are highly active catalysts in most atmospheric chemical chains, because they are involved in vital oxidizing reactions leading to ozone in the troposphere.

The sulfur content of petroleum varies widely and, as a result, so do the emissions of sulfur dioxide. In highly industrialized regions these emissions can be a major cause of acidic precipitation and the subsequent degradation of ecosystems and materials. Sulfur dioxide and sulfate are also important species in aerosol and cloud chemistry.

Petroleum burned in inefficient combustors can be a prolific producer of carbon soot, which is involved in visibility degradation and in condensed-phase atmospheric chemistry. Carbon soot may also influence Earth's radiation budget by absorption of sunlight, a process especially noticeable during the early spring in the Arctic.

Negligible amounts of petroleum were used as an energy source prior to the twentieth century, but petroleum use since then has grown rapidly (Figure 12.1), especially by motor vehicles. The most dramatic feature of the figure is that in about 1975 petroleum combustion passed coal combustion as a source of carbon dioxide. This pattern will probably continue for the next century, until world reserves of petroleum become sufficiently depleted that it ceases to be a major energy source.

Natural Gas Production and Use

The third energy source whose emissions are quantified in Figure 12.1 is natural gas. This gas, which can also be recovered as a by-product of the production of petroleum, saw little use prior to the first quarter of the twentieth century, but its utilization for the production of energy has risen rapidly since that time. Natural gas tends to burn much more cleanly than coal or petroleum, with only carbon dioxide, and perhaps methane, being an emittant of great concern. Per unit of energy produced, the CO_2 emissions from natural gas are smaller than those from oil, which in turn are smaller than those from coal.

Like the supply of petroleum, Earth's supply of recoverable natural gas is quite limited. It may be largely used up by the middle of the next century, thereafter ceasing to be an important factor in atmospheric impact assessments.

Waste Disposal

Methane is produced by anaerobic decay of organic municipal and industrial wastes in landfills. Worldwide emission estimates from this source have high uncertainty but suggest methane releases to the atmosphere of the order of 0.5 g CH_4 per gram of biodegradable carbon disposed of on land. Counting all organic waste disposal worldwide, this source of methane may be between 6 and 15% of the global CH_4 source. By far the greatest part of this emission comes from the industrialized world. Release rates from industrialized countries have been increasing steadily during the past few decades, but growth rates are now gradually stagnating. In the future, the contribution from developing nations is expected to grow rapidly because of strongly increasing population and urbanization. Consequently, methane release from landfills may become one of the main sources of atmospheric methane in the next century, and it will be important to develop methods to use this methane as an energy source rather than to permit its escape into the atmosphere. Because most of the carbon in landfills originates from plant material (such as paper, food wastes, and wood wastes), the emissions of CO_2 and CH_4 from landfills do not represent new sources of atmospheric carbon; they are only a recycling of a portion of the carbon pool.

A variety of more complex organic and halogenated compounds are released from landfills in addition to CO_2 and CH_4, but the total emissions of all of them are thought to be small when evaluated on a global scale.

An alternative to burying waste is to incinerate it. Modern incinerators have low rates of emission of trace species, but carbon dioxide is emitted, of course, as are somewhat uncertain amounts of hydrogen chloride (the latter arising from the combustion of chlorine-containing plastics) and small amounts of a number of other chemicals. The emissions of CO_2 and HCl vary widely with the mix of incinerated materials and with the age and type of incinerator facility.

Industrial Processes

The most diverse group of emissions into the atmosphere comes from the wide variety of industrial processes and subsequent product use where fuel combustion is not the primary source. Three classes of emittants can be singled out for special attention. The first is the family of chlorofluorocarbon compounds that was discussed in Chapter 8. The CFCs have been the subject of recent international agreements that are intended to prohibit their emissions by the end of the century. A second emittant of note is CO_2 from the manufacture of cement, i.e., the conversion of limestone into lime:

$$CaCO_3 + heat \rightarrow CO_2\uparrow + CaO. \qquad (12.1)$$

Cement manufacture currently is responsible for about 2% of all anthropogenic CO_2 emissions.

A third industrial emittant of importance is atmospheric particles, which play major roles in reducing visibility, in some health-related impacts, and as a platform for various transformations of gaseous pollutants. Transition metals, produced by many of the materials-processing industries, generally reside on emitted particles, although some are emitted as gases. For several of the metals the emission rates are quite large, and soil- and water-quality impacts may be anticipated for species such as mercury and cadmium. Emission rates of trace metals are widely variable depending on the industrial process, the raw materials used, and the degree of emission controls incorporated into the industrial process. Reduction in the emission rates of larger particles can often be accomplished without great expense. In the case of the smaller particles, however, emissions control can be difficult and costly. This situation is unfortunate, because it is these small particles that cause most of the particle-related decreases in visibility.

Industrial emissions include many other species, some rather poorly quantified. Many hydrocarbons, used as solvents, paints, and so on, are major emittants. A number of these compounds are highly volatile and are factors in the formation of photochemical smog in and downwind of urban areas.

THE ICE CORE RECORD

Climate

Climate information, particularly temperature and amount of precipitation, can readily be derived from the study of midlatitude glacier ice cores, the former parameter being derived from $^{18}O/^{16}O$ isotopic analysis, the latter by measurements of ice accumulation with time. We show in Figure 12.2 temperature data from 1600 to the present, derived from a Peruvian ice core. Temperature histories for the

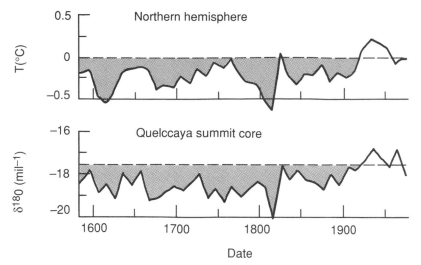

Figure 12.2 Decadal temperature departures (from the 1881–1975 mean) in the northern hemisphere as determined by a number of long record reconstructions and variations in the temperature indicator $\delta^{18}O$ for the Quelccaya ice cores from southern Peru. (Reproduced with permission from L. G. Thompson, E. Mosley-Thompson, W. Dansgaard, and P. M. Grootes, The Little Ice Age as recorded in the stratigraphy of the tropical Quelccaya ice cap, *Science, 234,* 361–364. Copyright 1986 by the AAAS.)

northern hemisphere derived by independent techniques are shown as well. The remarkable agreement confirms that the Little Ice Age and the climatic impact in 1815 of the eruption of the Tambora volcano were global in extent and that midlatitude glacial records can play important roles in studies of historical climate and air quality.

Gases

Most estimates of historical atmospheric CO_2 concentrations have been derived from ice core analyses that give values of about 270–280 ppmv for samples from the midnineteenth and preceding centuries. At the end of the past century, some atmospheric measurements were made directly on air samples; for example, the CO_2 concentrations of about 285–290 ppmv measured then are in excellent agreement with the ice core data. Some exciting new ice core analyses have shown substantial variability of atmospheric CO_2 over millennia, with low values of about 180–200 ppmv during ice ages and 260–280 ppmv during interglacials (see Figure 1.4). Many of the more recent CO_2 measurements are collected in Figure 12.3a; they show that the CO_2 concentrations in the atmosphere have increased steadily since the middle of the eighteenth century. It is of interest to compare these concentrations with the emission rates, as done in Figure 12.3b. The figure shows that until as recently as 1960, agricultural expansion and deforestation had a greater impact on atmospheric CO_2 than did fossil fuel combustion.

Methane has many sources, a number of which are influenced by human activity. The historical methane content of the atmosphere provides evidence that those sources caused little change in the atmospheric methane concentration until about two centuries ago. In Figure 12.4 methane concentrations for several hundred years are shown; they are derived from ice core studies by a number of different

Figure 12.3 (a) Carbon dioxide concentrations at remote global sites, 1750–1989. Filled circles represent data from an ice core at Siple Station, Antarctica; open circles represent data acquired in situ by modern analytical techniques at Mauna Loa, Hawaii. (Reproduced with permission from H. Friedli, H. Lötscher, H. Oeschger, U. Siegenthaler, and B. Stauffer, Ice core record of the ^{13}C/^{12}C ratio of atmospheric CO_2 in the past two centuries, *Nature, 324,* 237–238. Copyright 1986 by Macmillan Magazines Limited.) (b) Net emission rates of carbon to the atmosphere from the terrestrial biosphere (dotted line) and from fossil fuel consumption (dashed line) for the period 1860–1970. The solid line represents the sum of the two emissions. (Reproduced with permission from B. Moore et al., in *SCOPE 16: Carbon Cycle Modelling,* B. Bolin, ed., pp. 365–385, John Wiley & Sons, Chichester, UK. Copyright 1981 by Scientific Committee on Problems of the Environment.)

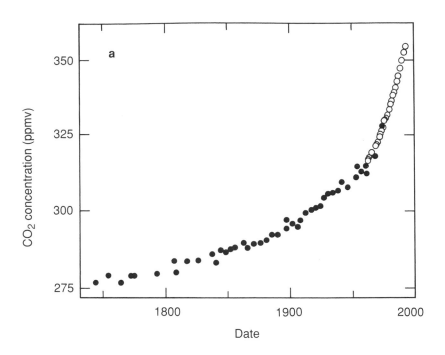

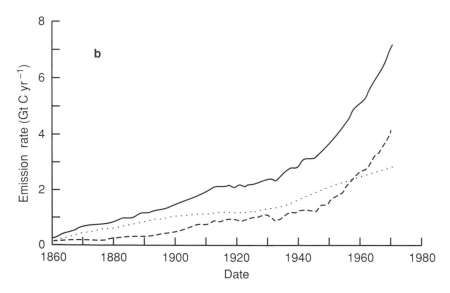

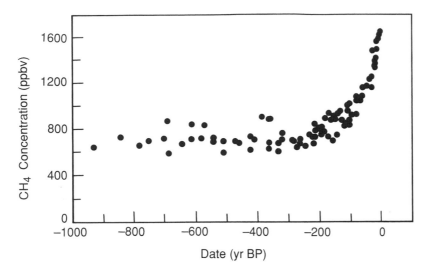

Figure 12.4 The atmospheric concentrations of CH_4 over the last millennium. The figure shows the time when CH_4 concentrations began to increase rapidly and the doubling of the CH_4 concentration over the last 200 yr. (Reproduced with permission from M. A. K. Khalil and R. A. Rasmussen, Atmospheric methane: Trends over the last 10,000 years, *Atmospheric Environment, 21*, 2445–2452. Copyright 1987 by Pergamon Press, plc.)

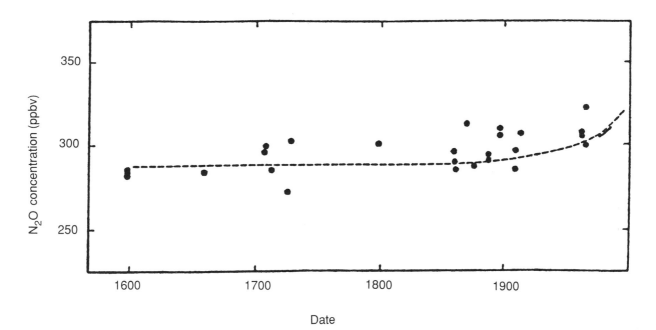

Figure 12.5 Historical record of atmospheric N_2O concentrations based on measurements of air trapped in an ice core from Antarctica. The dashed line is the global tropospheric average concentration from a model incorporating constant noncombustion and increasing combustion sources. The short solid line is modern data from the South Pole and from Tasmania. (Reproduced with permission from G. I. Pearman, D. Etheridge, F. de Silva, and P. J. Fraser, Evidence of changing concentrations of atmospheric CO_2, N_2O, and CH_4 from air bubbles in Antarctic ice, *Nature, 320*, 248–250. Copyright 1986 by Macmillan Magazines Limited.)

research teams. The concentrations are nearly stable, at about 700 ppbv, until after the middle of the seventeenth century. After that date, atmospheric methane began a relentless increase, which will be illustrated in more detail in the following chapter.

Nitrous oxide is a natural product of bacterial activity in ocean waters and soils. Its historical concentrations have recently been examined by ice core analyses; the results are shown in Figure 12.5. They indicate that N_2O concentrations have not changed dramatically except during the present century. Recent data show a definite increase at a rate of between 0.2 and 0.3% per year. Nitrous oxide averaged 285 ppbv between 1700 and 1800, compared with modern global values of between 305 and 310 ppbv. The data from ice cores for relatively modern periods agree with modern measurements, thus confirming the validity of the ice core record.

Ions

As noted in Chapter 11, snow and ice cores collected from glaciers provide an opportunity to determine the chemistry of previously deposited precipitation and hence allow the production of reliable historical data bases for areas remote from local anthropogenic influences. An example of such a study is that made on ice cores from a site in Greenland. In the previous chapter we demonstrated that the data extend backward in time some 1500 yr. This data record, and others like it, indicate that the acidity of high-latitude ice cores during this period was sporadically influenced by volcanic eruptions, particularly eruptions at high latitudes, and much less by anthropogenic emissions. Ice sheet acidity is thus an excellent tracer for historical volcanic data.

Anion concentration measurements for "excess sulfate" are shown in Figure 12.6. The value for excess sulfate (referred to hereafter simply as sulfate) is derived by subtracting the portion of sulfate due to seasalt from the total sulfate value to yield the component that is largely anthropogenic or volcanic in origin. The figure shows a number of

Figure 12.6 The sulfuric acid profile over the past 200 yr, as derived from an Antarctic ice core. Nine major events are numbered for identification and have been related to the following volcanic eruptions: 1, Tambora, Indonesia (1815); 2, Galunggung, Java (1823); 3, Coseguina, Nicaragua (1835); 4, Armagura Island, S. Pacific (1846); 5, Cotopaxi, Ecuador (1856), and/or Makjan, Molucca Island (1860); 6, Tarawera, New Zealand (1886), and/or Krakatoa, Java/Sumatra (1883); 7, Santa Maria, Guatemala (1902); 8, Nilahue, Argentina (1955); and 9, Agung, Bali (1963). (Adapted with permission from M. Legrand and R. J. Delmas, A 220 yr continuous record of volcanic H_2SO_4 in the Antarctic ice sheet, *Nature, 327,* 671–676. Copyright 1987 by Macmillan Magazines Limited.)

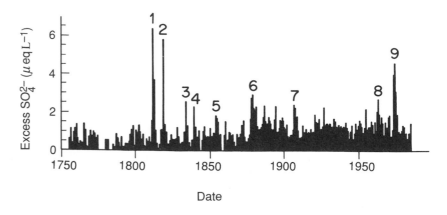

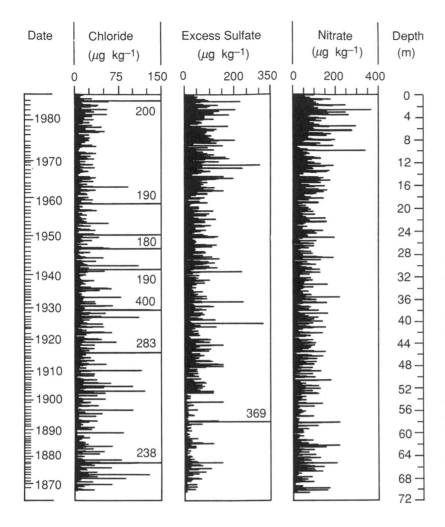

Figure 12.7 Chloride, "excess sulfate," and nitrate time series from a South Greenland ice core, 1869–1984. Chloride and excess sulfate values that are off scale appear as numbers. The time scale is compressed to take account of the nonuniform layer compaction with depth. Excess sulfate is the value given by subtracting seasalt-related sulfate (that correlated with Na^+ and K^+) from total sulfate. (Adapted with permission from P. A. Mayewski et al., Sulfate and nitrate concentrations from a South Greenland ice core, *Science, 232,* 975–977. Copyright 1986 by the AAAS.)

major events that can be attributed to volcanic eruptions. In addition, because it complements the H^+ identification of volcanic occurrences, it indicates that ice core acidity is largely due to the deposition of airborne sulfuric acid or its precursors.

Unlike sporadic events, which tend to be closely linked to natural processes, gradual trends are more likely to reflect the increasing emissions produced by global development. A Greenland ice core record for several ions for the past century appears in Figure 12.7. The ice core sulfate data are quite sporadic in character, but a clear increase in average concentration over the past century is indicated. The abrupt changes in sulfate mostly represent volcanic input. This work and that of others demonstrate that average nonvolcanic sulfate deposition in South Greenland precipitation now is approximately three times what it was at the turn of the century. This result is in rough agreement with increased SO_2 emissions from fossil fuel combustion, which in the northern hemisphere contributes about three to five times more sulfur than do natural processes.

Precipitation nitrate is also analyzed in ice cores from the glaciers of Greenland and elsewhere. The sources of the nitrate include natural emissions from soils and by lightning, as well as anthropogenic activities such as fossil fuel and biomass burning.

Changes in the nitrate record of the Greenland ice core (Figure 12.7) indicate a relative constancy in value from 1869 to 1955, followed by a period of general increase from 1955 to 1975 and an intensified increase after 1975. The magnitude of the increase is in good agreement with the results from other ice cores and, for the last several decades, with North American emission estimates for NO and NO_2, as shown in Figure 12.8.

Chloride in ice cores has known sources from the marine environment and from volcanoes. Anthropogenic sources are also a possibility,

Figure 12.8 Annual emissions of sulfur dioxide and oxides of nitrogen from the United States, 1950–1980. (The data are from G. Gschwandtner, K. Gschwandtner, K. Eldridge, C. Mann, and D. Mobley, Historic emissions of sulfur and nitrogen oxides in the United States from 1900 to 1980, *Journal of the Air Pollution Control Association, 36,* 139–149, 1986.)

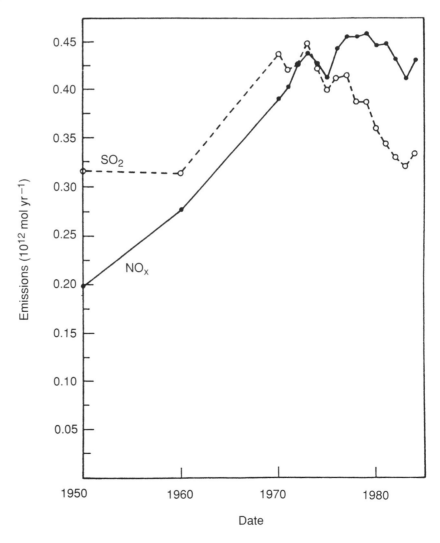

because HCl is emitted during the combustion of coal and of industrial and domestic waste. The influence of the marine source can be estimated by studying the ratio of chloride to sodium in the ice, whereas volcanically related signals are deduced by comparison with ice core sulfate records. For the Greenland ice core data shown in Figure 12.7, no significant anthropogenic trend is identifiable in the chloride concentrations for the period 1869 to 1984. The main reason may be the very short atmospheric lifetime (of the order of a day) of HCl, a gas that is highly soluble and therefore is removed extremely efficiently by precipitation and by deposition onto wet surfaces. Hydrogen chloride thus does not survive transport from the industrial sources to the Greenland glacier.

Total Particulate Matter

Historical concentrations of atmospheric insoluble aerosols are available from a number of high-latitude ice cores in both the Arctic and Antarctic. Like the ice core acidity data, the aerosol data are sporadically influenced by volcanic eruptions. Careful study suggests that the total aerosol loading does not show a significant anthropogenic influence. The relatively short lifetimes of most aerosol particles compared with the time scale for cross-latitude transport (see Chapters 4 and 5) come into play in explaining this absence of anthropogenic signatures in polar regions, but effects can readily be seen in ice cores from midlatitude glaciers.

THE SEDIMENT RECORD OF CHANGING CHEMISTRY

In some circumstances, sediment samples removed from lakes or bogs can be useful in establishing historical chemical compositions, at least in a relative way. As an example, Figure 12.9 shows data from sediment cores representing more than two centuries of deposition from the atmosphere to a peat bog in western Pennsylvania. Clearing and burning of the neighboring forest occurred between the mid-eighteenth and early twentieth centuries. The mass deposition was a maximum during the early twentieth century, presumably because atmospheric constituents were no longer being intercepted by the forests. (The total mass data are not shown here.) No local emissions sources of any significance are present, atmospheric deposition being the only route by which the elements enter. The region, although not the local area, is one in which heavy industry has been prevalent, and in which the manufacture of steel has been a feature of the last century. Coal is a major source of industrial and residential power generation.

Examine first the data for scandium in Figure 12.9. Scandium concentrations in the sediments were higher in the early twentieth century than before or after, a pattern consistent with the general mass deposition pattern. Such agreement is reasonable, because scandium is a trace constituent of soil, and this pattern appears to reflect the deposition in the bog of windblown soil dust. Look next at the deposition pattern for iron. Iron concentrations are also substantially higher than before the Industrial Revolution but have not decreased with the imposition of particle emission controls in the 1970s, probably because the sources of atmospheric iron include not only industrial emissions (which are most subject to emissions controls) but also windblown soil dust and fly ash from distant electrical power generation stations.

In the cases of sulfur and nitrogen, the patterns are quite different, showing very rapid increases over the past few decades from initially very low values. This pattern almost certainly reflects the very rapid increase in the use of coal, and, later, other fossil fuels, for energy generation, with a concomitant increase in the emissions of the gaseous combustion products SO_2 and NO.

Sediment chemistry records are indirect in that they cannot reflect actual atmospheric or hydrospheric concentrations of species

Figure 12.9 Deposition rates for four elements as functions of time and depth in a sediment core from a mountain top peat bog in western Pennsylvania during the period 1757–1981. (Adapted with permission from W. R. Schell, A historical perspective of atmospheric chemicals deposited on a mountain top peat bog in Pennsylvania, *International Journal of Coal Geology*, 8, 147–73. Copyright 1987 by Elsevier Science Publishers.)

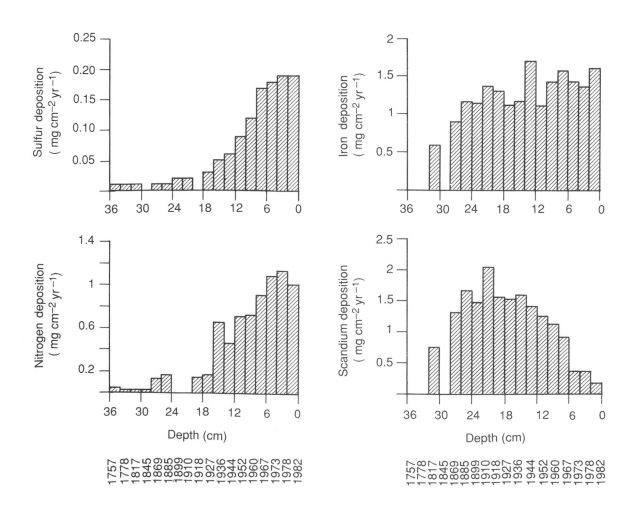

unless detailed calibrations can be carried out. The data are useful in a relative sense, however, for showing differences in deposition at different times and locations and for studying the impact of nutrient or toxic materials from the atmosphere on agricultural soils.

SUMMARY

Evaluated historical measurements provide a surprisingly informative picture of a few of the environment's trace species over the past few centuries. Depending on the residence time of the species, the information can be global in scope, regional, or local. In all cases, the general picture is the same: relative constancy in concentrations until a few hundred years ago, followed by increases at various rates in the concentrations of nearly all the measured constituents that have anthropogenic sources, first for agriculturally related species, then for species in industrial emissions. Although we can never hope to have a complete picture of the environmental chemistry of the past few centuries, its outlines can be inferred from the limited data available and are consistent with the known patterns of population growth, agricultural development, energy use, and industrial development.

EXERCISES

12.1 Construct a graph of total atmospheric weight of methane-related carbon as a function of year from 1770 to 1980. [Note: This project will require taking data points from a published graph (Figure 12.4), something scientists do all the time.]

12.2 Using the hydrostatic equation of Chapter 4, compute the approximate total mass present in the troposphere in 1850 of carbon dioxide, methane, and nitrous oxide. Assume that the data in Figures 12.3–12.5 are representative of the surface troposphere. Repeat the computation for 1750 and 1950 and comment on the results.

12.3 Peak 1 in Figure 12.6 is from the Tambora volcanic eruption of 1815. Its sulfate signal extends for a depth of 57 cm in the ice core at an average concentration of 2.6 μeq L^{-1}. (Recall that for sulfate 1 μeq = 2 μmol.) What is the total deposition to the ice of sulfate from the Tambora event, in units of kg SO_4^{2-} km^{-2}? If two-thirds of Earth's surface (both land and sea) received a similar deposition of sulfate, what was the approximate total sulfate emission of the Tambora volcano?

12.4 Figure 12.6 illustrates the sulfate signatures of several volcanoes in an ice core from Antarctica. Similar information has been published for the same time period for an ice core from Greenland, almost at the opposite pole of the world and in a different spatial relationship with anthropogenic and volcanic sources of sulfate (P. A. Mayewski, W. B. Lyons, M. J. Spencer, M. S. Twickler, C. F. Buck, and S. Whitlow, An ice-core record of atmospheric response to anthropogenic sulphate and nitrate, *Nature, 346*, 554–556, 1990; W. B. Lyons, P. A. Mayewski, M. J. Spencer, M. S. Twickler, and T. E. Graedel, A northern hemisphere volcanic chemistry record (1869–1984) and climatic implications using a South Greenland ice core, *Annals of Glaciology, 14*, 176–182, 1990.) Compare the information from the two ice cores and prepare a summary of what the sulfate data in the cores have to say about the atmospheric circulation, the importance of volcanoes to the global sulfur cycle, and the importance of anthropogenic emissions to the global sulfur cycle.

12.5 As Figure 12.9 shows, elemental analysis of deposited airborne material can sometimes be used to trace the origins of emissions having an impact in a particular region. A recent example of such a study is that for air quality in and air transport to the Arctic. Prepare a report on those studies, drawing on the following references and others of your choosing: G. E. Shaw and M. A. K. Khalil, *The Handbook of Environmental Chemistry*, Vol. 4, Part B, pp. 69-111, Springer, Berlin, 1989; *Atmospheric Environment, 23*, No. 11, 1989; *Geophysical Research Letters, 11*, No. 5, 1984.

FURTHER READING

P. Brimblecombe, London air pollution, 1500–1900, *Atmospheric Environment, 11*, 1157–1162, 1977.

C. U. Hammer, H. B. Clausen, and W. Dansgaard, Greenland ice sheet evidence of post-glacial volcanism and its climatic impact, *Nature, 288*, 230–235, 1980.

B. L. Turner, II, W. C. Clark, R. W. Kates, J. F. Richards, J. T. Matthews, and W. B. Meyer, eds., *The Earth as Transformed by Human Action*, Cambridge University Press, Cambridge, UK, 1990.

CHAPTER 13

Global Change: The Last Several Decades

The last several decades have been times of accelerated change on Earth. The industrial capacity of the western free market economies and those of the former USSR and the other socialist nations increased markedly throughout most of the century. Subsequently, in the 1970s and 1980s, the "third world" transition into modern society and industrialization began to cause major changes on much of the remainder of Earth. Throughout this period, as we have seen, the human population increased dramatically, especially in the less developed world.

In the developed parts of the world, rapid improvements in manufacturing technology and in air and water quality legislation brought with them gradual emission reductions and improved energy efficiency. As a result, emissions of trace gases and particles from these processes into the atmosphere and hydrosphere have, on the whole, gradually decreased *on a unit basis*. For example, emission of sulfur dioxide per kilogram of combusted coal has decreased at many facilities. On a global basis, however, this technological improvement has been countered by a great increase in demand for energy, manufactured goods, and food. Overall, the twentieth century has been one of rapid change accompanied by degradation in environmental quality.

The period of creation of the modern industrial world coincides with that of the development of modern analytical chemistry. As a result, it has been possible in many regions, and in remote locations characteristic of the unperturbed planet, for changes in the atmosphere's and hydrosphere's composition and chemical reactions to be monitored in detail. In this chapter we present information on the trends revealed by such measurements during the twentieth century.

INTRODUCTION TO ANALYTICAL CHEMISTRY TECHNIQUES

Trace Gas Analysis

The first analytical measurements of an atmospheric trace gas were those developed for carbon dioxide and ozone in the mid-nineteenth century. The technique utilized for ozone was based on the rapid reaction of ozone with iodine compounds. In its earliest implementation, the technique involved the impregnation of filter paper with an iodine compound, the ozone concentration being determined by the paper's color change. Unfortunately, the early method was sensitive to wind speed and relative humidity, so many of the data taken with this simple technique are of little value. In later versions of the technique, air was passed through a solution of potassium iodide, and the amount of free iodine produced by the ozone was measured by titration or by electrolytic techniques. Other forms of "wet chemistry" have been used for a variety of atmospheric gases, particularly in the 1950s and 1960s. In almost all cases, the techniques suffered from insufficient sensitivity to the low gas concentrations common in the atmosphere and were subject to interferences from gases other than those whose concentrations were being sought.

Modern in situ methods of trace gas analysis involve the capture of a sample of air, followed by an analytical procedure based on the physical and/or chemical properties of the molecule to be studied. The capture itself is not a trivial task, and the generally small concentrations being determined require that every effort be made to avoid losses in the sampling system or contamination of the immediate environment of the sampling site by very local pollution sources, including the sampling vehicles themselves.

Once captured, the molecules may be detected in any of a number of ways. For example, optical techniques may involve high-resolution spectral absorption or emission, resonance fluorescence, resonance scattering, or stimulated fluorescence of the species to be measured. Chemical methods may utilize either reactions with a tracer species, followed by detection of the reaction product, or the response of a gas chromatograph or mass spectrometric detector to the compound. The modern techniques are much more sensitive than their classical analogues and much less prone to interferences from other atmospheric species. As a result, the data accumulated over the past decade or two are generally of much higher quality than those of the past and have been able to demonstrate extremely low levels of trace atmospheric species. Nevertheless, only very recently have major measurement programs been planned to start the process of determining the global distributions of many important atmospheric constituents. Considerable effort is also being devoted to comparing concentration measurements obtained by different analytical techniques and to developing international calibration standards. Furthermore, it is now customary to measure many chemical constituents at the same time, realizing that such an approach is critical to obtaining a complete picture of the chemical processes that are occurring.

Passive Optical Measurements

A number of different optical techniques are used in measuring atmospheric trace species. Among the most common is remote sensing by the analysis of spectroscopic data. As sunlight passes through the atmosphere, the molecules in the air absorb radiation at specific wavelengths. The amount of absorption is a function of molecular properties and abundance. Thus it is possible to determine the concentrations of atmospheric trace gases by examining the absorption lines in solar spectra. In the 1930s, reliable tropospheric ozone concentrations were determined by optical techniques. In this way it was confirmed that most of the atmosphere's ozone is located in the stratosphere. Mostly for astronomical purposes, solar spectra have been recorded and archived for nearly a century, and many of these old spectra have now been analyzed. The resulting information is of great interest, although it should be realized that this technique looks through the entire vertical column of air above the observation site at a particular instant and does not reveal either variations with altitude of the species measured or variations in its ground-level concentrations.

Aqueous Chemical Analysis

Performing a chemical analysis on aqueous environmental solutions is a two-step process. The first step is to acquire an uncontaminated sample, a process much more difficult than might be supposed. In the case of rain, samples can be gotten by relatively straightforward collection, provided carefully cleaned and acid-washed plasticware or similar containers are used. Samples must not remain exposed to the air between samplings, because gases may dissolve in the water, or bacterial or chemical action may alter the composition, or insects, bird droppings, and other contaminants may be added. Fog is much more difficult to collect than rain, but several advances in experimental design in the past few years have now provided numerous fog water samples. One device, a rotating arm collector, captures fog droplets by impaction on a slotted rod rotating at high velocity. Collected droplets flow by centrifugal force to bottles mounted at the ends of the support rod, a technique that immediately shelters the droplets in a quiescent environment and prevents sample evaporation. A similar approach, using stationary slotted rods or Teflon strands carried outside a moving airplane, is used to sample cloudwater. In the case of dew, the sample is collected by runoff from large Teflon surfaces, sometimes backed by cooling coils. For freshwater or seawater, collection of samples is accomplished by submerging precleaned and presterilized polymeric sampling bottles. The valves on the bottles are opened at the desired depths by pressure-sensing electronics or by remote control from shipboard or submersible research vessel.

Once the sample is collected, analyses of several types are generally performed. The ions in the sample, including Cl^-, NO_3^-, SO_4^{2-}, Na^+, NH_4^+, and K^+, are quantified by ion chromatography or a similar liquid-phase technique. Atomic absorption spectroscopy is used to determine concentrations of metals, including lead, calcium, magnesium, cadmium, iron, copper, manganese, and nickel. Detailed organic analyses may utilize mass spectroscopy, chromatography, or a variety of liquid-phase techniques.

GLOBAL AIR QUALITY DATA AND TRENDS

CO_2

Carbon dioxide was the first atmospheric gas to be routinely monitored, the remarkable effort being begun in Mauna Loa, Hawaii, in 1958 by David Keeling of the Scripps Institution of Oceanography. Data from that site for more than three decades are shown in Figure 13.1. The seasonal cycle produced by photosynthesis and respiration of vegetation at temperate and subtropical latitudes is clearly evident, as is the strong long-term upward trend.

The increase in atmospheric CO_2 since the beginning of the nineteenth century has been due to a variety of worldwide human activities. Until about 1950, the release of CO_2 was mainly caused by the oxidation of organic matter exposed by the tilling of agricultural soils. Currently, the largest net source of CO_2 to the atmosphere comes from fossil fuel combustion. Tropical deforestation also contributes

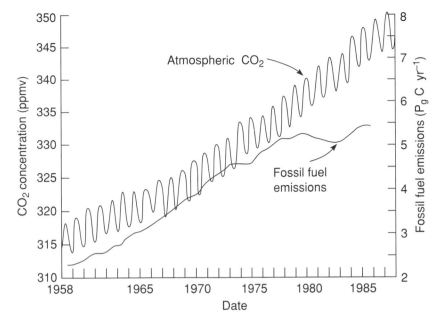

Figure 13.1 Concentration of atmospheric carbon dioxide at Mauna Loa Observatory, Hawaii, 19.5 °N, 155.6 °W, for the period 1958–1988. Also shown is the annual emissions of CO_2 from the burning of fossil fuels.

to the atmospheric increase; the net source is estimated to be 20 to 60% that of combustion. (The wide range indicates that there is a substantial uncertainty concerning the strength of the deforestation source term.) The rather strong deviation since about 1980 between the slopes of the atmospheric CO_2 concentration data and the fossil fuel CO_2 emissions (Figure 13.1) may indicate an expansion in tropical deforestation rates during the 1980s.

CO

Carbon monoxide is present in the atmosphere largely as a result of the atmospheric oxidation of methane and other alkanes, oxidation of organics emitted by vegetation, and the incomplete combustion of fossil fuels and biomass. Because the lifetime of CO is relatively short (a few months) and its land-based sources discontinuously spread all over the world, its concentrations vary substantially with time at any one location and from location to location; thus it is extremely difficult to observe a small systematic increase without frequent measurements spanning a relatively long period of time. Nonetheless, several sets of data appear to have established clear seasonal behavior and an upward long-term trend in carbon monoxide concentrations.

The first data set was collected in three urban New Jersey locations between 1968 and 1977. Studying the minima in these data as a function of time to look for changes in the "global background," Thomas Graedel and Jean McRae first suggested the possible increase. Later, studies by other groups confirmed this suggestion, at least in a preliminary way, for locations on several continents. It now appears that CO concentrations have shown an average annual increase of 0.7–

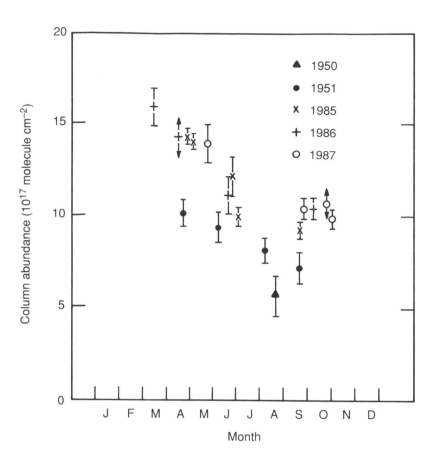

Figure 13.2 Monthly mean vertical column abundance of CO, derived from solar spectra recorded at an altitude of 3.6 km at the Jungfraujoch Station, Switzerland. Careful scrutiny reveals that the more recent data are the highest in concentration. The data show a strong seasonal cycle and an annual CO increase averaging 0.85%. (R. Zander, Ph. Demoulin, D. H. Ehhalt, U. Schmidt, and C. P. Rinsland, Secular increase of the total vertical column abundance of carbon monoxide above central Europe since 1950, *Journal of Geophysical Research, 94,* 11021–11028, 1989.)

1% between the mid-1950s and the present at midlatitudes in the northern hemisphere. Perhaps the best current demonstration of this trend is the concentration patterns derived from optical spectra taken at the Jungfraujoch Observatory in the Swiss mountains, shown in Figure 13.2.

It appears clear from Figure 13.2 that a steady increase in northern hemispheric CO is seen during all seasons of the year. This is of particular interest in view of the progress being made by the industrial nations in reducing their emissions of CO, as shown in Figure 13.3. The continuing global increase may be attributed to larger photochemical CO production from the oxidation of increasingly larger CH_4 concentrations, from increased CO emissions in the less developed countries, especially from biomass burning, and possibly also by gradual decreases in global hydroxyl radical concentrations.

CH_4

The same data set from New Jersey that first suggested an increase in atmospheric CO concentrations also did so for atmospheric methane (CH_4). Since that time, extensive and carefully calibrated measurements all over the world have confirmed that suggestion. The data from a number of experimenters for much of the past century are

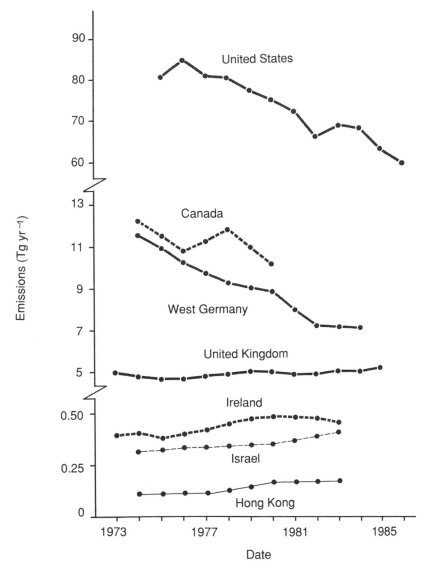

Figure 13.3 Trends in emission of carbon monoxide in different countries, 1973–1986. (After Global Environment Monitoring System, *Assessment of Urban Air Quality*, United Nations Environment Program, Geneva, 1988.)

pictured in Figure 13.4. The concentrations of methane as measured in ice cores for the years prior to about 1965 blend smoothly into the modern gas chromatographic measurements of the past two decades. The data show a steady increase over the entire period, although the most recent trend is at a higher rate than was the case earlier, amounting to about 0.6–1% (about 10–17 ppbv) per year. There are many CH_4 sources, most of which are related to human activity: coal mining, leaks in natural gas distribution systems, rice cropping, and fermentation in the guts of ruminants, especially cattle. The emissions from the undomesticated animal kingdom are small compared with other methane sources, whereas those from domestic ruminant animals are very significant, accounting for about 15% of the total methane source.

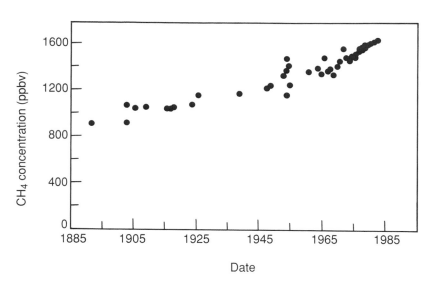

Figure 13.4 Atmospheric methane concentrations during the past century. Data up to 1965 is from ice cores; that for the most recent 20 yr are from measurements in air. (Reproduced with permission from M. A. K. Khalil and R. A. Rasmussen, Atmospheric methane: Trends over the last 10,000 years, *Atmospheric Environment, 21*, 2445-2452. Copyright 1987 by Pergamon Press, plc.)

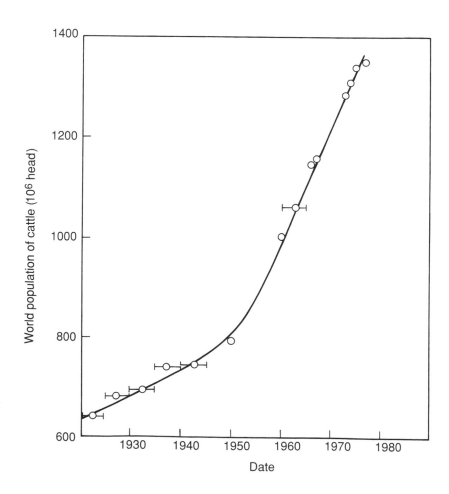

Figure 13.5 Estimated world population of cattle for the period 1920–1975.

The number and mass of natural animals on Earth have decreased in recent times because of the steady expansion of human influence. The effects of this decrease on atmospheric emissions of CH_4 are, however, overwhelmed by the growth in domestic animal populations and the concomitant conversion of natural lands, especially forests, to pasture and agricultural lands under intensive cultivation. The population of domestic cattle, the animal species responsible for most of the emissions, is shown in Figure 13.5. Since 1950 the number of domestic cattle has increased dramatically, with proportional increases in methane emissions.

Total Column Ozone

For several decades, systematic determinations have been made by spectroscopy of "column ozone," i.e., the total ozone in a column of air above the monitoring station. Most of this ozone is in the stratosphere. Complementary records have been provided for a little over a decade by satellite observations looking down through the same column of ozone. Recent concerns about possible stratospheric ozone destruction have stimulated careful examination of these historical records. A dramatic and recent example of these data appears in Figure 13.6. Two features of the data are worth noting. One is the seasonal pattern, showing larger ozone depletion during the winter months. This pattern is caused by the enhanced CFC–ozone chemistry that occurs in

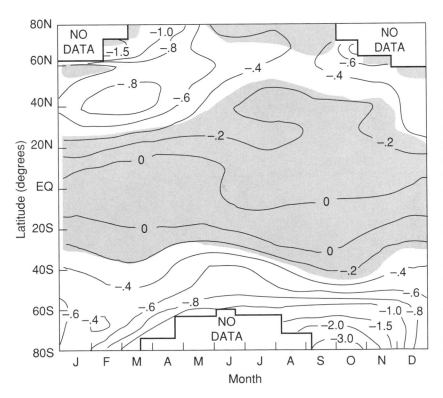

Figure 13.6 Trends in total column ozone (percentage per year) as a function of latitude and season for the period 1978–1990 as measured by the total ozone mapping spectrometer (TOMS) instrument on the *Nimbus 7* satellite. The data are averaged over all longitudes. Regions marked "no data" are those in the polar night, when no measurements are possible. The shaded area indicates where the trends are not significantly different from zero. The largest decreases occur near the South Pole in September and October, when the ozone hole forms. (R. S. Stolarski, P. Bloomfield, R. D. McPeters, and J. R. Herman, Total ozone trends deduced from *Nimbus 7* TOMS data, *Geophysical Research Letters, 18,* 1015–1018, 1991.)

the cold stratosphere over the poles, a process discussed in more detail in Chapter 8. The second feature is the tendency for diminished ozone depletion as one goes to successively lower latitudes. This pattern probably reflects the transport of ozone-poor air from the pole as a result of the formation and breakdown of the constraining polar vortex circulation. The further one gets from the pole, the less such processes influence the monthly average ozone concentration. These data indicate clearly that ozone destruction has occurred and is occurring, not only near the poles, but also at latitudes down to almost 20°.

The effects of changes in the concentration of stratospheric ozone are expected to be directly felt by biological systems. Figure 3.2 shows that the penetration of solar radiation to Earth's surface at wavelengths longer than about 210 nm is strongly limited by ozone absorption. This absorption extends to just past 300 nm, although with rapidly decreasing efficiency. As a result, the amount of radiation reaching the surface changes by a factor of about ten thousand from 290 nm to 320 nm, as shown in Figure 13.7.

The wavelength band 290–320 nm is termed the *UV-B region*, and it is this region to which biological organisms are most sensitive. Figure 13.7 gives the spectrum of radiation absorption by DNA, which changes by a factor of about 10^4, and in the opposite sense from the received surface radiation. (This is, of course, an evolutionary product, because organisms whose DNA was susceptible to solar

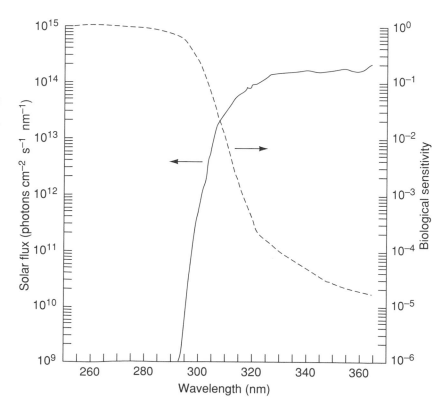

Figure 13.7 A comparison of the spectral flux of solar radiation reaching Earth's surface with the spectral sensitivity of biological organisms to radiation. The solid line shows the solar irradiance (scale at left) reaching Earth's surface when the Sun is at a zenith angle of 39°. The dashed line shows biological sensitivity (scale at right) to radiation at different wavelengths. (The sensitivity parameter is the inverse ratio of the efficiency of absorption of solar radiation at a specified wavelength by DNA compared with the absorption by DNA of radiation with a wavelength of 260 nm.) Decreases in stratospheric ozone result in increases in irradiance in the wavelength region between 290 and 320 nm. (Courtesy of C. Brühl, Max-Planck-Institut für Chemie, Mainz, Germany.)

radiation would not have been survivors over the long term.) As a consequence of the overlapping curves, any decrease in ozone increases the penetration of radiation exactly in the wavelength region where DNA is most sensitive. It is this circumstance that renders maintenance of the stratospheric ozone shield such a high priority.

We illustrated in Figure 1.1 the increasing loss of ozone in the Antarctic stratosphere in the austral spring. Recently, John Frederick and Amy Alberts of the University of Chicago have measured the impact of that loss on the amount of ultraviolet radiation that penetrates to the surface. Some of their results are shown in Figure 13.8, where the ground-level irradiance in the wavelength region near 300 nm is compared for the years 1988 and 1990. The Antarctic ozone

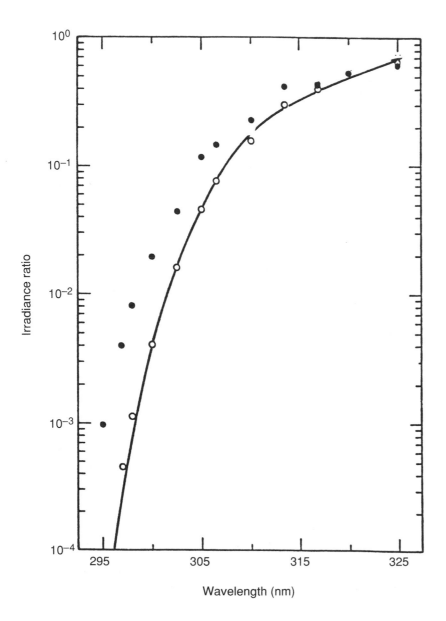

Figure 13.8 Ground-level irradiance ratio at Palmer Station, Antarctica. The data are for local noon on December 21, the austral summer solstice, in 1988 (o) and 1990 (•). The solid line is the result of a computer model. (J. E. Frederick and A. D. Alberts, Prolonged enhancement in surface ultraviolet radiation during the Antarctic spring of 1990, *Geophysical Research Letters, 18,* 1869–1871, 1991.)

hole in 1988 was deep, but not catastrophically so, and had nearly closed by December 21 of that year. In 1990, however, the ozone hole was deeper and more persistent, and on December 21 was still present. The relative impact on biologically active ultraviolet radiation was substantial, with noon irradiances at 298 nm about 10 times larger in 1990 than in 1988.

Effects of Ultraviolet Radiation on Biological Organisms

Humans. UV-B radiation causes cancer in light-skinned people. Studies have shown that a 1% reduction in the ozone layer increases the effective UV-B dose by about 2%. This increase in dose in turn leads to about a 4% increase in the incidence of basal cell carcinoma and about a 6% increase in squamous cell carcinoma. For a 10% decrease in stratospheric ozone, the latter two increases become about 50% and 90%. Thus in Germany, for example, a 10% reduction in stratospheric ozone would lead to approximately 20,000 additional cases of skin cancer each year. These carcinomas, unlike the melanoma skin cancers also caused by UV-B exposure, are curable with early treatment. Melanomas are much less frequent but much more deadly. Other effects of increased UV-B radiation on humans include a higher incidence of cataracts and a weakening of the immune system.

Plants. Approximately 200 plant species have thus far been examined for sensitivity to UV-B radiation. About half demonstrate significant adverse effects, including reductions in average leaf area, reduced shoot length, and decreases in the rate of photosynthesis. The most complete information has been gathered for agricultural crops; it suggests among other results that a 25% reduction in stratospheric ozone would lead to a 50% reduction in soybean yield. There is also preliminary evidence that natural nitrogen fixation may be negatively affected by increased exposure to UV-B.

Marine Ecosystems. Even under natural conditions, many species of plankton are sensitive to low-level doses of UV-B; the defense mechanism developed over evolutionary time has been to live at a depth that screens most of the UV-B but allows the longer wavelength radiation used for photosynthesis to penetrate. A loss in protective atmospheric ozone thus implies direct harm from the radiation or indirect damage from a decrease in photosynthetic activity as the plankton move to greater depths. Because plankton are low on the food chain, a reduction in their numbers is expected to have an influence on the higher level members of the ecosystem such as zooplankton and fish. Studies to examine these processes in more detail are being actively pursued, especially in Antarctica.

Heavy Metals

The ability of high-latitude ice cores to provide globally integrated atmospheric samples has been discussed in previous chapters. When the recently deposited layers are analyzed, the ice cores reveal recent air quality trends as well as those of the distant past. For example, Figure 13.9 illustrates concentrations of lead in Greenland ice cores. The data show a gradual increase during the early stages of the Industrial Revolution, followed by a large, rapid increase during the 1950s and 1960s as leaded gasoline use increased dramatically. Following 1970, as legislation reducing the amount of lead in gasoline was implemented, the ice core lead content decreased dramatically. These data constitute one of the best examples of the effectiveness of emissions controls in changing the global environment for the better.

REGIONAL AIR QUALITY DATA AND TRENDS

Ozone in Paris and Northern Europe

In 1845, Ch. Schoenbein in Germany introduced an "ozonometer," an iodized starch paper that changed color when exposed to ozone and was extensively used for ozone measurements at many sites throughout the world. By the last quarter of the nineteenth century, it had been clearly demonstrated that ozone was a normal constituent of atmospheric air.

Although further efforts continue to extract useful information from the old data, there is probably only one long series of measurements that may be taken as representative of ground-level ozone concentrations during the past century. Those measurements were

Figure 13.9 Changes in lead concentrations in Greenland ice and snow from 5500 yr BP to the present. The different symbols represent data from several ice cores in Greenland and show about a 200-fold increase between several thousand years ago and the early to mid-1950s. Over the most recent 20 yr, the lead in the Greenland ice has decreased rapidly. (Reproduced with permission from C. F. Boutron, U. Görlach, J. P. Candelone, M. A. Bolshov, and R. J. Delmas, Decrease in anthropogenic lead, cadmium and zinc in Greenland snows since the late 1960s, *Nature, 353*, 153–156. Copyright 1991 by Macmillan Magazines Limited.)

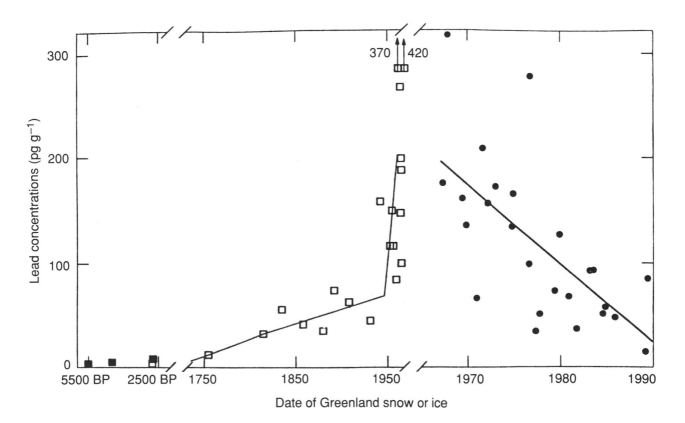

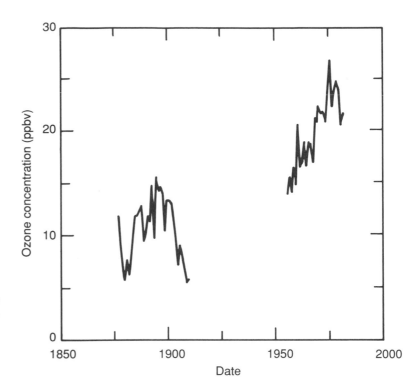

Figure 13.10 A comparison of 1876–1907 surface ozone volume mixing ratios at Montsouris (near Paris) with those at Arkona, Germany, 1956–1982. (Reproduced with permission from A. Volz and D. Kley, Evaluation of the Montsouris series of ozone measurements made in the nineteenth century, *Nature, 332*, 240–242. Copyright 1988 by Macmillan Magazines Limited.)

made by A. Levy and coworkers at the Montsouris observatory near Paris between 1876 and 1907 and were recently validated by Andreas Volz and Dieter Kley of the Nuclear Research Center at Jülich, Germany. The data indicate average concentrations three to four times lower than those presently measured at comparable sites. These early ozone observations are compared with modern data in Figure 13.10.

More extensive research into tropospheric ozone concentrations had to await the development of more reliable instrumentation, especially optical techniques. Spectrographic techniques for a long time were too cumbersome to use for long-term observations. Where measurements were made, however, all ozone concentrations determined by optical techniques from low-altitude (< 1 km) stations were substantially lower than those currently measured with modern chemical and optical techniques at representative stations in similar areas. Repetitive measurements at Hohenpeissenberg since 1967 clearly show tropospheric ozone increasing and stratospheric ozone decreasing (Figure 13.11).

Present-day average annual ozone volume mixing ratios at low-altitude stations in Germany are typically in the range 25–35 ppbv, with average summertime (April–September) values of 35–45 ppbv. It is not uncommon during summer smog episodes for ozone maximum volume mixing ratios to exceed 100 ppbv, values that were never reported in the old data sets. Clearly, there have been enormous increases in background ozone concentrations over the past half-

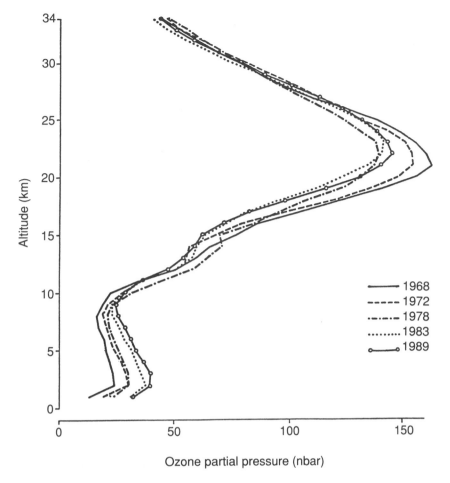

Figure 13.11 Tropospheric and stratospheric ozone annual average concentrations (expressed as partial pressure in nbar = 10^{-9} bar) measured at Hohenpeissenberg, Germany, during the period 1967–1988. (W. Attmannspacher, private communication, 1989.)

century, especially downwind of the heavily motorized urban and industrial areas. As discussed in Chapter 8, such behavior is the result of high simultaneous concentrations of oxides of nitrogen and reactive hydrocarbons.

Gases and Particles in the United States

The concentrations of a number of atmospheric trace species have been monitored regularly for more than a decade at state or federally supervised locations throughout the United States. The information from these sites has been collected and summarized on a national basis since 1975 and is presented in a form in which typical concentrations, their ranges throughout the differing monitoring sites, and their long-term trends can be readily seen. The "boxplot" technique that is used is shown in Figure 13.12a, where the 5th, 10th, and 25th percentiles of the data depict the "cleaner" monitoring sites, the 75th, 90th, and 95th percentiles depict the "dirtier" sites, and the median and average describe the "typical" sites. Although the average and median both characterize typical behavior, the median has the advantage of not being affected by a few extremely high or low observations.

Figure 13.12 (a) The plotting convention used for boxplots in subsequent figures for U.S. air quality trends. (b) Boxplot comparisons of trends in annual mean sulfur dioxide concentrations at U.S. sites, 1975–1987. On this figure and others of its type, NAAQS indicates the U.S. national ambient (i.e., outdoor) air quality standard for this species. The number of sites has varied over the years, but constitutes several hundred. (c) Boxplot comparisons of trends in annual mean nitrogen dioxide concentrations at U.S. sites, 1975–1987. (d) Boxplot comparisons of trends in maximum 3-month average lead levels at U.S. sites, 1975–1987. (Environmental Protection Agency, *National Air Quality and Emission Trends Report*, EPA-450/4-84-029, Research Triangle Park, NC, 1985.)

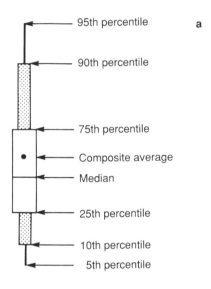

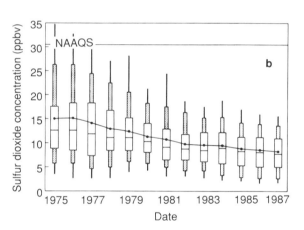

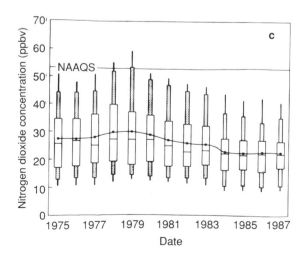

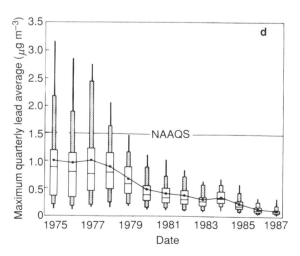

The U.S. urban concentrations of sulfur dioxide over the past decade are shown in Figure 13.12b. The emission of SO_2 is dominated by fossil fuel combustion, with industrial activity, motor vehicles, and other sources being much less important. As increased controls and cleaner fuels have been used over the past decade, SO_2 concentrations have steadily decreased, and continued improvement is expected over the next decade. In a world in which deteriorating air quality is the norm, this marked improvement is well worth noting. As of 1987, the typical annual mean concentration of U.S. urban SO_2 was about 8 ppbv.

Boxplots of annual mean nitrogen dioxide concentrations are given in Figure 13.12c. For the upper percentiles, they show a concentration increase from 1975–1979, followed by a decrease from 1979–1983 and relatively stable concentration distributions since. This behavior mirrors the trend in total NO_x emissions, which increased slightly in the late 1970s, decreased slightly in the early 1980s, and has now become relatively stable, with removal of emittants by catalytic converters being balanced by increased numbers of vehicles. The trend is only slightly evident in the mean values of the annual concentrations and disappears entirely from the lower percentiles of the data. Most sites have annual mean concentrations of NO_2 in the range of 20–30 ppbv, and all are below the national ambient air quality standard of 53 ppbv.

As is the case for Los Angeles, the composite data for the United States show a decrease over the past decade in CO concentrations and a slight downward trend for O_3. Significant decreases have occurred in the concentrations of total suspended particulates. Perhaps the most dramatic evidence for improvement in an air quality component as a result of legislation is that for airborne lead in U.S. urban areas. Figure 13.12d shows the 12-yr trend in ambient lead concentrations. It is easy to see that the substantial reductions in leaded gasoline use in the United States that have occurred during this period have been reflected in sharply decreasing atmospheric lead concentrations. As of 1983, the mean lead level was about 0.5 $\mu g\ m^{-3}$. Air from the United States is an important part of the input to the Greenland ice cores, and the data in Figure 13.12d are quite consistent with the ice core lead data of Figure 13.9.

Ozone in the Tropics and Subtropics

Few atmospheric measurements have been made in tropical Brazil, but the region is sufficiently undeveloped that the measurements that have been made provide a unique look at a continental region whose air quality is becoming increasingly influenced by human activity. One feature of the region of particular interest is large-scale biomass burning. This activity not only involves the much-discussed deforestation but even more the extensive practice of burning savanna grassland during the dry season to remove weeds and allow vigorous growth of grass during the rainy growing season that follows. In

Figure 13.13 Vertical profiles of ozone in the tropical troposphere. The profile over the equatorial Pacific shows no influence from biomass burning, whereas the profile over the Pacific off South America suggests ozone en-hancement due to long-range transport from the tropical continents. The ozone profiles over Brazil and the Congo region of Zaire show high ozone con-centrations between 1 and 4 km altitude due to photochemical pro-duction in biomass burning plumes. At higher altitudes, ozone concentra-tions are also substantially enhanced, perhaps as a consequence of ozone production by reactions in biomass burning effluent plumes. (P. J. Crutzen and M. O. Andreae, Biomass burning in the tropics: impact on atmospheric chemistry and biogeochemical cycles, *Science, 250*, 1669–1678, 1990.)

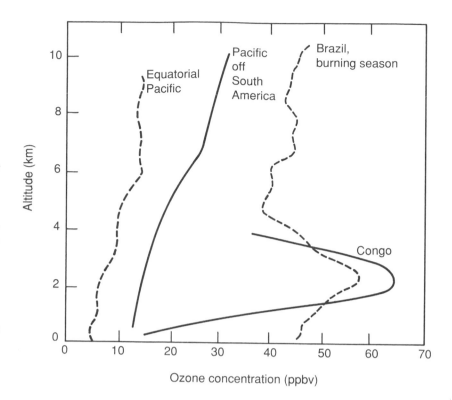

addition to the major impacts such practices are certain to have on the region's biota, biomass burning causes marked changes in air quality. An example is shown in Figure 13.13, which compares ozone profiles over and downwind of tropical forests with those over the ocean at the same latitude. High ozone concentrations are produced in the plumes, often exceeding values typical of polluted regions of the northern hemisphere. The highest concentrations are usually found in discrete layers at altitudes between 1 and 4 km, in accordance with transport mechanisms of burning plumes.

URBAN AIR QUALITY DATA AND TRENDS

Reactive Gases in Los Angeles

Photochemical smog was first recognized as a product of emissions into the atmosphere in Los Angeles, and observations of several atmospheric trace species in the Los Angeles Basin date from the mid-1950s. Many of the early measurements suffered from inaccuracies, especially sensitivity to compounds other than that for which the measurement was designed. These difficulties were gradually resolved by extensive analytical effort. To assess trends in the region over relatively long periods of time, G. Kuntasal and Tai Chang of the Ford Motor Company have recently summarized the data for the period

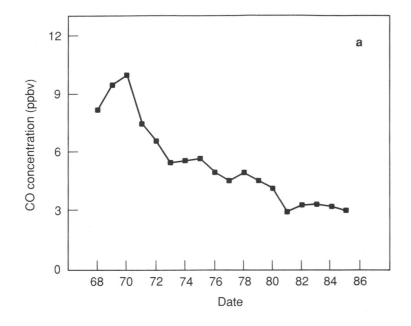

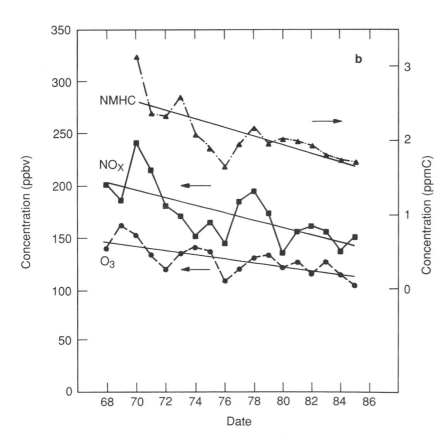

Figure 13.14 (a) Trends for third-quarter daily maximum hourly CO concentrations in the South Coast Air Basin (Los Angeles and environs) for the period 1968–1985. (b) Observational data and linearized trends of nine-station third-quarter daily maximum hourly averages for nonmethane hydrocarbons (NMHC), oxides of nitrogen (NO_x), and ozone in the South Coast Air Basin, California. The arrows indicate the appropriate ordinate for each species' concentrations. (Reproduced with permission from G. Kuntesal and T. Y. Chang, Trends and relationships of O_3, NO_x, and HC in the South Coast Air Basin of California, *Journal of the Air Pollution Control Association*, 37, 1158–1163. Copyright 1987 by Air Pollution Control Association.)

1968–1985. In urban areas, trace gas concentrations are measured continuously and averaged over each hour. The highest of the hourly averages is then selected for trend analysis. For carbon monoxide, the data show a definite downward trend consistent with vehicular emission control measures (Figure 13.14a). Basin-wide hydrocarbons, NO_x, and O_3 show downward trends as well (Figure 13.14b), although less dramatic than that for CO.

Urban Particulate Matter

By far the most comprehensive worldwide assessment of atmospheric air quality is that published by the United Nations Environment Program for the period 1980–1984. Data from that report for annual averages of total particulate matter in major urban areas are shown in Figure 13.15. More meaningful data would be that for the submicron fraction of the particulate matter, as that fraction is the primary particle influence on visibility and health, but the total particulate matter data that are available still give a sense of relative concentrations and trends. More than half of the cities providing data exceed the United Nations Environment Program guidelines, and these cities are located in nearly every continent. The annual mean levels range from a low of about 35 μg m^{-3} to a high of about 800 m^{-3} μg, a range of a factor of about 25.

Figure 13.15 The range of annual averages of total particulate matter concentrations measured at multiple sites within cities throughout the world for the period 1980–1984. Each bar represents a city, coded as follows: 1, Frankfurt; 2, Copenhagen; 3, Cali; 4, Osaka; 5, Tokyo; 6, New York; 7, Vancouver; 8, Montreal; 9, Fairfield; 10, Chattanooga; 11, Medellin; 12, Melbourne; 13, Toronto; 14, Craiova; 15, Houston; 16, Sydney; 17, Hamilton; 18, Helsinki; 19, Birmingham; 20, Caracas; 21, Chicago; 22, Manila; 23, Lisbon; 24, Accra; 25, Bucharest; 26, Rio de Janeiro; 27, Zagreb; 28, Kuala Lumpur; 29, Bombay; 30, Bangkok; 31, Illigan City; 32, Guangzhou; 33, Shanghai; 34, Jakarta; 35, Tehran; 36, Calcutta; 37, Beijing; 38, New Delhi; 39, Xian; 40, Shenyang; 41, Kuwait City. Few cities in the developing world have been extensively monitored and therefore do not appear in this figure; many would be expected to have high levels of particulate matter. The gray shading indicates the concentration range recommended by the United Nations Environment Program as a reasonable target for preserving human health. (Global Environment Monitoring System, *Assessment of Urban Air Quality*, United Nations Environment Program, Geneva, 1988.)

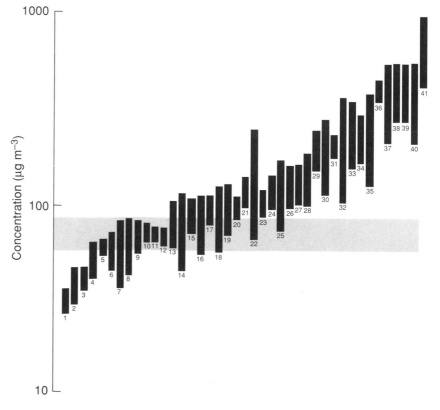

The UNEP also studied trends in total particulate matter air quality data. Of the 37 cities surveyed for 1980–1984, 6 had major upward trends, including urban areas in South America, Asia, and Europe. It is clear that regional trends are sometimes masked by local emissions and that efforts to control emissions are countered by steady increases in global emission fluxes. Many developing countries are not yet monitoring their air quality, but it seems likely that most of them have increasing levels of emissions and decreasing air quality.

Urban Sulfur Dioxide

Figure 8.14 showed data for sulfur dioxide in a display similar to that of Figure 13.15. In more than a quarter of the cities, the annual average SO_2 concentrations exceed that thought to have a high potential for severe corrosion impact. The difference between the combined site averages in the 54 cities surveyed is nearly a factor of 100. The half-dozen cities with the highest annual averages include locations in Europe, Asia, and South America.

Trends in the ambient annual average levels of sulfur dioxide for 33 cities in 1980–1984 showed that 27 have downward or stationary trends, reflecting switches to lower-sulfur fuels and efforts to reduce emissions of sulfur gases from industrial sources. However, six cities have significant upward trends, including locations in Asia, Australasia, and Europe.

TRENDS IN PRECIPITATION CHEMISTRY

Ammonium and Nitrate Ions in Northern Europe

Historical data for nitrate and ammonium ions in precipitation began to be collected in 1853 in England and several decades later on the European continent. They show that the deposition of nitrate ion to the ground has increased throughout the duration of the record (by a factor of about 4). The amount of deposited ammonium ion has increased as well, but by a factor of between 1.5 and 2. These results are approximately consistent with the flux of oxides of nitrogen thought to have been released during the century 1880–1980 as a consequence of increased combustion in the western European region. During the same period, ammonia emissions have increased as a result of enhanced emissions from fertilizer application and animal husbandry.

Acidity and Anions at Hubbard Brook, NH

At Hubbard Brook in the mountains of the northeastern United States there exists the longest continuous, scientifically rigorous record of precipitation ion concentrations in North America, from 1964 to the present time. The Hubbard Brook concentrations have been extended further back in time with the use of a computer model linked to the emission fluxes of species of interest. Trends in the sulfate, nitrate, and hydrogen ions for the period 1900-1980 are shown in Figure 13.16.

Figure 13.16 Estimated concentrations (dotted lines) and measured concentrations (solid lines) of sulfate (a), nitrate (b), and acidity (c) in precipitation at Hubbard Brook, NH. (Reproduced with permission from J. A. Fay, D. Golomb, and S. Kumar, Modeling of the 1900–1980 trend of precipitation acidity at Hubbard Brook, New Hampshire, *Atmospheric Environment, 20*, 1825–1828. Copyright 1986 by Pergamon Press, plc.)

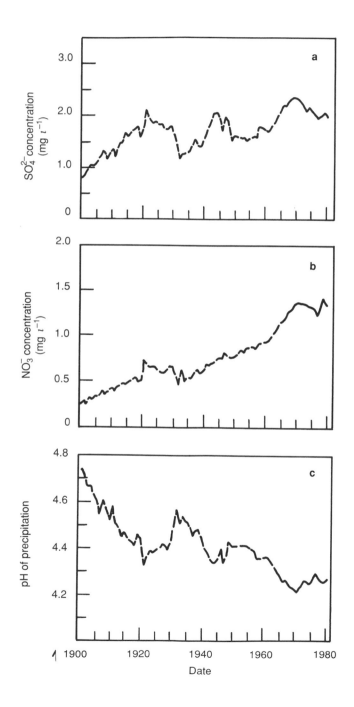

Historical Hubbard Brook values for sulfate appear in Figure 13.16a. They mirror to some degree the large sulfur emissions during the two world wars and the decrease during the depression years. The marked decrease in SO_2 concentrations that has occurred is not mimicked by the SO_4^{2-} data, however. This nonlinearity is a consequence of the fact that the conversion of SO_2 to SO_4^{2-} requires an oxidizer, generally H_2O_2. During the winter or during periods of overcast skies, however, the production of H_2O_2 is severely inhibited. As a result, predictions of the amount of sulfate that will result from emission of a specific amount of sulfur dioxide require the use of complex computer models that simultaneously investigate meteorological transport and chemical conversion.

Nitrate ion concentrations are shown in Figure 13.16b. The near-monotonic increase is substantially different from that for SO_4^{2-}, because it reflects the more gradual and more recent increase in high-temperature combustion of fossil fuels in power plants and in automobiles.

The precipitation pH is related to both the sulfate and nitrate concentrations and is shown for the 1900–1980 period in Figure 13.16c. The figure shows a reasonably steady decrease in pH throughout the period, interrupted chiefly by the depression years of the mid-1930s. Since about 1970, the pH has become much more stable as a consequence of the increasing limitations being placed on emissions to the atmosphere.

TRENDS IN SURFACE WATER CHEMISTRY

Surface water chemistry is influenced by three factors: the natural environment in the environs of the body of water and its drainage basin, the effects of humanity's activities in altering that environment (such as applying fertilizer or discharging industrial waste streams), and the effects of humanity's activities as communicated through the atmospheric transport and deposition of anthropogenic emittants. In any particular situation, all of the above may contribute and any one may be dominant.

As an example of industrial waste stream influences on surface water, we show in Figure 13.17 the flux of chloride ion in the Rhine River over the past century. The chloride ion load has increased by a factor of about 8 during that period. Especially notable are the precipitous drop during the few years after World War II and the equally rapid increase as the countries through which the Rhine flows rebuilt their industry and resumed their discharges to the river in the late 1940s and early 1950s.

Opposite trends in river water chemistry within a single country are shown in Figure 13.18, where nitrate concentrations for two rivers are shown for nearly simultaneous 10-yr periods. The Grand River has

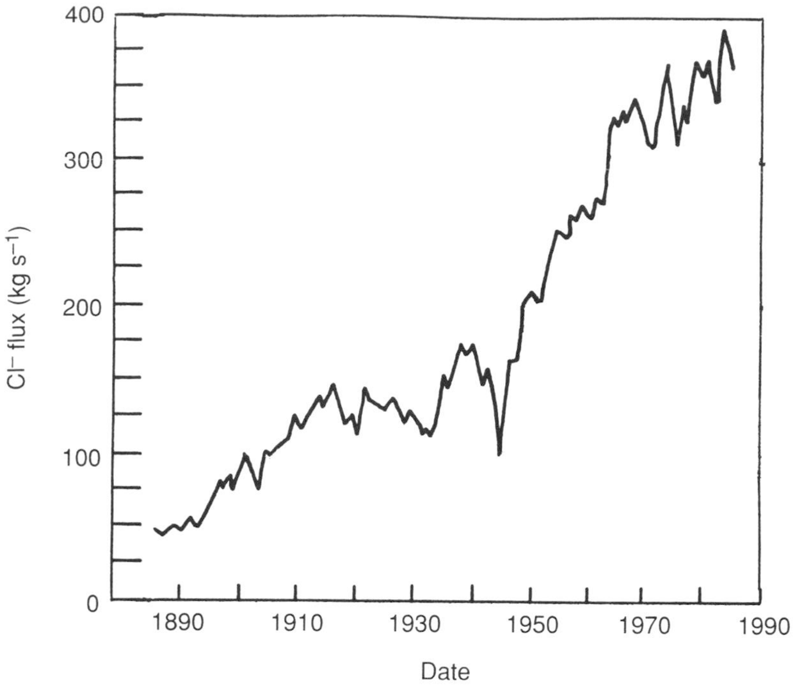

Figure 13.17 The flux of chloride ion in the Rhine River as measured at the Netherlands–German border throughout the last century. (Adapted from a figure by T. S. Uiterkamp in *Annual Report of the Cooperating Rhine and Meuse Drinking Water Companies*, 1985.)

intensively cultivated lands in its drainage basin. These lands were receiving high and increasing applications of nitrogen fertilizer during most of the period illustrated. The fertilizer-related impact on the North Platte River was less intense, and improved municipal waste treatment is credited for lowering the riverine nitrate concentration. Airborne nitrate doubtless contributed to the nitrate concentrations of both rivers but was not dominant.

The final surface water chemistry example is derived from a sediment core taken from the mouth of the Mississippi River. An analysis was performed separately for "pollutant lead" and total lead, the former being based on the difference between the measured level and that attributable to natural sediment lead concentrations. The data show (Figure 13.19) that pollutant lead contributions were at a maximum in about 1970, coincident with the maximum in the consumption of lead in gasoline. This figure, together with Figures 13.9 and 13.12d, demonstrate that a change in the emission rate of a single emittant may be detected near the source and far from it, and in the same medium into which the species was emitted or in a different medium into which it was deposited.

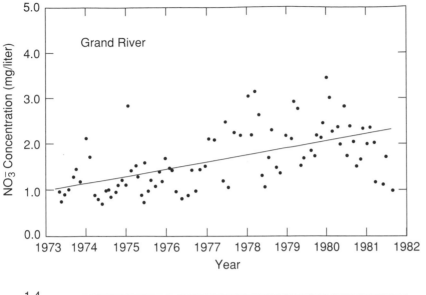

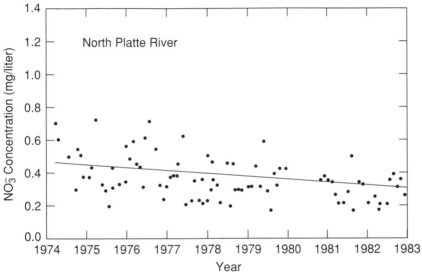

Figure 13.18 Trends in total nitrate concentration in two United States rivers: (a) Grand River, Michigan; (b) North Platte River, Nebraska. (Reproduced with permission from R. A. Smith, R. B. Alexander, and M. G. Wolman, Water-quality trends in the nation's rivers, *Science, 235*, 1607–1615. Copyright 1987 by the AAAS.)

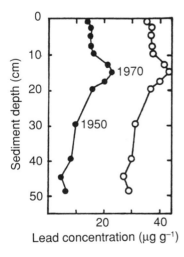

Figure 13.19 Lead concentration as a function of depth in sediments near the river mouth. The closed circles refer to "pollutant lead" (see text), the open circles to total lead. The top of this sediment record was formed during 1982, the bottom in about 1850. Two intermediate calibrated dates are indicated: 1950, when lead emissions from gasoline were rather low following the Second World War, and 1970, when lead use in gasoline was high and when the lead peak in the sediments occurred. (Adapted with permission from J. H. Trefry, S. Metz, R. P. Trocine, and T. A. Nelson, A decline in lead transport by the Mississippi River, *Science, 230*, 439–441. Copyright 1985 by the AAAS.)

CONCLUDING COMMENTS

Much of the record of environmental chemistry over the past several centuries is contained in the ice cores that also provide the only direct samples of ancient atmospheres and hydrospheres. For parts of the past century, however, and for a few specific species, actual analytical measurements begin to reveal states and trends. In some cases, surrogate information is available to supplement the analytical data. The information is sometimes local in scope, sometimes global. In either situation, examples of which are discussed herein, the general picture is the same: gradual increases in concentrations from the dates of the earliest reliable measurements, followed by dramatic increases in concentration in the twentieth century. The few exceptions to this picture are the result of vigorous controls of specific emissions by some of the developed nations.

EXERCISES

20.4	23.7	25.9	20.9	15.8	10.8	10.0	10.9	12.1	18.8	25.6	31.7	34.9	44.0	37.2
23.6	17.7	25.5	31.0	34.1	37.2	41.0	45.5	55.4	41.2	57.8	44.7	32.0	26.2	21.5
18.4	13.8	12.4	17.2	18.5	20.2	26.4	31.3	37.7	24.2	27.8	26.0	31.5	39.2	47.6
49.7	72.2	61.1	57.1	41.3	40.0	35.6	51.7	68.6	75.6	51.2	41.2	36.2	31.0	25.0
21.2	24.8	27.2	35.7	40.5	32.5	26.2	17.4	21.5	33.3	40.7	29.6	23.2	18.6	13.9
21.1	23.4	31.8	45.2	21.9	18.7	12.8	13.7	25.7	34.2	37.8	64.4	27.5	18.7	34.6
40.7	32.3	18.9	25.0	32.7	11.1	10.7	19.6	27.8	30.1					

13.1 Refer to Figure 13.12a, where the convention for making a boxplot is shown. Boxplots are efficient ways of summarizing large amounts of data. Display the data above (for NO_x in ppbv in a U.S. city in 1984) as a boxplot. Are the 25th and 75th percentiles equidistant from the median? What does that mean? Are the 5th and 95th percentiles equidistant from the median? What does that mean?

13.2 Refer to Figure 13.15. A typical adult respiration rate is 14 breaths per minute and each breath has an intake volume of about 500 cm^3. Compute the mass of particles with which the body must deal in a day if one is breathing air containing the midpoint concentration of particles in Montreal. How much mass is this in a 50-yr adult lifetime? Repeat the calculations for Calcutta and Kuwait? (Note that a few milligrams is a typical medication dose.)

13.3 Compute the mass of lead ingested in a year by an adult in a U.S. city with median levels equal to the national median in 1975 and in 1983. Repeat the calculation for a city at the 95th percentile. (Refer to Figure 13.12d and use information from Exercise 13.2.)

13.4 Rain chemistry information from Hubbard Brook, New Hampshire, is shown in Figure 13.16. If the annual rainfall at Hubbard Brook is 86 cm yr^{-1}, compute the annual sulfate and nitrate ion deposition to the ground in 1900 and 1980 in units of μM cm^{-2}. Compute the hydrogen ion deposition for the same years. Are the concentrations explained if it is assumed that the sulfate and nitrate are in the forms H_2SO_4 and HNO_3, or are other precipitation constituents indicated?

FURTHER READING

Environmental Protection Agency, *National Air Quality and Emission Trends Report, 1985*, EPA-450/4-87-001, Research Triangle Park, NC, 1987.

Global Environment Monitoring System, *Assessment of Urban Air Quality*, United Nations Environment Program, Geneva, 1988.

J. S. Levine, ed., *Global Biomass Burning: Atmospheric, Climatic, and Biospheric Implications*, MIT Press, Cambridge, MA, 1991.

Monitoring the global environment: An assessment of urban air quality, *Environment, 31* (8), 6–13, 26–37, 1989.

CHAPTER 14

Budgets and Cycles

SYNTHETIC ENVIRONMENTAL CHEMISTRY

Most of the information that has been presented and discussed thus far is what would be termed *experimental environmental chemistry*. In most of the remainder of the book, we will be discussing *synthetic environmental chemistry*. This latter activity is usually performed in the office rather than in the laboratory or the field and generally uses the computer as the principal analytical instrument, in order to combine the information that has been gathered about the working of the various processes that influence Earth system chemistry.

The synthesist first attempts to take the vast amounts of information produced by his or her experimentally oriented colleagues and form from it coherent overviews of physical and chemical processes. In Earth system chemistry one important method for doing so is the construction of a *budget*, an attempt to understand in a quantitative way the flows into, the chemical transformations within, and the removal from the environment of a specific constituent or group of constituents. The synthesist may construct budgets for the interaction of the atmosphere with specific ecosystems or for the entire planet's solid, liquid, and gaseous regimes, rather than restricting the analysis to any one regime. Budgets of various kinds are the subject of this chapter.

The second synthetic tool is the *model*, which attempts to understand the dynamics of component and process interactions in the Earth system, usually with the aid of a computer. We discuss models in the following chapter. It is often true that inconsistencies in the results of budget analyses or model calculations suggest new projects for experimentalists, just as experimental results are used to form and refine budgets and models. In Earth system chemistry, as in all of science, experimental and synthetic approaches work hand in hand in improving knowledge.

THE BUDGET CONCEPT

Most people are familiar with the concept of a financial budget and balance sheet. On such an accounting, one records estimates or actual values of income and expense for a selected period. If income and expense are equal, then the budget is in balance. If income exceeds expense, then the monetary reserves increase. If expense exceeds income, then the monetary reserves decrease.

An approach very similar to that of financial budgeting is used to fashion budgets in Earth system science. An important difference, however, is that in an Earth system budget all of the input or output processes important for the system may not have been identified. Even if they have, their magnitudes may be uncertain. The situation can be appreciated with the aid of the diagram in Figure 14.1, which shows a tub receiving water from several faucets and having a number of drains of different sizes. When the water is supplied at constant (but probably different) rates by all the faucets and is removed at an equal total rate by drains with probably different flows, the water level remains constant. When the tank is very large, however, and has some wave motion that makes it difficult to tell whether the absolute level is changing, an observer may have difficulty telling whether the system is in balance or not. In that case, he or she may try instead to measure the rate of supply from each of the faucets and the rate of removal in each of the drains over a period of time to see whether the sums are equivalent. A part of this technique involves the determination of the *pool size* (the quantity of water in the tank) and either the rate of supply or the rate of removal. Determination of changes in the pool size then gives information about rates that are difficult to measure. The process of estimating or measuring the supply and removal fluxes and checking the overall balance by measuring the amount present in the reservoir constitutes the Earth system budget analysis.

Suppose that the input from one of the sources is increased; i.e., in our analogy, the flow from one of the faucets increases. Will the water level keep increasing? The answer depends on whether one of the drains can accommodate the additional supply, as can the "trough drain" at the right side of the tank in the figure. If no such drain is present, then the water level will indeed increase. Conversely, if the flow into a drain is enhanced for some reason, such as the removal of an obstruction, then the water level will decrease in the absence of a corresponding increase in the supply. Such a process will continue in this manner unless the flows through the drains adjust themselves to this new factor or unless the new factor results in other drains changing their functioning. The system is inherently nonlinear, as is often the case with Earth system budgets, a characteristic that severely complicates analysis.

All of the circumstances mentioned above occur in budgets devised for Earth system studies, and all budgets involve the same concepts some of which we have used earlier. One is that of the *reservoir*, which is an entity defined by characteristic physical, chemical, or biological properties that are relatively uniformly distributed. Ex-

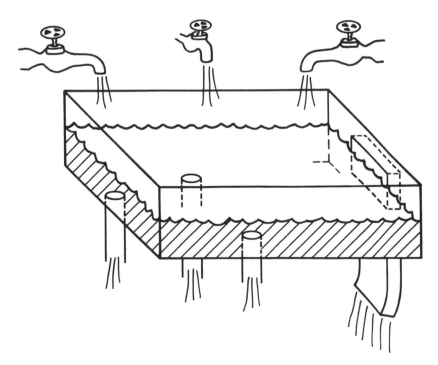

Figure 14.1 A simple conceptual system for budget calculations. The water level in the tub is determined by the water flows in and out, as discussed in the text.

amples include the atmosphere as a whole, spatial sections of the atmosphere, or chemically defined subsets such as the oxygen or water vapor pools. A second concept is that of *flux*, which is the amount of a specific material entering or leaving a reservoir per unit time. Examples include the rate of evaporation of water from Earth's surface, the conversion rate of methane to carbon monoxide in the atmosphere, and the rate of transfer of ozone from the stratosphere to the troposphere. Third, we have *sources* and *sinks*, which are rates of creation or destruction of a specific material within a reservoir per unit time. Examples include the photochemical production or destruction of ozone, or the deposition of nitric acid to Earth's surface.

A system of connected reservoirs that transfer and conserve a specific material is termed a *cycle*. Over the longest of time scales, the entire universe is the only such system, but for our purposes many Earth system cycles can be considered closed (i.e., no material is lost). An example is the planetary water cycle, the reservoirs being the oceans, lakes, groundwater, soils, glaciers, ice sheets, the biosphere, clouds, and atmospheric water vapor.

Earth system budgets have the same three basic components as those for the tank in Figure 14.1: determination of the present level (the concentration of a single species or a group of species), a measurement or estimate of sources, and a measurement or estimate of sinks. A perfect determination of any two of these three components determines the other. Because any species of interest in the environment is likely to have several sources and sinks, however, each source and sink must generally be studied individually. There are some significant similarities between the chemical budgets and cycles discussed in this chapter and the radiative energy budget presented in Chapter 3, and it is worth reviewing that budget presentation in the context of the present chapter.

Time Scales in Budget and Cycle Analysis

It is convenient to define a number of different time scales as part of the budget and cycle process. The first is the turnover time τ_0. This parameter is the ratio between the content [C] and the flux of a specific material into (F_i) or out of (F_o) a reservoir that is in steady state:

$$\tau_0 = \frac{[C]}{F_o} = \frac{[C]}{F_i} . \qquad (14.1)$$

The turnover time reflects the spatial or temporal variability of a property within a reservoir, with a small variability indicating a long turnover time and a large variability indicating a short turnover time compared with the mixing time in the reservoir, mixing time being related to the molecular and eddy diffusivities of the medium. If material enters or leaves the reservoir by several paths, then the overall turnover time of the reservoir is related to the individual turnover times of the pathways $\tau_{0,i}$ by

$$\tau_0 = \frac{\sum_i F_i}{[C]} = \frac{\sum_o F_o}{[C]} = \frac{1}{\sum_i \frac{1}{\tau_{0,i}}} . \qquad (14.2)$$

A second useful parameter is the *residence time*, τ_r, which is the average time spent in the reservoir by a specific material. If physical rather than chemical processes are involved, then the term *transit time* may be used as an alternative. The average residence time is composed of those of all appropriate molecules, weighted with appropriate probability factors. For example, when one evaluates water flow into and out of a lake, one finds that some of the water rapidly flows from its inlet to the outlet whereas other water follows much less direct paths. We can define the average residence time in a situation with a variety of residence times as

$$\tau_r = \int_0^\infty \tau\, \Psi(\tau)\, d\tau\ , \quad (14.3)$$

where $\Psi(\tau)$ indicates the fraction of the constituent having a residence time between τ and $\tau + d\tau$. The probability fraction $\Psi(\tau)$ is a function of the reservoir processes. In the case of radioactive decay, for example, it can be shown that $\tau_r = \tau_\alpha$, the time constant for the exponential decay of the radionuclide.

The age is the time elapsed since a particle entered a reservoir. The average age of all particles of a specific kind within a reservoir is thus given by

$$\tau_a = \int_0^\infty \tau\, \Psi(\tau)\, d\tau\ , \quad (14.4)$$

where $\Psi(\tau)$ is the age probability function.

For a reservoir in steady state, the turnover time τ_0 and the average residence time τ_r are equal. They may, however, be significantly different from the average age of the particles in the reservoir, depending on the properties of the functions $\psi(\tau)$ and $\psi(\tau)$.

An obvious example showing that $\tau_r \neq \tau_a$ is the human population: In the developed world, the average age is between 35 and 40 yr, whereas the average residence time (life expectancy) is more than 70 yr.

Another useful concept is that of *response time* τ_e, which is the time needed to reduce the effect of a disturbance from equilibrium in a reservoir to $1/e$ of the initial perturbation value. For example, in many chemical applications the equilibrium of species i in a system is expressed by

$$0 = P_i - \Lambda_i[i]\ , \quad (14.5)$$

where P_i is the production rate of species i and $\Lambda_i[i]$ its loss rate, [i] being the reservoir content. If we now introduce a disturbance from equilibrium [i'], the change in [i] with time becomes

$$\frac{d[i]}{dt} = P_i - \Lambda_i([i] + [i']) = -\Lambda_i[i']\ , \quad (14.6)$$

so

$$[i'](t) = [i'] \exp(-\Lambda_i t)\ . \quad (14.7)$$

In this case, $\tau_e = \Lambda^{-1}$.

A final parameter to discuss, the *transport time*, applies in the absence of well-defined reservoirs and is a critical factor in many environmental assessments. If advection is the only transport mode under consideration, the transfer of a property from one location to another is simply given by the ratio of distance L and flow speed V:

$$\tau_{adv} = L/V . \qquad (14.8)$$

For example, in midlatitude westerly winds the average velocity is $V = 20$ m s^{-1} and the transport time around the globe is about 2 wk. If we consider processes transverse to the flow rather than with it, then a transport time under diffusive conditions can be defined as

$$\tau_{diff} = L^2/K . \qquad (14.9)$$

where K is the diffusivity, a space- and time-dependent property of the regime in which diffusion occurs. A summary of applicable diffusivities for regimes of interest in environmental chemistry is shown in Figure 14.2. The atmosphere is clearly the region of most rapid vertical mixing, particularly between the surface and the top of the planetary boundary layer (usually 1–2 km in altitude during the day). Mixing across the tropopause is slow, but not as slow as mixing across the sea surface, with its microlayer skin of organic detritus. Once below the sea surface microlayer, mixing in the upper 100 meters of the oceans is vigorous, although still some two orders of magnitude slower than in the troposphere. Within the benthic sediments, diffusion is very slow and is ultimately determined not by molecular diffusion in sediment pore waters but by *bioturbation*, the disturbance of the sediment surface by benthic organisms.

The application of the diffusivities of Figure 14.2 to Equation (14.9) allows the construction of a chart of characteristic transport times for exchanges of chemical species within the atmosphere and oceans. The result, shown in Figure 14.3, provides a quick summary of the time scale over which various reservoirs are altered by an anthropogenic or natural disturbance.

COMPUTING BUDGET INPUTS

The physicochemical processes that tend to maintain the atmosphere and climate are to a large extent controlled or modified by the emissions of trace constituents to the environment. Historically, these trace constituents have been supplied by the natural processes of Earth: chemical and photochemical processes in the atmosphere, terrestrial and marine biological processes, volcanic eruptions, tectonic activity, and so forth. More recently, the activities of humanity on Earth have provided substantial sources of trace constituents to the atmosphere as well.

The computation of emissions to the environment is conceptually simple. The flux of emittant i is the product of the magnitude α_i of the activity of a given process and the emission factor ϵ_i:

$$\Phi_i = \alpha_i \cdot \epsilon_i . \qquad (14.10)$$

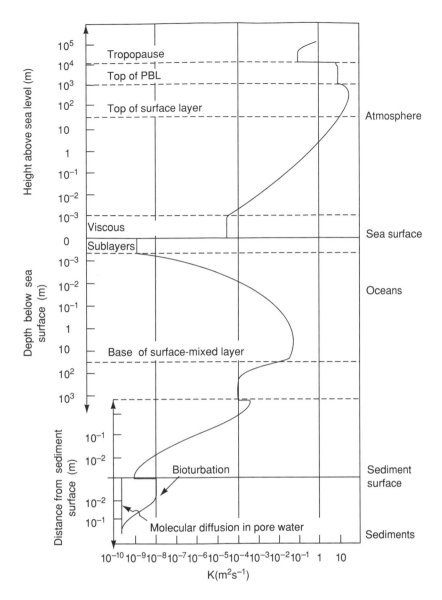

Figure 14.2 A schematic summary of the average vertical molecular or turbulent diffusivity (whichever is larger) through the lower portion of the atmosphere, the oceans, and the upper portion of the ocean sediments. PBL, planetary boundary layer. (Reproduced with permission from H. Rodhe, Modeling biogeochemical cycles, in *Global Biogeochemical Cycles*, S. S. Butcher, R. J. Charlson, G. H. Orians, and G. V. Wolfe, eds., pp. 55–72. Copyright 1992 by Academic Press, Inc.)

For example, the emission of CO_2 from the combustion of coal is computed by expressing ϵ_{CO_2} in mass of CO_2 per mass of combusted coal, with α_{CO_2} being the total mass of coal combusted per unit time. Emission factors are average values and may differ substantially on local or regional scales because of such things as (in the case of CO_2) boiler type, boiler temperature, and other engineering parameters. Fluxes generally vary by season, time of day, and weather; flux computations designed to take these variations into account are therefore quite complex. Emission factors must be separately determined for each emittant and each source. In addition, many sources emit the same species, so computations must sum over all relevant sources. An example for the fossil fuel component of anthropogenic CO_2 emissions was shown in Figure 12.1. This figure, which treats the

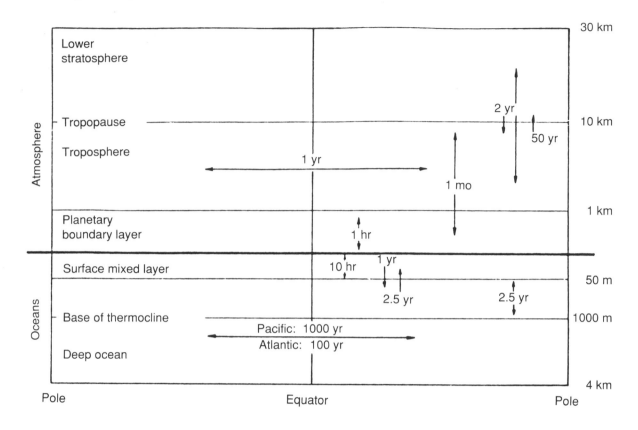

Figure 14.3 Characteristic transport times for the exchange of chemical species in the atmosphere and oceans. The thermocline is the region of the ocean in which density variations are determined largely by temperature variations. (Reproduced with permission from H. Rodhe, Modeling biogeochemical cycles, in *Global Biogeochemical Cycles*, S. S. Butcher, R. J. Charlson, G. H. Orians, and G. V. Wolfe, eds., pp. 55–72. Copyright 1992 by Academic Press, Inc.)

time period 1860 to the present, shows that petroleum emissions were insignificant until the beginning of the twentieth century, and natural gas emissions began only in about 1930.

Difficult as it is to accurately determine anthropogenic emission fluxes, natural emission fluxes are more difficult still. The sources are generally uncontrolled and uncontrollable, the emissions complicated to measure, and the source regions highly variable and often in geographically remote regions. A modest example of the difficulties is illustrated in Figure 14.4, which shows the dependence of the emission of α-pinene from different groups of trees as a function of temperature. (α-Pinene is the molecule that makes an evergreen forest smell the way it does.) The ordinate on Figure 14.4 is logarithmic, so it can be seen that experiments at 30 °C vary by a factor of 100 or more in emission rate, probably because no consideration of other important variables such as soil nutrients, soil moisture, and ambient air pollutants has been taken into account.

A rational response to this discussion of emissions estimation is to wonder if it ever can be done. The answer is yes, provided one is willing to grant that emissions inventories always have built-in uncertainties, sometimes quite high ones. Many inventories have been

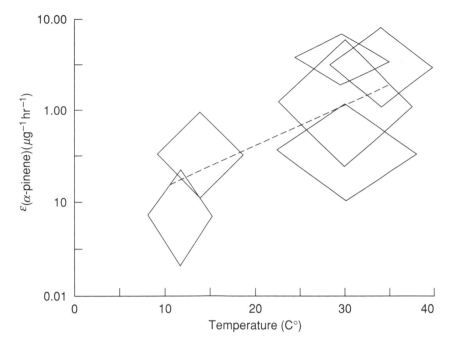

Figure 14.4 Emission fluxes of α-pinene from evergreen vegetation as a function of ambient temperature, as measured by several different research groups at different locations. The dashed line is the best fit to all the observations. (Adapted with permission from B. Lamb, A. Guenther, D. Gay, and H. Westberg, A national inventory of biogenic hydrocarbon emissions, *Atmospheric Environment, 21*, 1695–1705. Copyright 1987 by Pergamon Press, plc.)

constructed, many scientists have been involved, and the inventories being used for budget and model studies are continually improving (but never finished).

REGIONAL SCALE BUDGETS

Approaches to Regional Budgets

Earth system budgets may be derived for systems as simple or as complex as one wishes. It is often the case that one is concerned about a particular regime, the atmosphere, for instance, and builds a budget restricted to that regime. Budgets are also often restricted to a small or moderate sized spatial region. Each of these approaches simplifies the problems involved in budget preparation, but overlooks the fact that input fluxes in a budget always come from some other regime and the output fluxes go into another regime, or, alternatively, they come from and go to adjacent geographical areas. Restricted budgets are therefore sometimes accused of "moving problems around" rather than solving them. Often, however, one cannot approach an understanding of a particular problem unless one has a good picture of its budget components. As a result, each type of budget has its uses and its limitations.

Sulfur in the Western North Atlantic

The starting point for any regional budget is the assessment of emission fluxes within the region for species of interest. A second crucial parameter is the rate of loss of species of interest within the region. For a budget restricted to the atmosphere, loss processes for reactive species include chemical transformation and deposition to the surface, either under dry conditions or during periods of active precipitation. Transport upward into the global troposphere must also be considered, especially if the species of interest have long lifetimes. An additional aspect of a regional budget that is not encountered in a global budget is the advection of species into and out of the region as a consequence of atmospheric motions.

These budget elements are illustrated by a sulfur budget constructed for the western North Atlantic Ocean, as far east as Bermuda. The aim was to assess how much of the sulfur emitted in the highly industrialized regions of eastern North America was deposited within the region and how much was lost by advection to the marine atmosphere and subsequently to the ocean surface. The project involved a large number of measurements on land, in the air, and at island locations. The results are summarized in Figure 14.5.

Figure 14.5 A sulfur budget for the western North Atlantic Ocean. The numbers represent transfer rates in teragrams S per year (Tg S yr^{-1}). (J. N. Galloway and D. M. Whelpdale, WATOX-86 overview and western North Atlantic Ocean S and N atmospheric budgets, *Global Biogeochemical Cycles, 1*, 261–281, 1987.)

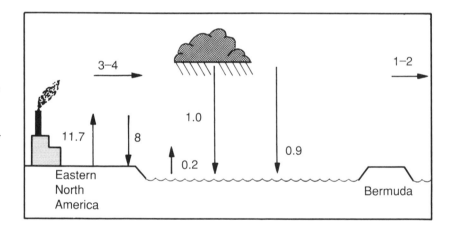

As the figure shows, about 11–12 Tg S are emitted each year into the atmosphere from eastern North America, mostly as sulfur dioxide from coal and oil combustion. Of this amount, about 8 Tg S are deposited on the North American continent by direct uptake of SO_2 from the atmosphere or by acid precipitation following conversion of sulfur dioxide to sulfuric acid in the atmosphere. The remainder, 3–4 Tg S, is exported over the ocean by the westerly winds. Loss to the ocean surface removes an estimated 1.9 Tg S yr^{-1}, approximately equally divided between wet and dry deposition. A small amount of reduced sulfur (mostly dimethyl sulfide), about 0.2 Tg S yr^{-1}, is injected into the marine atmosphere from biochemical processes in the surface ocean. Between 1 and 2 Tg S yr^{-1} is advected eastward, out of the region of analysis. Most of that amount is thought to be lost to the ocean surface east of Bermuda, with only a few tenths of a teragram of North American sulfur reaching Europe annually.

Multiregime Budgets

Regional multiregime budgets are among the most challenging to construct, because they require an understanding of processes that encompass a number of different scientific specialties. To demonstrate, consider Figure 14.6. This figure shows a budget for the annual cycle of nutrient and atmospheric nitrogen in the Hubbard Brook ecosystem in New Hampshire, and includes total reservoir contents, transfer rates, and biological production and loss rates. Note that the hardwood forest ecosystem is a reservoir like that in Figure 14.1, but there are several subreservoirs, each of which must be evaluated and interconnected with the others. Determining some of the fluxes, such

Figure 14.6 An example of a regional budget: the annual nitrogen budget for an undisturbed northern hardwood forest ecosystem at Hubbard Brook, New Hampshire. Bound nitrogen denotes organically bound nitrogen not readily available for plant growth; mineralization is the microbiological transformation of bound to inorganic nitrogen which can be utilized by plants, especially ammonium (NH_4^+), nitrite (NO_2^-), and nitrate (NO_3^-). Nitrogen fixation, the dissociation of atmospheric N_2 into fixed nitrogen (NH_3), is accomplished in the roots of leguminous plants by symbiotic microorganisms. The reverse process, denitrification, is accomplished by denitrifying bacteria which, under anaerobic conditions, convert nitrate to molecular nitrogen and nitrous oxide. The reservoir values in boxes are in kilograms of nitrogen per hectare. The rate of accretion of each pool (numbers in parentheses) and all transfer rates are expressed in kilograms of nitrogen per hectare per year. (Reproduced with permission from F. H. Bormann, G. E. Likens, and J. M. Melillo, Nitrogen budget for an aggrading northern hardwood forest ecosystem, *Science, 196*, 981–983. Copyright 1977 by the AAAS.)

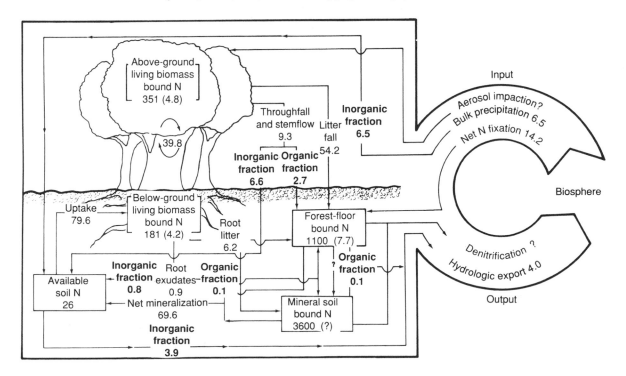

as nitrate or ammonium in precipitation, is relatively straightforward. Other factors, such as the amount and transformation rates of mineral soil-bound nitrogen, can be estimated only with difficulty.

At the right of the diagram are the input and output fluxes. Just as the subreservoir budget construction required expertise in botany and soil science, so the assessment of inputs and outputs requires expertise in atmospheric chemistry and hydrology. Several question marks appear in the figure, indicating fluxes or reservoir contents that the budget authors were unable to calculate. The diagram looks comprehensive and elegant, but in such a complex system one cannot be sure that all relevant subreservoirs and processes are included. Another important limitation is the lack of an attempt to treat spatial variations within the ecosystem. (Such variations are termed *sub-grid scale variability*.) Despite these problems, the budget in Figure 14.6 is one of the better multiregime budgets yet derived.

GLOBAL SCALE BUDGETS

Approaches to Global Budgets

More challenging even than the construction of budgets of the type in Figure 14.6 is the derivation of global budgets. In such budgets, each factor of significance must be assessed, not transferred to an out-of-budget reservoir.

The stringent requirements for accurate analyses of many factors that span the scientific disciplines result in budgets that are not always completely successful. The mere operation of constructing a budget, however, even if it proves to be unsatisfactory, often serves to organize thinking and as an impetus for field measurements to help quantify the uncertain parameters. To illustrate the techniques and problems involved, we present here the most intensively studied global budget in Earth science: that for carbon.

The Global Carbon Budget

The global carbon budget can be constructed in many different ways, for a variety of purposes, and at many levels of sophistication. In the simplest version of this budget, one considers only the atmosphere and attempts to quantify the flows of carbon into and from it. Although many carbon-bearing constituents are present, the most abundant by far and the species that dominates the atmospheric budget is carbon dioxide. A potential budget for the most important terms describing the fluxes of carbon dioxide in and out of the atmosphere for the beginning of the 1980s is shown in Figure 14.7. Of the five different fluxes identified in the figure, three (photosynthesis, detritus decomposition, and ocean cycling) are wholly or predominantly natural, whereas two (fossil fuel combustion and deforestation) are wholly or predominantly anthropogenic.

Constructing the budget requires that accurate figures on each flux be available. For fossil fuel combustion, commercial records are sufficient to estimate CO_2 fluxes. Also required is the rate of deforestation, a process that releases CO_2 into the atmosphere because the standing biomass in forests is greater than that of the successional vegetation. This number is harder to come by, because much of the deforestation occurs in less developed countries with insufficient record-keeping. Recent efforts to derive such information by analyses of data from scientific satellites will help substantially in making assessments. Perhaps most difficult of all to assess in the budget operation are the natural components of the carbon cycle. These processes occur over large and poorly accessible geographical areas and have uncertain variations due to changes in temperature, local environment, plant species, and a host of other factors.

Figure 14.7 shows that an average annual net emission flux of about 5.5 Pg of carbon (C) entered the atmosphere as a result of the combustion of fossil fuels in the 1980s. The historical rate of emissions by fuel type was shown in Figure 12.1; it featured a rapid rise from 1950 to about 1975, followed by a modest leveling off due largely to more efficient energy use in the developed world. Future fluxes will be heavily influenced by the fuels used by the developing world as they strive to raise the standards of living for their populations. In many of these countries, the predominant current practice resulting in CO_2 emissions is deforestation and wood combustion for heating and cooking. As Figure 14.7 shows, deforestation is estimated to have made a significant impact of 1–3 Pg C per year to the carbon budget

Figure 14.7 A simplified annual average atmospheric carbon dioxide budget for the 1980s, with fluxes given in units of petagrams C per year (Pg C yr^{-1}). From left to right the processes are anthropogenic fossil fuel combustion, respiration, photosynthesis, detritus decomposition, deforestation, respiration from ocean surface waters, dissolution into ocean surface waters. Various degrees of uncertainty are associated with all numbers in all budgets; we omit them here in the interest of clarity. (Data from B. Bolin in SCOPE 21, *The Major Biogeochemical Cycles and Their Interactions*, B. Bolin and R. B. Cook, eds., John Wiley & Sons, Chichester, UK, 1983.)

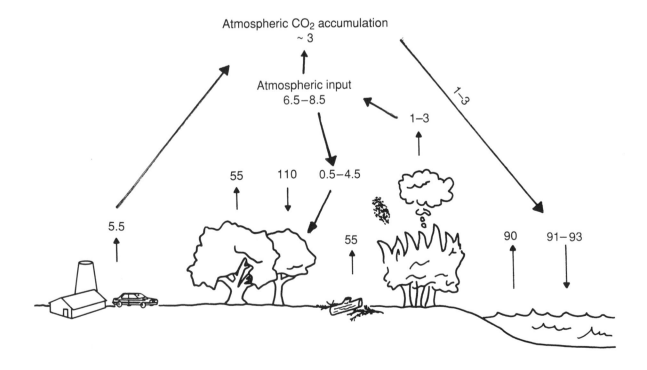

during the 1980s. Altogether, therefore, some 6.5–8.5 Pg C was added annually to the atmosphere by anthropogenic activities. However, the annual increase of CO_2 in the atmospheric reservoir was only about 3 Pg C. The difference of 3.5–5.5 Pg C (half of the emissions) is thought to have been taken up by the oceans and by terrestrial ecosystems.

Figure 14.7, simple and potentially inaccurate as it is, contains some extremely interesting information, such as the implication that the natural fluxes of carbon through the vegetative cycle dominate the atmospheric cycling of carbon. It is worth noting that the carbon incorporated by vegetation during photosynthesis is roughly in balance with that released during detrital decomposition. (Think of these fluxes as the faucets in Figure 14.1 with the dominantly large input flow balanced by the largest of the flows into the outflow trough.) This balance does not mean that vegetative fluxes are inconsequential. For instance, that fraction that is due to seasonality of CO_2 uptake and detrital release by nontropical vegetation also explains the annual behavior (but not the long-term trend) of the atmospheric CO_2 content, shown in Figures 12.3a and 13.1.

Because the gross exchange fluxes between the atmosphere and the terrestrial and marine biosphere are so large and difficult to

Figure 14.8 Size of reservoirs (in units of 10^{15} g C) and fluxes (in units of 10^{15} g C yr^{-1}) for the part of the global carbon cycle that is in a state of comparatively rapid turnover, i.e., characteristic turnover times less than about a millennium. The assessment was made for the year 1980. (B. Bolin in SCOPE 21, *The Major Biogeochemical Cycles and Their Interactions*, B. Bolin and R. B. Cook, eds., John Wiley & Sons, Chichester, UK. Copyright 1983 by Scientific Committee on Problems of the Environment.)

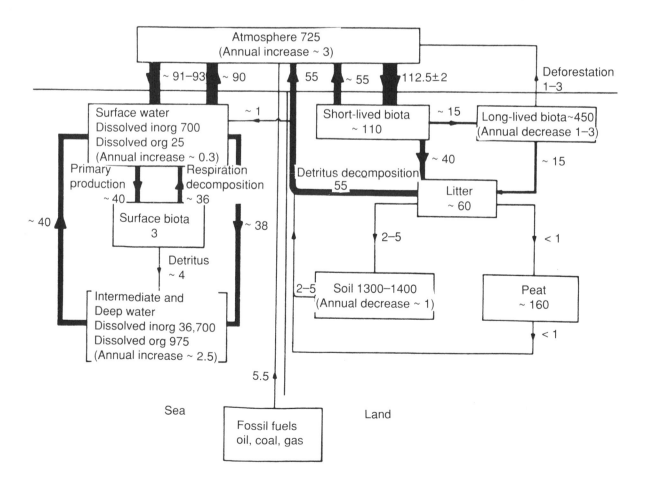

estimate, there remain large uncertainties in the net fluxes shown in Figure 14.7. In fact, Pieter Tans of the U.S. National Oceanic and Atmospheric Administration in Boulder, who (with colleagues) studied the budget of CO_2 isotopes with a general circulation model, suggests that the role of the oceans as a net sink for atmospheric CO_2 may have been overestimated. His idea is that some net sink of carbon to soil organic matter or to that portion of the world's vegetation that is not affected by deforestation may be taking place. (In other words, the flux, assumed to balance that added to vegetation by photosynthesis, may be out of balance.) A suggestion for why such a process might be occurring is that increased atmospheric CO_2 may fix more CO_2 in plant matter, a process termed CO_2 fertilization. The increased input of atmospheric SO_2 and NO by combustion and the increased use of nitrogen- and phosphorus-containing fertilizers may likewise stimulate plant growth and carbon storage. These are important and unresolved problems, severely hampering reliable predictions of future atmospheric CO_2 concentrations.

Figure 14.9 Fluxes (in units of 10^{15} g C yr^{-1}) for the carbon cycle in the Earth's crust, where the characteristic turnover times are of the order 10^8 yr; inorg, inorganic; org, organic; diss, dissolved; part, particulate. (B. Bolin in SCOPE 21, *The Major Biogeochemical Cycles and Their Interactions*, B. Bolin and R. B. Cook, eds., John Wiley & Sons, Chichester, UK. Copyright 1983 by Scientific Committee on Problems of the Environment.)

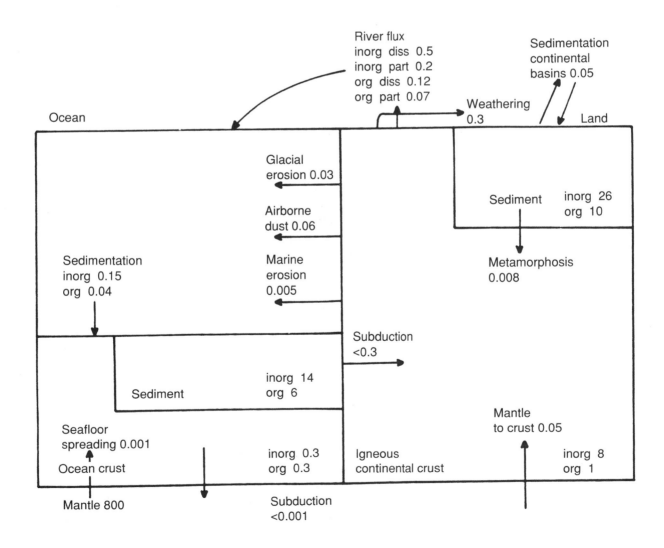

A comprehensive look at the carbon cycle is provided if the processes occurring on land and in the sea are considered in detail. Such an analysis is shown in Figure 14.8 for those processes which, on a geophysical scale, have relatively rapid turnover. This figure reveals the comparative sizes of the carbon reservoirs and gives some idea of the fluxes that join them. It shows that the carbon mass in soil and peat is much larger than that in leaf litter, yet the latter is much more active in cycling carbon. In the oceans, the flows between surface water and deeper waters are indicated. Uncertainties in these rates are highly significant, because the atmospheric accumulation rate is well within the uncertainties of the various flows.

The final component of the carbon cycle is provided by analyses of the longest-term aspects shown in Figure 14.9. These processes determine the natural background CO_2 levels over geological time. The fluxes in this portion of the cycle are thought to be quite small and to involve plate tectonic motions, erosion, sedimentation, and other processes moving at truly lethargic rates. Nonetheless, the total carbon cycle can be considered well in hand only if the variables in Figures 14.7 to 14.9 are all well understood and well quantified. The fluxes in Figure 14.8 are, of course, those of most concern for the next century or so.

Global Budgets for Selected Nitrogen Species

The nitrogen cycle in the atmosphere is far from well quantified. It is known that nitrogen, N_2, constitutes more than 99.9999% of the nitrogen present, and that nitrous oxide (N_2O) makes up more than 99% of the remainder. The other nitrogen-containing species are present in extremely small and highly variable amounts, yet they are crucial constituents in several aspects of atmospheric chemistry. Ammonia, for example, is the only basic gas and is the principal neutralizing agent of the acidic atmospheric aerosol. Nitric oxide (NO) and nitrogen dioxide (NO_2) are far from abundant, yet we saw in Chapter 8 that through participation in several catalytic cycles they are very important to the chemistry of ozone in the troposphere and stratosphere. All of the nitrogen-containing gases are involved in the complex processes of biological nitrogen fixation and denitrification. Nitrogen fixation is mostly accomplished by symbiotic bacteria living in the roots of leguminous plants; the usual product is ammonia. Denitrification, which generally occurs under anaerobic (oxygen poor) conditions, involves denitrifying bacteria that transform nitrate to N_2 and some N_2O, the latter two species being returned to the atmosphere. An alternative process is *pyrodenitrification*, in which some 40% of the fixed nitrogen in plant material is released to the atmosphere as N_2 during biomass burning.

Although some of the components of the atmospheric N_2O cycle are not well understood, we can estimate the N_2O budget by making use of a number of diverse sources of information. We first draw up a list of sources, including bacterial processes in soils, bacterial processes in the oceans, and a variety of industrial activities. The sinks are thought to be limited to chemical breakdown of N_2O in the stratosphere. We begin with ice core measurements of the preindustrial atmospheric N_2O concentration, 285 ppbv. The present concentration is about 310 ppbv and the growth rate is 0.2–0.3% per year (3.0–4.5 Tg N yr^{-1}). Because the photochemistry of atmospheric N_2O is very simple and its distribution in the atmosphere rather well determined, estimates can be made of the preindustrial and current stratospheric destruction rates of N_2O; these turn out to be about 8.1 and 8.7 Tg N yr^{-1} and imply an atmospheric lifetime for N_2O of about 170 yr.

This information enables us to construct a tentative budget for N_2O, shown in Table 14.1. To begin, we set the natural source at a total of 8.1 Tg N yr^{-1}. Of this amount, recent oceanic measurements suggest a flux of 1.4–2.6 Tg N yr^{-1}, leaving 5.5–6.7 Tg N yr^{-1} to come from soils, with most emissions probably emanating from tropical forest regions. Because the current stratospheric loss rate is estimated and the buildup (the imbalance between source and sink) measured, we can estimate that 11.7–13.2 Tg N yr^{-1} are emitted to the atmosphere from all sources. The anthropogenic contribution is thus the difference between this range and the estimated natural flux, or 3.6–5.1 Tg N yr^{-1}, about 50–65% of the natural source. In order to achieve a budget in balance, this flux must be reduced by about 3.0–4.5 Tg N yr^{-1}, approximately 85% of its magnitude. The anthropogenic sources are not well understood, however, consisting of a combination of poorly quantified emissions from cultivated soils following nitrogen fertilizer application, nylon and nitric acid production and other industrial activities, and certain combustion processes. Until these are better determined, it will be difficult to devise emission reduction plans for N_2O, yet its long atmospheric lifetime implies that the response to corrective measures will be very slow.

TABLE 14.1. Estimated Sources and Sinks of Atmospheric N_2O

PROCESS	FLUX (Tg N yr^{-1})
Natural sources	8.1
Oceans	1.4–2.6
Continents	5.5–6.7 (by difference)
Current sources	11.7–13.2 (sum of below)
Breakdown in stratosphere	8.7
Annual increase	3.0–4.5
Anthropogenic sources	3.6–5.1 (by difference)

In a somewhat analogous fashion we can proceed to derive fluxes for NO, recognizing, however, that the much shorter lifetimes of NO and its oxidation products NO_2 and HNO_3 make global budgeting a problematical exercise. Complicating the problem are the many chemical and physical sinks for the nitrogen compounds. In combination, these factors make it impossible to derive emissions fluxes from observations scattered in space and time and to prepare a meaningful global budget. For some perspective, however, we list in Table 14.2 the estimated source fluxes:

TABLE 14.2. Estimated Source Fluxes for Atmospheric NO	
PROCESS	FLUX (Tg N yr^{-1})
Natural sources	13–29
Soils	10–20
Lightning	2–8
Transport from stratosphere	0.5
Current anthropogenic sources	24–30
Fossil fuel combustion	21
Biomass burning	2.5–8.5
Aircraft	0.6

The wide ranges given for most of the fluxes are a measure of the degree of uncertainty. Indeed, only for fossil fuel combustion, emissions from which are well monitored, can reasonably accurate flux estimates be made on a global basis. Nonetheless, it appears from the estimates that anthropogenic sources of NO now exceed the natural sources.

A further point with respect to Table 14.2 is that one should not uncritically equate the magnitude of the emission flux with the chemical importance. Atmospheric residence times are also of significance, and those times are longer for emissions at higher altitudes. This fact gives emissions from lightning and aircraft greater weight. In fact, aircraft emissions of NO continue to be regarded as potentially of great significance for stratospheric ozone depletion and climate change and will be a focus of attention by atmospheric chemists over the next decade.

A Global Atmospheric Sulfur Budget

A current assessment of the global sulfur cycle is pictured in Figure 14.10. It is generally conceded that the cycle is dominated by industrial activities, which result in the emission of some 70 Tg S yr^{-1} to the atmosphere, almost entirely as SO_2. It is estimated that of this amount about 25% is deposited as sulfate over the oceans and that the

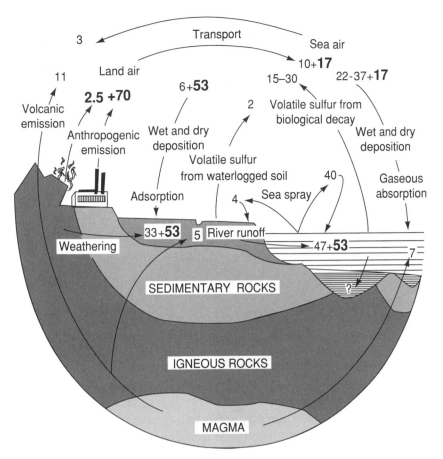

Figure 14.10 A global atmospheric sulfur cycle. Numbers in small type denote the estimated natural contribution; numbers in larger bold type denote the anthropogenic contribution. The units of the fluxes are teragrams S per year. (Modified from L. Granat, H. Rodhe, and R. O. Hallberg, The global sulfur cycle, in *Nitrogen, Phosphorus, and Sulfur: Global Cycles*, B. H. Svensson and R. Söderlund, eds., *Ecological Bulletins (Stockholm)*, *32*, 1976.)

remainder reaches the surface over land. Natural biological emissions of sulfur over the continents amount to only about 2 Tg S yr⁻¹. The emissions of sulfur from the oceans are much larger. This sulfur is produced by microbial processes and is manifested as the gas dimethyl sulfide (CH_3SCH_3), which is emitted at the rate of 15–30 Tg S yr⁻¹. Most of this sulfur is returned to the oceans as sulfate in precipitation, but perhaps 10% is transported to the continents by surface winds. Another natural source of atmospheric sulfur is volcanoes and emissions from fumaroles. This source is quite erratic, and the estimated average annual flux of 11 Tg S is quite uncertain. Because most of the volcanically active regions are on islands or at oceanic margins, much of the sulfur from volcanic eruptions will be deposited over the oceans.

Anthropogenic sulfur, almost entirely produced on land, has a substantial net flux to the oceans. The flow to the oceans also has a large geological term, as a result of the weathering of rock and the transport of sulfur to the ocean in river flow. Much of the river sulfur flow also is strongly anthropogenic, however. On a global basis, natural and anthropogenic sulfur contributions to the oceans are roughly equal.

Finally, we note that the sulfur budget includes carbonyl sulfide (COS), which is relatively stable chemically and is present in the lower atmosphere at concentrations everywhere near 0.6 ppbv. Carbonyl sulfide is broken down photochemically in the stratosphere, eventually providing the main source of sulfur for the stratospheric sulfate layer.

A Global Atmospheric Budget for Chlorine

The atmosphere contains substantial amounts of chlorine and smaller amounts of fluorine, bromine, and iodine. In the lower atmosphere, important chemical reactions involving the halogens have not been established, except for corrosion reactions on surfaces. In the stratosphere, however, chlorine and bromine are of vital importance in catalytic cycles that lead to lower concentrations of ozone. Particularly for this latter reason, the fluxes involved in the atmospheric halogen cycles are of interest and importance.

An attempt to construct a budget for atmospheric chlorine is shown in Figure 14.11. By far the largest flux indicated in the figure is that from seasalt spray over the oceans. This flux is very uncertain, as are the percentage of seasalt chlorine that is converted to HCl by reactions with H_2SO_4 and HNO_3 and subsequent vaporization and the percentage returned to the surface by sedimentation and precipitation. Because these numbers dominate the tropospheric chlorine cycle, their uncertainty renders the entire cycle of only academic interest and of no use in assessing trends for total atmospheric chlorine. Moreover, seasalt and HCl are efficiently returned to Earth's water and land surfaces, minimizing their long-range transport through the atmosphere and confining them to the troposphere.

Industrial processes of various kinds (including the combustion of coal, which has a modest chlorine content) produce HCl, organic reactive chlorine gases, and the chlorofluorocarbons, which are, depending on the species, slightly reactive or unreactive in the troposphere. The reactive chlorine gases are quite soluble or are converted into soluble compounds, and most of the supply is lost to precipitation. The entire emissions of the CFCs, currently about 0.8 Tg Cl yr^{-1}, will

Figure 14.11 A global atmospheric chlorine cycle. The units of the fluxes are 10^{12} g Cl yr^{-1}.

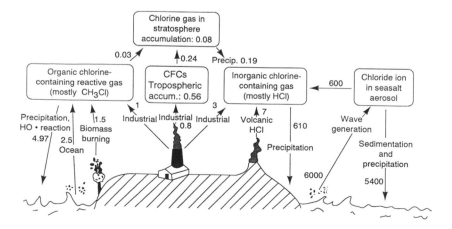

ultimately reach the stratosphere. However, that process takes about a decade for the average CFC molecule, and in the interim about 0.56 Tg Cl yr^{-1} is added to the tropospheric load and 0.24 Tg Cl yr^{-1} is transferred to the stratosphere. Of this latter amount, about 0.16 Tg Cl yr^{-1} is broken down by photochemical processes to yield inorganic Cl, while about 0.08 Tg Cl yr^{-1} adds to the stratospheric accumulation. The natural supply of chlorine to the stratosphere (as CH$_3$Cl) is only about 0.03 Tg Cl yr^{-1}. Some 0.2 Tg Cl yr^{-1} are removed from the stratosphere to the troposphere, mostly as soluble, readily removable HCl.

The main natural source of organic chlorine is the generation of CH$_3$Cl by kelp. The chlorine does not accumulate in the stratosphere, however, because it is balanced by a roughly equivalent return flow in precipitation and air exchange.

The relative magnitudes of the fluxes in Figure 14.11 are striking, the natural inorganic flux being some 2000–3000 times that of industrial components or of the natural organic chlorine fluxes. A marked distinction between reactivity, solubility, and (hence) lifetime makes all the difference, however. It is quite impressive to note how a relatively minor branch of the chlorine cycle can have such a prominent influence on stratospheric ozone, and thereby on atmospheric chemistry and on the biosphere. This situation emphasizes that although budget studies are of great value in atmospheric science they provide only part of the total picture.

If the atmospheric budget for chlorine is poorly determined, those for the other halogens cannot be specified at all. The concentrations are known to be small, and most atmospheric injection is thought to occur as sea spray or as a consequence of biological activity. Industrial sources are modest for Br, I, and F. In the case of anthropogenic organic bromine, however, much of the emission is in the form of "halons" (bromine analogues of the CFCs), which are used as fire extinguishing gases in areas where water is undesirable. Molecule for molecule, the halons are more destructive to stratospheric ozone by about an order of magnitude than are the CFCs (see Chapter 17). The reason is that the Br• and BrO• radicals are far less efficiently converted into the less reactive acid form (HBr) and thus are much more effective ozone depleters.

Global Budgets for Carbon Monoxide and Methane

The cycle of carbon monoxide in the atmosphere is closely coupled with those of methane and of the nonmethane hydrocarbons. For methane, the coupling occurs as a consequence of the chemical destruction of CH$_4$ in the lower atmosphere by Reaction (14.11),

$$CH_4 + HO\cdot \rightarrow CH_3\cdot + H_2O , \quad (14.11)$$

followed by

$$CH_3\cdot + O_2 + M \rightarrow CH_3O_2\cdot + M . \quad (14.12)$$

Most of the $CH_3O_2 \cdot$ then proceeds through reaction chains ending in the production of carbon monoxide. In the background atmosphere where low concentrations of NO are the rule, an important continuation in the chain process is

$$CH_3O_2 \cdot + HO_2 \cdot \rightarrow CH_3OOH + O_2 \quad (14.13)$$

$$CH_3OOH + h\nu \, (\lambda < 360 \, nm) \rightarrow CH_3O \cdot + HO \cdot \quad (14.14)$$

$$CH_3O \cdot + O_2 \rightarrow HCHO + HO_2 \cdot \quad (14.15)$$

$$HCHO + h\nu \, (\lambda < 360 \, nm) \rightarrow CO + H_2 \, . \quad (14.16)$$

More complex chemical chains along similar lines produce carbon monoxide from naturally emitted hydrocarbons such as isoprene (C_5H_8) and the terpenes ($C_{10}H_{16}$), and anthropogenically emitted hydrocarbons such as propane (C_3H_8). As with methane, the principal loss mechanism for atmospheric carbon monoxide is chemical in nature, the reaction with the hydroxyl radical:

$$CO + HO \cdot \rightarrow CO_2 + H \cdot \, . \quad (14.17)$$

As noted earlier, increasing concentrations of carbon monoxide and methane are thought to be depleting hydroxyl radical concentrations. If so, the rate of reaction (14.17) will decline as well, and concentrations of CO will increase more than would be suggested by merely looking at changes in input fluxes. The same applies to CH_4. Whether such processes are indeed occurring cannot be determined from measurements of the hydroxyl radical concentrations, as these are highly variable and exceedingly small. A more promising approach being attempted is to determine instead the trends in long-lived trace gases for which the source flux is very well known and which are removed from the atmosphere only by reaction with HO·. An example of such a gas is methyl chloroform (CH_3CCl_3), which is produced in well-quantified amounts by the chemical industry. Other examples are the hydrogen-containing replacements for the CFCs, such as CHF_2Cl.

It is of interest here to emphasize distinctions among the budgets of carbon dioxide, methane, and carbon monoxide. That for carbon dioxide encompasses only physical and biological processes, only small amounts of CO_2 being created by oxidation of CO and hydrocarbon gases and none being destroyed in the atmosphere by chemical reactions. For methane, none is created in the atmosphere, but atmospheric reactions are the principal sink. For carbon monoxide, the budget contains chemical reaction terms on both the source and sink sides of the ledger.

The high levels of ultraviolet radiation in the tropics produce more HO· there, as we will see in the computer model results of later chapters. As a consequence, the chemical loss of many species is maximized there. For CO, those large losses are balanced in the tropics by substantial sources, as shown in Figure 14.12. Biomass burning and the breakdown in the atmosphere of methane and vegetative emittants such as isoprene are the primary source processes.

Currently established source and sink rates of carbon monoxide are given in Table 14.3. Many of the sources are related to anthropogenic activities, but significantly more than half of the flux from the oxidation of nonmethane hydrocarbons is thought to be related to natural emissions of hydrocarbons from vegetation. Table 14.3 shows clearly that there are substantial uncertainties in the various individual source and sink terms. For this reason, the apparent imbalance between the total source and sink is not significant. Nevertheless, long-term observations of carbon monoxide at a midlatitude station in the Swiss Alps have indicated an annual growth in concentration of 0.7–1% over the past several decades, an increase suggesting that the budget is correct in indicating that the source strength of CO may be larger than the sinks, at least in the northern hemisphere.

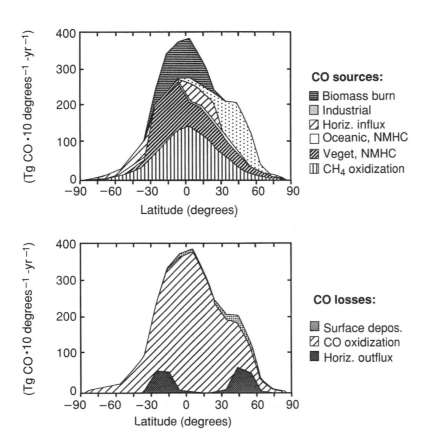

Figure 14.12 Annual carbon monoxide budget estimate as a function of latitude. The source terms are defined as *Biomass burn*, CO emissions from biomass burning in the tropics; *Industrial*, CO emissions from fossil fuel use and industrial processes; *Horiz. Influx*, net gain of CO in the atmospheric columns extending to 25 km as a consequence of horizontal transport; *Oceanic, NMHC*, oceanic emissions of CO plus CO produced in the boundary layer over the oceans through photooxidation of marine nonmethane hydrocarbons; *Veget, NMHC*, vegetative emissions of CO plus CO produced in the boundary layer over land through photooxidation of nonmethane hydrocarbons emitted by vegetation; *Methane oxidation*, production of CO by the atmospheric oxidation of methane. The sink terms are defined as *Surface depos.*, surface deposition of CO; *CO oxidation*, loss of CO through oxidation by radicals HO•; *Horiz. outflux*, net loss of CO in the atmospheric columns extending to 25 km as a consequence of horizontal transport. (Courtesy of K. M. Valentin, Max-Planck-Institut für Chemie, 1990).

TABLE 14.3. Global and Tropical Budget of Atmospheric Carbon Monoxide[a]

BUDGET ITEM	GLOBAL[b]	TROPICS[b] (30°S–30°N)
Sources		
Technological sources	440 ± 150	—
Biomass burning	700 ± 200	600 ± 200
Vegetation	75 ± 25	60 ± 20
Ocean	50 ± 40	25 ± 20
CH$_4$ oxidation	600 ± 200	400 ± 150
NMHC oxidation	800 ± 400	500 ± 200
Total production	2700 ± 1000	1600 ± 600
Sinks		
Oxidation by OH•	2000 ± 600	1200 ± 400
Uptake by soils	250 ± 100	70 ± 35
Flux into stratosphere	110 ± 30	80 ± 20
Total destruction	2400 ± 750	1400 ± 450

a. Adapted from R. J. Cicerone, in *The Changing Atmosphere*, F. S. Rowland and I. S. A. Isaksen, eds., John Wiley & Sons, New York, p. 49, 1988, and World Meteorological Organization, *Atmospheric Ozone, 1985*, Report No. 16, Geneva.
b. In Tg CO yr^{-1}.

Because CO is the usual end product in the methane oxidation chain, the two budgets are closely linked. That for methane is given in Table 14.4. The most important natural source of CH$_4$ is emission from wetlands, with termites and the oceans being of modest import. There are several anthropogenic sources, all of similar magnitude. In assessing biomass burning and fossil fuel combustion, measurements of the ^{14}CH$_4$ and ^{13}CH$_4$ isotopes have proved to be of great value, as biomass burning emits ^{13}CH$_4$-rich methane and fossil fuels emit ^{14}CH$_4$-free methane. The other sources emit methane with isotopic compositions typical of the recent atmospheric CO$_2$ content.

The most important sink for CH$_4$ is reaction with HO•, a flux that can be estimated with greater reliance than that of any of the sources except the CH$_4$ production by ruminants. Atmospheric measurements clearly indicate an increase in concentrations (see Figures 12.4 and 13.4), but the uncertainties in the fluxes and the great number of sources of similar magnitude indicate that much work remains to be done on the global methane budget.

TABLE 14.4. Estimated Sources and Sinks of Atmospheric CH$_4$

BUDGET ITEM	FLUX (Tg C yr^{-1})
Sources	
Natural	
Wetlands	120 (100–200)
Termites	20 (10–50)
Ocean	10 (5–20)
Hydrates	0 (0–5)
Anthropogenic	
Coal, gas mining	100 (70–120)
Rice fields	60 (20–100)
Ruminants	80 (65–100)
Waste treatment	80 (60–100)
Biomass burning	40 (20–80)
Sinks	
Reaction with HO$^\bullet$	430 (350–510)
Removal by soils	30 (15–45)
Atmospheric increase	37 (34–40)

BUDGETS OVER TIME: LEAD IN THE HUDSON-RARITAN BASIN

An extension of the process of constructing a budget for a particular species in a particular regime at a particular time (the initial effort almost always referring to the present) is constructing a budget for an extended time period. In such a process, changes in sources and sinks must be assessed, in addition to the current source and sink magnitudes. As an example of such an effort, we present briefly a one-century budget for lead in a river basin.

The region chosen for analysis by Robert Ayres and coworkers at Carnegie-Mellon University is the Hudson-Raritan Basin, which has its outflow to the sea at New York Harbor. Over the period 1880–1980, a number of different processes have contributed to lead levels in its atmosphere, surface waters, and soils, as shown in Figure 14.13. The figure emphasizes that different sources can have impacts on lead levels in a single regime and that several regimes (atmosphere, soil, water, etc.) can be influenced by a single source.

We will not reproduce the entire budget analysis but show in Figure 14.14 the magnitudes of the most influential emissions sources over the 1880–1980 period. The principal source in the first half of this period was emission, largely into waterways, from paint pigments. In the last half of the period, emission of lead additives from

Figure 14.13 Emission and transport processes for trace metals in river basins. (R. U. Ayres, L. W. Ayres, J. A. Tarr, and R. C. Widgery, *A Historical Reconstruction of Major Pollutant Levels in the Hudson-Raritan Basin: 1880–1980*, NOAA Tech. Memo. NOSOMA43, National Oceanic and Atmospheric Administration, Rockville, MD, 1988.)

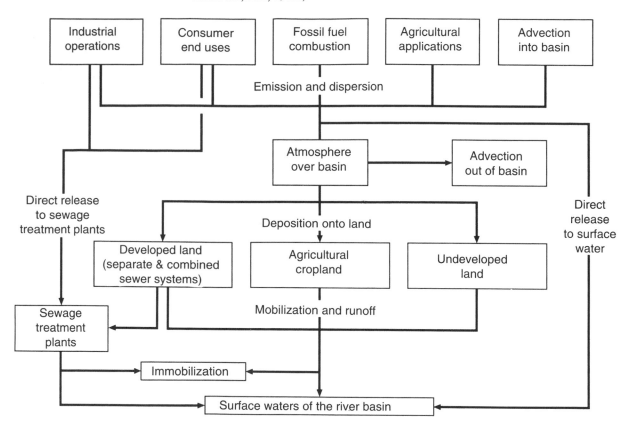

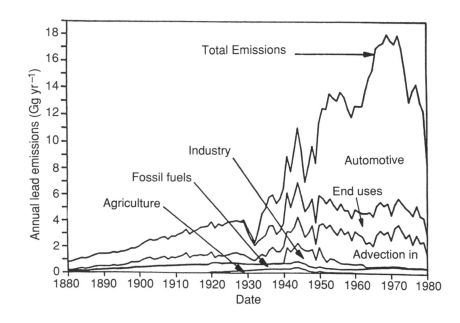

Figure 14.14 Annual lead emissions in the Hudson-Raritan Basin, 1880–1980. (R. U. Ayres, L. W. Ayres, J. A. Tarr, and R. C. Widgery, *A Historical Reconstruction of Major Pollutant Levels in the Hudson-Raritan Basin: 1880–1980*, NOAA Tech. Memo. NOSOMA43, National Oceanic and Atmospheric Administration, Rockville, MD, 1988.)

automotive combustion was dominant. This latter emission was almost entirely into the atmosphere, so the total emissions of lead were dispersed over a wide area as their concentrations rose.

Beginning in the 1970s, several industrialized countries began to eliminate the use of lead as a constituent of gasoline. The decline in airborne lead emissions that resulted in the United States was dramatic. Most of the rest of the industrialized world has also acted to limit automotive lead emissions. The same is not generally true for the developing countries, where lead concentrations in urban air may become an increasing problem.

The portion of the 1880–1980 Hudson-Raritan Basin lead emissions ending up in surface waters is indicated in Figure 14.15. The diagram is clearly related to Figure 14.14, yet the input from soil erosion indicates that deposition to soil has been assessed, and then the fraction of that input lost to erosion determined. Such an assessment is difficult enough when one can go into the field and attempt to measure the parameters of interest. It is considerably more difficult when one attempts to quantify the emissions from technologies and practices of the past. Nonetheless, it is only through such attempts that one can assess trends in environmental quality over extended time periods.

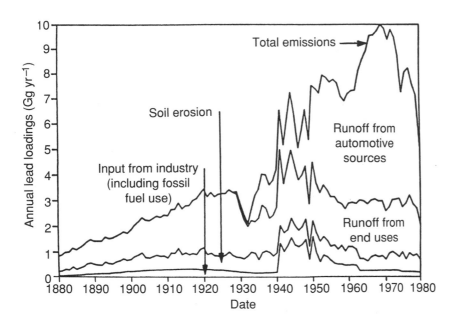

Figure 14.15 Annual lead inputs into the surface waters of the Hudson-Raritan Basin, 1880–1980. (R. U. Ayres, L. W. Ayres, J. A. Tarr, and R. C. Widgery, *A Historical Reconstruction of Major Pollutant Levels in the Hudson-Raritan Basin: 1880–1980*, NOAA Tech. Memo. NOSOMA43, National Oceanic and Atmospheric Administration, Rockville, MD, 1988.)

SUMMARY: BUDGETS IN AND OUT OF BALANCE

An important fact that should be evident from this chapter and from those preceding it is that in no case that has been studied extensively is an Earth system budget known to be in balance. Budget imbalance is not a central issue in regional budgets, because spatial variability is always present and some regional budgets may be out of balance to the extent and direction needed to balance other regional budgets. Another consideration is that some budget elements are so difficult to quantify that it may be hard to tell whether a budget is in balance or not. When sufficient information is available, however, a well-validated but out of balance global scale budget is an indication of global change.

The lifetime of a species is an important factor in establishing the distance scale for which a budget is appropriate. For atmospheric particles or trace metals, for example, no detailed global budgets are available. Such budgets would be of limited use in any case, because those species have lifetimes of only a few days to a few weeks; too short to make them of global significance. On a regional scale, however, they can have important effects on soil toxicity and visibility, and regional budgets and models are quite suitable to their study.

In Table 14.5, we collect information on various global atmospheric budgets, including some of those assessed above but also several others of interest. The degree of imbalance, often a rather uncertain number in a budget, varies from a very small percentage of the total to a very large percentage. Given the indisputable fact that *every* budget we have mentioned is out of balance in the positive direction, we have a clear signal that many different stresses are being placed on the environment.

The overall message from the study of budgets is that increasing portions of the planet are in danger of becoming more and more contaminated with the passage of time, especially on a local and regional scale in the developing world, in contrast to some of the situations that exist today in the developed world. These latter trends may only be temporary; on a global scale, budget imbalance is the rule rather than the exception.

TABLE 14.5. The Assessment of Imbalance in Budgets of Long-Lived Atmospheric Species			
SPECIES	1990 CONCENTRATION[a]	ESTIMATED ANNUAL CONCENTRATION CHANGE	ESTIMATED ANNUAL PERCENTAGE CHANGE
CO_2	354 ppmv	1.8 ppmv	0.5
CH_4	1.72 ppmv	15 ppbv	0.9
CO	90 ppbv	0.6 ppbv	0.7
N_2O	310 ppbv	0.8 ppbv	0.3
$CFCl_3$	0.28 ppbv	0.01 ppbv	4
CF_2Cl_2	0.48 ppbv	0.02 ppbv	4
CH_3CCl_3	0.14 ppbv	0.01 ppbv	4

a. Remote boundary layer.

EXERCISES

14.1 Table 14.5 lists the current annual concentration changes estimated for a number of atmospheric trace species. Assume that these changes occur uniformly throughout the troposphere and compute the mass of each species added to or subtracted from the troposphere each year. For purposes of this problem, assume that 90% of the atmosphere's mass is in the troposphere.

14.2 Atmospheric sulfur budgets have been constructed for some twenty years, with perhaps half a dozen of those efforts receiving wide recognition: W. W. Kellogg, R. D. Cadle, E. R. Allen, A. L. Lazrus, and E. A. Martell, *Science, 175*, 587–596, 1972; R.D. Cadle, *J. Geophys. Res., 80*, 1650–1652, 1975; L. Granat, H. Rodhe, and R. O. Hallberg, in *Nitrogen, Phosphorus, and Sulphur: Global Cycles*, SCOPE Report No. 7, Ecol. Bull. (Stockholm), *22*, 89–134, 1976; C. F. Cullis and M. M. Hirschler, *Atmos. Environ., 14*, 1263–1278, 1980; A. G. Ryaboshapko and C. H. Williams, in *The Global Biogeochemical Sulfur Cycle*, M. V. Ivanov and J. R. Freney, eds., John Wiley & Sons, Chichester, 203–296, 1983.

Select three of these sulfur budgets and write a comparative assessment, examining the processes treated, the sources of data used, and the degree of certainty assigned by the authors to the estimates. How do the budgets differ? What does each include? If the sulfur fluxes for a particular process are not the same, why not?

14.3 Information on flows of trace metals into air, water, and soils throughout the globe in 1983 is presented by J. O. Nriagu and J. M. Pacyna, *Nature, 333*, 134–139, 1988. Select two metals and draw global budget diagrams for them. Comment on the relative importance of the different sources and sinks.

14.4 Using information from the reference in Exercise 14.3 and from all other relevant literature, select an industrial area of the world (the size of a small country or less) and construct regional budgets for copper, mercury, and zinc in 1983.

FURTHER READING

B. Bolin and R. B. Cook, eds., *The Major Biogeochemical Cycles and Their Interactions*, SCOPE 21, John Wiley & Sons, Chichester, 1983.

J. N. Galloway and D. M. Whelpdale, WATOX-86 overview and western North Atlantic Ocean S and N atmospheric budgets, *Global Biogeochemical Cycles, 1*, 261–281, 1987.

W. Meyer and B. L. Turner, eds., *The Earth as Transformed by Human Action*, Cambridge University Press, Cambridge, UK, 1990.

P. P. Tans, I. Y. Fung, and T. Takahashi, Observational constraints on the global atmospheric CO_2 budget, *Science, 247*, 1431–1438, 1990.

CHAPTER 15

Building Environmental Chemical Models

THE PHILOSOPHY OF SCIENTIFIC MODELS

It is becoming more and more commonplace in the environmental sciences to describe systems of complex, interacting, nonlinear processes involving physics, chemistry, and biology through the design of what are termed numerical models. Although we can study individual system interactions in detail by laboratory simulations or, under favorable conditions, directly in nature, because of the many possible combinations of system processes, we must use numerical mathematical models when we try to comprehend system behavior as a whole. These models may be thought of as attempts to create computer replicas of natural system behavior so that causes and effects may be better understood.

From a scientific point of view, the most interesting stage of model development may well exist when knowledge about the basic system processes has advanced enough to allow the investigators to design a reasonable model, but not so far that all variables influencing the system are well understood. At this stage, model results often reveal deviations from what is measured, thus leading to the search for processes not yet considered. This intermediate, innovative stage of model development is approximately the one in which Earth system chemical models exist at the present time. As a result, the conclusions of model calculations have the potential to greatly improve our understanding of chemical systems. Model results must, of course, be carefully checked against appropriate simultaneous measurements of as many model variables as possible to discover deficiencies of the model that signal insufficient understanding or the absence of key processes and interactions. In just such a way it was discovered that photochemical processes in the gas phase alone could not explain the rapid loss of ozone under ozone hole conditions, a finding that led to the discovery of several important reactions that take place on ice particles. (Ice particles do not exist only in the atmosphere within the "ozone hole," so this finding stimulates research progress in other atmospheric regions as well.) Models can also be very useful in studying the effects of changes in individual processes on the total model system, thus gaining insight into the most critical sensitivities and feedbacks.

Once sufficient confidence in a model has been acquired, that model can then be used to predict future conditions arising from changes in relevant processes: natural ones, such as solar activity or volcanic eruptions, or anthropogenic ones, such as emissions of industrially produced trace gases. A current example of the significant role of numerical models is their simulation of changes that might result from different regulatory or political actions, such as the phaseout of chlorofluorocarbons, or from a change in the balance among the current mix of modifiable energy sources: coal, oil, natural gas, and nuclear. In these applications, models become policy tools, and much responsibility is put on the shoulders of the scientific community to critically test the models and to convey the results with the necessary caveats. Because of the urgency for advice, some chemical models are already used in this mode, although scientific knowledge is still incomplete. Models may even make incorrect predictions, just as sometimes occurs with weather prediction models. Much research is needed to improve the scientific basis and the performance of models, and skepticism concerning details of the results is clearly legitimate. Nonetheless, current models constitute some of the best available tools for incorporating large amounts of information and putting it into a form that can be critically appraised and used. Their results and implications clearly deserve to be taken seriously.

In the research mode, models can also be applied to studies of the past, providing additional means for testing the models or for learning more about the changing nature of the forcing functions. It is this ability of models to explore situations unavailable in reality, that is, to conduct "numerical experiments," that makes them of such potentially great value.

Models differ greatly in their degree of complexity. We will discuss below some of these differences, perhaps the most important of which are the approaches taken in how models handle time and space. For all modeling efforts, computer resources ultimately provide a trade-off among the amount of spatial resolution and time resolution that can be achieved and the degree of detail of the physical, chemical, and meteorological processes that can be considered. For example, when the spatial and process detail is highly restricted, it is often possible to use short time resolution. However, in such a system, interesting physical and chemical changes may take place over times that are long relative to the total time simulated by the computation. Figure 15.1 illustrates the point. To construct the figure, we have taken mean atmospheric lifetimes for a few species of concern, allowed the species to be advected at a typical wind speed of 5 m s^{-1}, and computed the mean distance that each species travels before it is removed from the atmosphere by the predominant chemical or physical loss process. On the figure we indicate as well a few distances of interest: the Tokyo, Japan urban area, an example of the local scale; Paris, France to Frankfurt, Germany, an example of the regional scale; Los Angeles to Boston, an indicator of continental scale distances; and

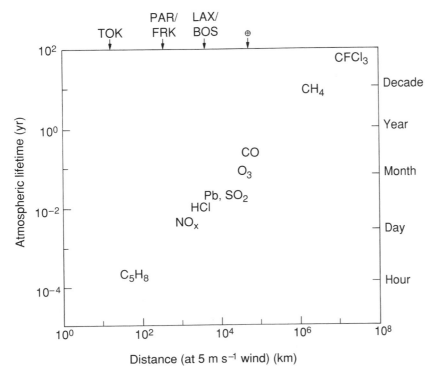

Figure 15.1 Time and source-to-sink space scales for selected species emitted from or generated by anthropogenic activities. TOK, Tokyo; PAR/FRK, Paris to Frankfurt; LAX/BOS, Los Angeles to Boston; ⊕, Earth's diameter.

Earth's diameter, the ultimate global scale indicator. Isoprene (C_5H_8), which reacts rapidly, is clearly best characterized as a locally acting trace constituent, while NO_x is regional in spatial scale. Sulfur dioxide, lead, and hydrogen chloride are continental in scale, but the scale for O_3 and CO is of the order of the hemispheric scale and CH_4 (methane) and $CFCl_3$ (CFC-11) clearly exceed the global scale (that is, they can circle the planet without being lost from the atmosphere). The atmospheric concentration patterns of the species are consistent with their typical spatial scales, the concentration of isoprene being the most variable of the species in Figure 15.1 and that of $CFCl_3$ the least.

Earth system chemistry models are sufficiently complicated that limitations in computer resources inevitably force compromises. For instance, models may emphasize the details of the chemistry and minimize those of transport, or vice versa. In the following, we will briefly discuss several types of models that have been used for Earth system studies.

BOX MODELS FOR FRESHWATER CHEMISTRY

The simplest of the chemical models is termed the *box model*; it has the form shown in Figure 15.2. This diagram is drawn for streamwater chemistry, a common application, and deserves detailed study because

Figure 15.2 A schematic diagram showing the features of a "zero-dimensional" box model of streamwater chemistry.

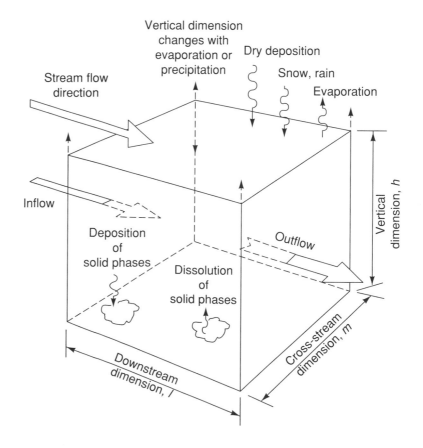

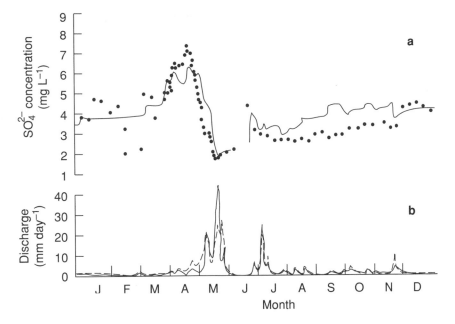

Figure 15.3 (a) The observed (bullets) and calculated (solid line) sulfate concentrations in stream-water at Storgama, Norway for the year 1978. (b) The observed (dashed line) and calculated (solid line) streamwater discharge rate. (Adapted with permission from N. Christophersen, L. H. Dymbe, M. Johannssen, and H. M. Seip, A model for sulphate in streamwater at Storgama, Southern Norway, *Ecological Modelling, 21,* 35–61. Copyright 1983/84 by Elsevier Science Publishers BV.)

it includes many features that show up in models of increasing complexity. Within the box, the size of which is defined by the model builder, chemical species enter in two ways: from the atmosphere and from upstream or runoff flows. Chemical species within the box are assumed to be well mixed. Downstream flow out of the box represents a loss of the chemical species. The changes in composition that occur result from ions and molecules in the incoming and outgoing flow streams and deposition and/or dissolution between the liquid and solid phases. Approaches vary, but it is common in freshwater models to assume that solid and liquid phases are in chemical equilibrium and to then apply solubility product and ion pairing constants as discussed in Chapter 7.

When the environmental conditions related to streamwater chemistry are precisely monitored, box models can produce results that quite accurately reproduce both the hydrologic flow and the hydrologic chemistry of specific streams. An example is shown in Figure 15.3. You may recall that the experimental data pictured in this figure was shown in Figure 9.1 to demonstrate the pulse of sulfate released to the stream by spring snow melt. Here those data are compared with the results predicted by a computer model. The model is successful in reproducing the essence of the hydrologic and chemical history of the stream, the correspondance indicating that the measurement data are accurate and that such parameters as the ion balance, the phase transformations between solid and liquid, and the water and ion absorption capacity of the soil in the catchment basin are well captured by the model.

A box model for atmospheric applications is pictured in Figure 15.4. We illustrate here an application that is somewhat more complex than the streamwater chemistry example of Figure 15.2. In this atmospheric perspective chemical species enter in two ways: from source emissions and as a consequence of atmospheric motions. The latter includes both *advection* (the transport of a chemical species by the mean motion of air parcels) and *entrainment* (the subgrid vertical movement of air parcels as a consequence of turbulent mixing). Advection out of the box represents a loss of the chemical species, as does detrainment due to upward motions. Often a box model for atmospheric chemistry is placed at ground level, so changes with time of the vertical dimension of the box reflect diurnal variations in atmospheric mixing or height of the boundary layer. In addition, the presence of the surface results in the loss of chemical species to the ground.

The choice of the dimensions and placement of the box is often dictated by the particular problem of interest. For instance, if one

Figure 15.4 A schematic diagram showing the features of a "zero-dimensional" box model of atmospheric chemistry.

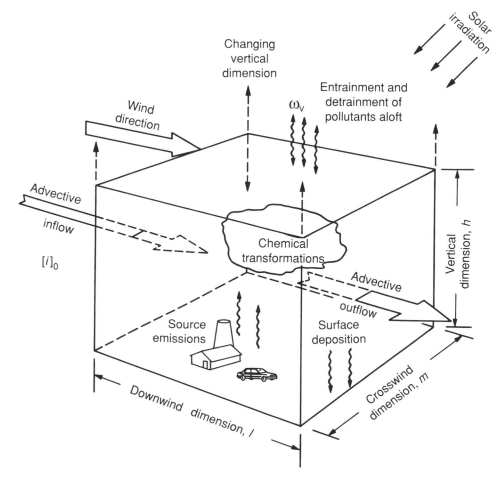

wishes to study the influence of urban emissions on the chemical composition of the air leaving an urban area, then as a first approximation the box may be designed to cover the urban area and emissions may be assumed to be mixed thoroughly throughout the box. Chemical species from the clean air side are "blown" into the box. The chemical composition of the air within and advected out of the box is then determined by the model. Chemical reactions must, of course, be taken into account. Because a few of the reactions are driven by solar radiation, appropriate wavelength-dependent solar fluxes must be incorporated into the calculations. Some reactions will be sources of specific chemical species, others will be sinks. Not only gas-phase reactions, but also chemical reactions on the surface of aerosol particles, in cloud or fog droplets, or in raindrops must be included. The period over which chemical reactions take place is approximated by the average residence time of air in the box, a quantity given by the horizontal dimension of the box in the direction of the flow divided by the advection velocity.

As with streamwater models, most atmospheric chemical models are time dependent. For example, a box model such as that in Figure 15.4 might attempt to compute air quality for a 24-hr day in an urban area. In that case, any variations in time in any of the terms need to be accurately supplied to the model; these variables would certainly include daily variations in traffic and other sources, diurnal wind speed patterns, variations in the height of the mixed layer, and the (wavelength-dependent) variation of solar radiation during the day. In some cases, one is more interested in the products of the chemical reactions than in the primary emissions themselves. Examples are the generation of ozone and the formation of sulfate particles.

Equations for Chemical Box Models

In the mathematical formulation of a box model, a separate equation is written for each chemical species to be represented. Consider the atmospheric situation pictured in Figure 15.4. For species i, present at concentration $[i]$, in a well-mixed box of length l and height h and having a fluid velocity of u, the time dependence of the concentration of species i is given by

$$\frac{\partial [i]}{\partial t} = \frac{u([i]_0 - [i])}{l} + S_i + C_i$$

$$+ \frac{w_v([i]_0^+ - [i]) - w_{\alpha_i}[i]}{h} \qquad (15.1)$$

where S_i is the emission source term, w_{α_i} is the surface *deposition velocity*, w_v the *ventilation velocity* (positive downwind) that describes the exchange with air above the box, $[i]_0^+$ the concentration of species i above the box, C_i the chemical loss or production rate, and $[i]_0$ the upwind concentration of species i.

The chemical terms that are included in this basic equation will have a degree of complexity that is dependent on the level of detailed desired by the model builder. The net chemical production terms are given by

$$C_i = P_i - I_i[i] \qquad (15.2)$$

in which the first term represents chemical reactions producing species i and the second term represents reactions removing it. The parameter I_i^{-1} is the chemical lifetime for species i, and the terms are defined by

$$P_i = \sum_{\substack{j \neq i \\ l \neq i}} k_{jl}[j][l] + \sum_{j \neq i} k_j[i] \qquad (15.3)$$

$$I_i = \sum_i k_{ij}[i] + k_i . \qquad (15.4)$$

In these equations the double-indexed k terms describe the reaction coefficients, the single-indexed terms the photolysis rates due to the absorption of solar radiation, and the terms in brackets are the concentrations of the various species. The determination of C_i often implies the simultaneous solution of up to hundreds of coupled differential equations, containing many nonlinear interactions. Because the various chemical compounds have widely different chemical lifetimes, the time steps for the integration should, in principle, be determined by the species with the shortest lifetimes. These can be very short, for some reactions as short as a microsecond. Thus, the solution of the system of equations can easily become such a formidable computational task that the development of techniques to solve the system of equations becomes a specialty in itself.

The model building and application proceed as follows. The builder first defines the dimensions of the box and the approach to time variations. Next, mathematical representations are devised for the factors involved: source fluxes, surface depositions, advection, entrainment, and so forth. Initial and boundary flux conditions must be specified. Suitable descriptions of each of these factors has been and is a continuing topic of research, so defining their representations is not a simple task. The set of chemical reactions is then decided upon. This necessary step is also a potential difficulty, because Earth systems are so chemically complex that not all reactions can be included. Sometimes the rate constants or absorption coefficients for each reaction have not been determined in the laboratory, in which case estimates are required. In addition, there is always the risk that some relevant processes are not even known.

Two versions of atmospheric box models have been devised. One is known as *Eulerian* and computes chemical species concentrations at a fixed geographical location (as done in Figure 15.4). Comparison with observations is relatively straightforward in this case, because most monitoring sites are at fixed locations. The alternative approach

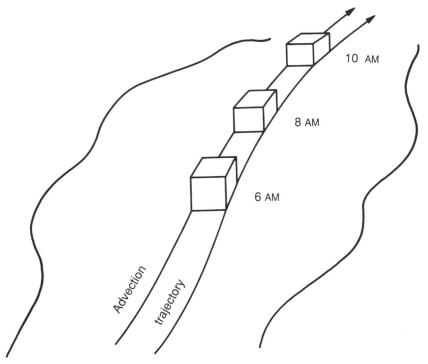

10 AM

8 AM

6 AM

Advection

trajectory

Figure 15.5 Movement of the reactive parcel of air as treated in a Lagrangian model.

is termed *Lagrangian* and is pictured schematically in Figure 15.5. In this approach the box is allowed to move with air or water motions, thus allowing one to follow the chemistry of an individual parcel. Advection terms are eliminated from the calculation, but source and sink terms vary as the parcel moves over different source regions. Comparison with observational data is difficult; a possible but seldom used technique has been to equip a balloon as a flying chemical laboratory and have on-board scientists measuring atmospheric constituents as they float among them. Lagrangian models are often employed for studies on local and regional scales.

Box models are perhaps best suited for taking first looks at a particular problem. In general, they have only limited use because they rely on assumptions that are not fulfilled in practice (such as rapid vertical and horizontal mixing or uniformity of surface sources). Also, in many circumstances important chemical reactions are still taking place in the air that leaves the urban center, such as in photochemical smog formation situations. To study that kind of problem, one would like to model the transport and chemical transformations taking place over space scales of hundreds of kilometers and time scales of hours to days. This can be done by dividing the study area into many boxes with appropriate dimensions determined by meteorological, orographical, and emission patterns. Such a technique leads to increasingly complex transport and photochemistry models, the so-called regional air or water pollution models, in which the details of the three-dimensional flow patterns are combined with chemical transformation calculations.

Chemical models are not designed only for local or regional pollution studies. In fact, some 20 yr ago, when computer resources were less than 1% of those of today, the earliest models were already successfully studying global pollution problems such as the effects of increasing concentrations of NO_x and chlorofluorocarbons on stratospheric ozone. We will next discuss some of the approaches that were taken for those kinds of studies.

ONE-DIMENSIONAL (1D) MODELS

The simplest improvement over the box model is to allow concentrations of chemical species to be computed as a function of one of the spatial dimensions. The earliest models were of the chemistry of the middle atmosphere (the stratosphere and mesosphere), where height was chosen as the variable dimension. In later years the troposphere was also treated with 1D models, so all three atmospheric regions were sometimes simultaneously considered. Conceptually, the picture is that of Figure 15.6, which may be thought of as a series of box models one atop the other. Equal-depth layers may be used, but in practice the models are usually designed to have more and thinner layers near the ground, where changes are rapid, and fewer and thicker layers for model regimes farther from the planetary surface.

The builder of a 1D model confronts several additional requirements compared with those facing the box model builder. Initial

Figure 15.6 A schematic diagram showing the features of a one-dimensional (1D) model of atmospheric chemistry. Elevated emissions are those coming from tall exhaust stacks, aircraft, volcanoes, or surface emissions drawn upward in convective clouds.

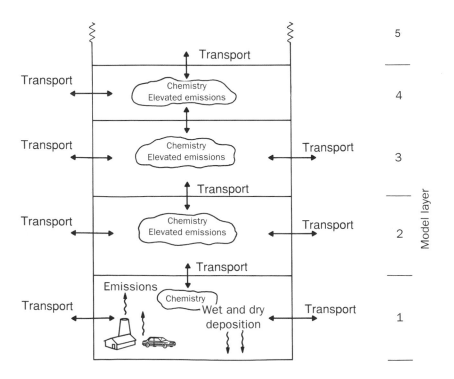

conditions must now be defined for each of the layers, and transport between layers must be appropriately represented. In the usual case of a 1D troposphere-stratosphere model, the layers are designed to represent global averages and advection is thus omitted by definition, as there is no net mass flux through a horizontal surface (or, more correctly, through a surface of constant atmospheric pressure). There is thus no lateral transport of chemical compounds, and vertical transport at specific altitudes is calculated for average planetary conditions.

Vertical Transport in 1D Models

In 1D models, vertical transport is approximated by an eddy diffusion parameterization, in which the net upward flux (e.g., in molecules per unit area per unit time), in analogy to molecular diffusion, is defined by

$$\phi_i = K_z [M] \frac{\partial \mu_i}{\partial z}, \qquad (15.5)$$

in which z is the height coordinate, K_z is the vertical eddy diffusion coefficient (in area per unit time), $[M]$ is the concentration of air (molecules per unit volume), and μ_i the volume mixing ratio (or mole fraction) of species i:

$$\mu_i = \frac{[i]}{[M]}. \qquad (15.6)$$

A typical vertical profile of K_z was given in Figure 14.2. That profile cannot be derived from physical principles but was determined empirically from atmospheric observations of trace gases, such as CO_2 for the troposphere and N_2O for the stratosphere. In the latter case, for instance, the net upward flux of N_2O through a horizontal surface may be calculated from global satellite observations of its distributions, because the flux is equal to the number of molecules that are lost above the horizontal surface by the photolysis of N_2O and by reactions with electronically excited $O(^1D)$ atoms. The sharp drop of K_z values between the troposphere and the stratosphere reflects the temperature increase with height just above the tropopause (Figure 3.5). This temperature gradient sets up stable meteorological conditions and inhibits mixing substantially.

Mathematical Structure of 1D Models

From a mathematical standpoint, the 1D model modifies the set of ordinary differential equations into a set of partial differential equations of the form

$$\frac{\partial [i]}{\partial t} = C_i - \frac{\partial \Phi_i}{\partial t}, \qquad (15.7)$$

where C_i is the net chemical production rate of species i (in molecules per unit volume per unit time). If the lower boundary of the model is located at Earth's surface, the net upward flux may be given by the expression

$$\Phi_{i|z=0} = S_i - w_{\alpha_i} [i], \qquad (15.8)$$

where S_i is the emission rate of gas i at Earth's surface and w_{α_i} the surface deposition velocity (in distance per unit time). Alternatively, emission and uptake at Earth's surface may be incorporated in the definition of the chemical source term C_i by adding two additional terms in the lower grid of the model:

$$C_i' = C_i + \frac{1}{\Delta z}(S_i - w_{\alpha_i}[i]), \qquad (15.9)$$

where Δz is the thickness of the lowest grid element.

More elaborate lower boundary layer formulations are often developed, in which vertical exchange processes in the boundary layer near the Earth's surface due to solar radiative heating of Earth's surface and its radiative cooling by emission of infrared radiation are taken into account.

Figure 15.7 Predicted changes in ozone concentrations by the year 2050 as a function of altitude for emissions of the chlorofluorocarbons CFCl$_3$ and Cl$_2$CF$_2$ under two different emission scenarios in the Brühl and Crutzen model. (C. Brühl and P. J. Crutzen, Scenarios of possible changes in atmospheric temperatures and ozone concentrations due to man's activities, estimated with a one-dimensional coupled photochemical climate model, *Climate Dynamics, 2*, 173–203, 1988.)

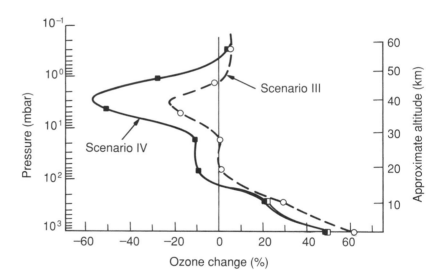

A typical example of a 1D model for the Earth's troposphere and stratosphere is that of Christof Brühl and Paul Crutzen of the Max-Planck-Institute for Chemistry in Mainz, Germany. A sample of the results is shown in Figure 15.7. The model from which these results were derived is an example of a "time-dependent" model, in that it follows the concentrations of chemical species and temperatures over particular time periods. A somewhat simpler and often employed alternative is the "time-independent" model, which is run for a specific set of boundary conditions with the time derivative being put equal to zero; it therefore gives the eventual result of a permanent change in conditions without attempting to reproduce results at different intermediate times.

To produce the results shown in Figure 15.7, computations were performed for two different assumptions of chlorofluorocarbon emissions. The results for predicted ozone reduction are typical of those of the many models that have studied the interactions between CFCs and stratospheric ozone. This model did not include reactions on aerosol particle surfaces, however, so could not predict the ozone hole.

TWO-DIMENSIONAL (2D) MODELS

Two-dimensional models of atmospheric chemistry move a step closer to reality as they add one more spatial dimension to the model. A schematic diagram of a typical 2D model for an urban area (that of one Thomas Graedel and James Schiavone at AT&T Bell Laboratories) is shown in Figure 15.8. This model calculates variations in chemical species concentrations as functions of both height and downwind distance. Such a model is, in some senses, a combination of the Eulerian and Lagrangian approaches, because chemical variations at a fixed location can be studied, but so can those of a parcel moving with the wind.

In its actual implementation, the model that is pictured might be called "$1^{1}/_{2}$", because its policy of constraining one dimension to be parallel to the downwind vector allows successive 1D computations to be performed in the individual boxes, proceeding from left to right in the direction of the airflow. The advection from one box to the next represents the integrated effects of the chemistry and emissions in the downwind regions. Typical results of this model are shown in Figure 15.9, which depicts lines of constant concentrations (*isoquots*) of ozone in two dimensions at three selected times. Among the features of the results that are worth noting are that ozone concentrations are quite low at night, when the lack of solar photons shuts off photochemical ozone generation, while ozone is being lost from the atmosphere by reaction with freshly emitted NO from automobiles, home heating, and industrial energy generation:

$$NO + O_3 \rightarrow NO_2 + O_2 . \qquad (15.10)$$

During the early morning and especially at midday, the highest ozone concentrations may form away from the regions of the heaviest emissions, as a consequence of atmospheric motions and the delay time involved in the photochemical generation process from source species to products.

Another common type of 2D atmospheric chemistry model is used to study global scale problems. The approach is to use altitude as one of the variable dimensions and latitude (from pole to pole) as the other.

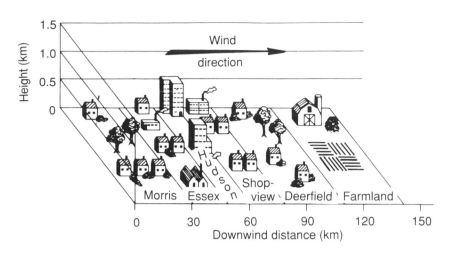

Figure 15.8 A schematic diagram of the two-dimensional (2D) urban atmospheric chemistry model of Graedel and Schiavone. (T. E. Graedel and J. A. Schiavone, 2-D studies of the kinetic photochemistry of the troposphere-II. Normal convective conditions, *Atmospheric Environment, 15*, 353–361, 1982.)

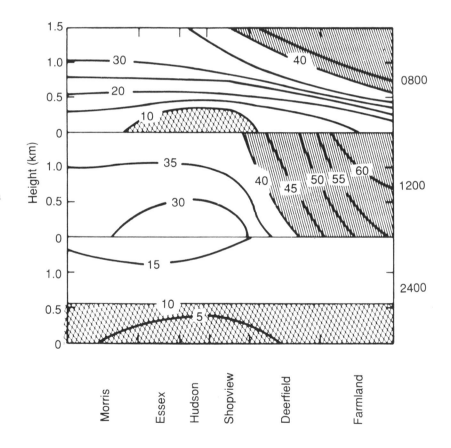

Figure 15.9 Calculated isoquots of ozone (ppbv) at three times of day (given on the right ordinate) on a downwind distance—height display of the 2D urban model results of Graedel and Schiavone; units on the isoquots are parts per billion by volume. The morning and midday increases in ozone concentrations downwind of the urban area are the result of ozone formation from urban precursor molecules as the air parcels are transported. (T. E. Graedel and J. A. Schiavone, 2-D studies of the kinetic photochemistry of the troposphere-II. Normal convective conditions, *Atmospheric Environment, 15*, 353–361, 1982.)

Diffusion of Chemical Constituents in 2D Atmospheric Models

The additional dimension in 2D models alters the mathematical description in such a way that the time rate of change of the concentration of a chemical constituent in a Cartesian coordinate system with coordinates y (south to north) and z (upward) is written as

$$\frac{\partial [i]}{\partial t} = C_i - \frac{\partial}{\partial y}\Phi_y - \frac{\partial}{\partial z}\Phi_z , \qquad (15.11)$$

where

$$\Phi_y = v[i] - K_y[M]\frac{\partial \mu_i}{\partial y} . \qquad (15.12)$$

and

$$\Phi_z = w[i] - K_z[M]\frac{\partial \mu_i}{\partial z} . \qquad (15.13)$$

In Equations (15.12) and (15.13), v denotes the average northward wind speed and w is the average vertical wind speed; K_y and K_z represent the meridional and vertical eddy diffusion coefficients. These parameters simulate the transport of chemical species as a consequence of mixing processes that are not described by the mean winds. In the exact treatment of these processes, corrections must also be made for the deviation of Earth from sphericity.

An implicit assumption in 2D models is that chemical species composition variations in the zonal (west–east) directions are much smaller than those in the vertical and latitudinal directions. Although such an approach is clearly an improvement over 1D global models, which take into account only variations in air composition in the vertical direction, the neglect of longitudinal variations will introduce substantial deviations from reality, especially at lower altitudes where the influence of chemical and biological processes as sources and sinks of trace gases at the Earth's surface is large. Two-dimensional meridional models have, therefore, mainly been used for stratospheric studies. For the troposphere, it may also be appropriate to develop 2D zonal models, sometimes termed channel models, with height and distance from west to east, the direction of the main airflow, chosen as the spatial coordinates. Such models can be instructive because there are considerable variations in the emission characteristics in zonal directions that are due to the locations of the oceans and the continents.

An example of a trace gas suitable for study with a stratospheric 2D model is N_2O, which is produced at the Earth's surface, principally by microbiological action in soils and waters. In the troposphere, nitrous oxide has no important sinks. In the stratosphere, at altitudes above about 25 km, solar ultraviolet radiation in the atmospheric "window region" of 190–230 nm penetrates fairly deeply because of relatively low absorption by O_2 and O_3. Nitrous oxide is dissociated by this radiation:

$$N_2O + h\nu\,(\lambda < 240\text{ nm}) \rightarrow N_2 + O\ . \qquad (15.14)$$

Nitrous oxide is also lost by reaction with electronically excited oxygen atoms that are formed by the photolysis of ozone at wavelengths shorter than 310 nm. Two reactions are possible:

$$N_2O + O(^1D) \rightarrow N_2 + O_2 \quad \text{or} \quad 2\,NO\ . \qquad (15.15)$$

The latter pathway is the most important source for NO in the stratosphere. As we have already discussed in Chapter 8, catalytic reactions involving NO are most important in controlling the abundance of ozone in the stratosphere, and, because most of the N_2O appears to be soil-derived, the reactions demonstrate the linkage between surface biogenic processes and stratospheric ozone. Altogether, the photochemistry of N_2O is very simple, because it has no sources within the atmosphere and only the aforementioned sink reactions, which can easily be calculated given the concentration of nitrous oxide and the measured ozone concentration distribution.

Figure 15.10 A comparison of 2D model results (solid lines) and satellite observations (dotted lines) for N_2O for February 1979. The numbers indicate computed or measured concentrations in ppbv. (R. L. Jones and J. A. Pyle, Observations of CH_4 and N_2O by the NIMBUS 7 SAMS: A comparison with in situ data and two-dimensional numerical model calculations, *Journal of Geophysical Research*, 89, 5263–5279, 1984.)

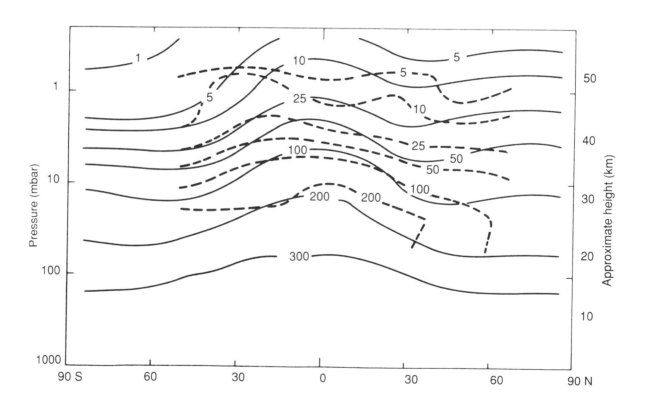

Latitude (degrees)

Comparison of satellite data for N_2O with the N_2O distribution calculated from a 2D model by R. Jones and J. Pyle of Oxford University is shown in Figure 15.10. The comparison reveals good agreement, a result suggesting that not only the simple photochemistry of N_2O loss but also the complex dynamics of the stratosphere may be represented well by the model. Areas in which the data and model results differ are being used to improve knowledge of chemistry and transport in the atmosphere.

Typical results of another 2D tropospheric global model are shown in Figure 15.11, in which Paul Crutzen and coworkers have computed the latitudinal distribution of the concentrations of the hydroxyl radical with height for the four seasons. Among the features of the results that are readily noted are the decrease of HO• as one goes to higher altitudes or (especially) toward the poles. This pattern reflects the higher intensities of solar ultraviolet radiation in equatorial regions due to higher Sun angles and minimum protection from ultraviolet radiation by low equatorial amounts of stratospheric ozone, as well as the larger concentrations of water vapor in equatorial regions and near the Earth's surface. The tendency for higher concentrations of HO• to occur over northern rather than southern midlatitude regions reflects the prevalence of higher concentrations of tropospheric ozone and nitric oxide in the northern hemisphere.

Figure 15.11 Calculated isoquots of the hydroxyl radical (in units of number of radicals per cubic centimeter) as a function of latitude and altitude: (a) January; (b) July; (c) April; (d) October. The results are given for cloud-free conditions and are from the 2D global model results of Crutzen and Gidel. (P. J. Crutzen and L. Gidel, A two-dimensional photochemical model of the atmosphere. 2: The tropospheric budgets of the anthropogenic chlorocarbons, CO, CH_4, CH_3Cl, and the effect of various NO_x sources on tropospheric ozone, *Journal of Geophysical Research, 88,* 6641–6661, 1983.)

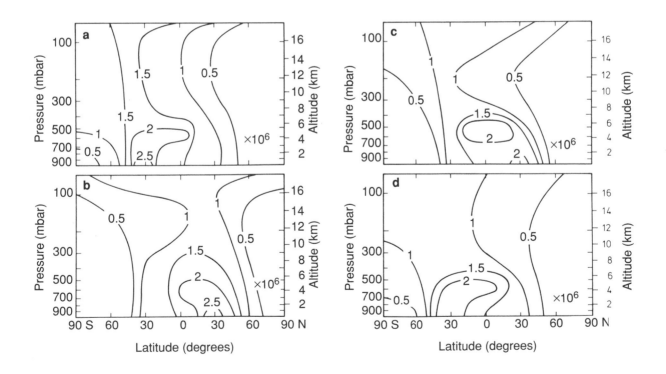

THREE-DIMENSIONAL (3D) MODELS

The most comprehensive of the atmospheric chemical models are those that treat variations in the concentrations of one or more chemical species in three spatial dimensions and one time dimension. The resulting equations look and are extremely complex, and their solution is more complex still. Recall first that one set of equations must be solved for *each* chemical species treated in the model. Even more daunting is the need to specify the wind velocity fields in three dimensions. The vertical component w is especially problematic, as it is generally too small to measure and must be inferred. The generation of the wind fields is one of the principal goals of synoptic and dynamic meteorology and is often accomplished through the design of weather prediction models and *general circulation models* (GCMs), which attempt to simulate global weather and climate, or of the more spatially limited regional or mesoscale models. The satisfactory achievement of that goal is only now coming within the grasp of the theoretical meteorological community, and then only with extensive use of today's largest computers. As a consequence of these computational complexities, a coupled global 3D chemical-transport model unfortunately has precious little room left for atmospheric chemistry computations. An example of a simple application of this kind of model is that of Jerry Mahlman, Hiram Levy, and coworkers at NOAA's Geophysical Fluid Dynamics Laboratory in Princeton, New Jersey; some of their results are shown in Figure 15.12. In this computation, the chemical involvement is limited to treatment of a single species, again the chemically simple nitrous oxide gas (N_2O), the sources and sinks of which were discussed earlier in this chapter.

The overall average lifetime of N_2O in the atmosphere is about 170 yr, so this gas is rather evenly distributed in the troposphere. In the model, N_2O is merely transported from its source at the Earth's surface to the stratosphere, where it is destroyed by chemical reactions. These processes determine its atmospheric distribution. Figure 15.12a shows that for the 10 mb pressure level the highest N_2O volume mixing ratios of about 110 ppbv occur in equatorial regions. This may appear surprising, because the most rapid destruction of N_2O occurs in the equatorial stratosphere. However, as is clearly indicated by observations of chemical tracers, transport from the troposphere to the stratosphere occurs largely at low latitudes through the tropical tropopause. The return flow at higher latitudes will thus be depleted in N_2O.

Figure 15.12a is analogous to the output of the 2D model shown in Figure 15.10. The 3D model produces results for one additional spatial dimension, of course, so one can also construct latitude–longitude figures of the type shown in Figure 15.12b at a specified pressure surface or height. This figure indicates that N_2O concentrations predicted for the middle stratosphere have only minor variations with longitude (a result that helps give confidence in 2D model results

a

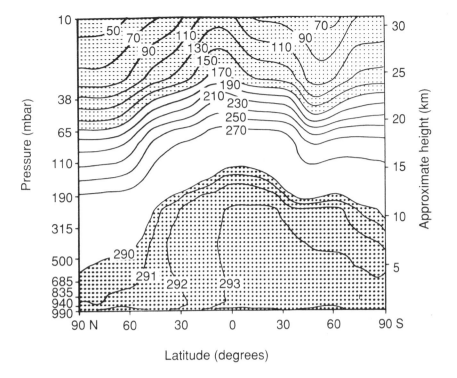

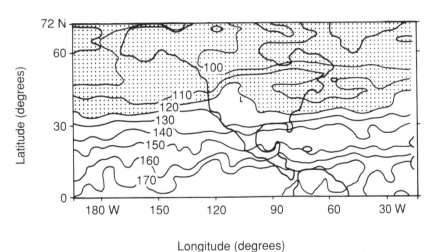

Figure 15.12 (a) Zonal mean mixing ratio computed for October as a function of latitude and altitude. (b) Calculated isoquots of N_2O mixing ratio for September 14 as functions of latitude and longitude for an atmospheric pressure height of 10 millibars (that is, an altitude of about 31 km). Units on the isoquots are parts per billion by volume. (J. D. Mahlman, H. Levy III, and W. J. Moxim, Three-dimensional simulations of stratospheric N_2O: Prediction for other trace constituents, *Journal of Geophysical Research, 91,* 2687–2707, 1986.)

in which longitudinal constancy is assumed); the variations that are seen reflect the atmospheric circulation more than the distribution of sources of N_2O.

The model by Mahlman, Levy, and coworkers was the first attempt to couple a simple chemical process with the meteorological output of a general circulation model. More ambitious attempts in this direction are now being tried by a number of research groups in the United States and Europe. None of the models can simultaneously treat the full panoply of major atmospheric chemical and meteorological processes. It will take a new generation of larger and faster computers before a detailed atmospheric chemical model can be combined with a GCM. Until then, simpler approaches must be followed. A few of these simplifications are discussed in the following paragraphs.

The 3D model described above is a fully coupled model (if an incomplete one) of atmospheric motions and chemistry, that is, it computes both processes simultaneously, although thus far the chemistry is not allowed to feed back into the meteorology. A variation of this approach that has proved useful for atmospheric chemistry studies is the *decoupled* or *off-line* 3D model. In this approach, a meteorological model is run for a chosen geographical region (in some cases, the entire globe) and time of year, and the calculated winds, temperature, water vapor, and cloud distributions are stored to be used as input variables in a 3D chemical model. Such a tactic simplifies the computations and allows more time and computer storage space to do the chemical computations. The approach is suitable for situations in which the changing chemical constituent concentrations will not change the meteorological result. This is the case for many tropospheric applications in which calculations are performed for limited time periods of perhaps a few years or less. It would not be the case, however, if enough greenhouse gases were added to alter the atmospheric wind and temperature structure. For the stratosphere, such decoupled models are, at best, approximate descriptions of reality, because the most important output of such models, the ozone concentrations, determines the heating and thus the circulation of the stratosphere.

Mathematical Structure of 3D Models

In 3D models, the equations to be solved to determine the chemical concentrations at a specific location in three-dimensional space and as a function of time are written (here again in simplified Cartesian coordinates) as

$$\frac{\partial [i]}{\partial t} = C_i + S_i - \frac{\partial}{\partial x} F_x - \frac{\partial}{\partial y} F_y - \frac{\partial}{\partial z} F_z \qquad (15.16)$$

with x, y, and z denoting the coordinate system, normally chosen to increase from west to east, from south to north, and upward from Earth's surface. F_x, F_y, F_z and are the corresponding fluxes of gas i, which by introducing the mean motions u, v, and w and the eddy diffusion coefficients K_x, K_y, and K_z in the corresponding directions can be written as

$$F_x[i] = u[i] - \left(K_x[M] \frac{\partial \mu_i}{\partial x}\right) \qquad (15.17)$$

$$F_y[i] = u[i] - \left(K_y[M] \frac{\partial \mu_i}{\partial y}\right) \qquad (15.18)$$

$$F_z[i] = u[i] - \left(K_z[M] \frac{\partial \mu_i}{\partial z}\right) \qquad (15.19)$$

These models often allow u, v, and w to be time-varying and, as in other models, continue to require mass continuity.

An example of a successful decoupled 3D chemical model, shown in Figure 15.13, is derived from the Acid Deposition Modeling Project (ADMP), begun by Julius Chang and colleagues at the National Center for Atmospheric Research in Boulder, Colorado and currently being further developed at the State University of New York in Albany and at the University of Cologne, Germany, in collaboration with other institutions. The model includes a meteorological segment that treats either all of the United States and lower Canada or Europe; this segment feeds air motion and cloud and precipitation data to the chemical segment of the model, which is applied to a chosen portion of the model area. Figure 15.13 shows the results of a tracer experiment designed to test the meteorological segment. Similarly successful comparisons have been made with chemically active species, and the model is designed to be readily usable for continental and subcontinental scale air quality studies.

A simplified attempt to design a global 3D chemical transport model has been tried by Paul Crutzen and Peter Zimmermann at the Max-Planck-Institute for Chemistry in Mainz, Germany. In their

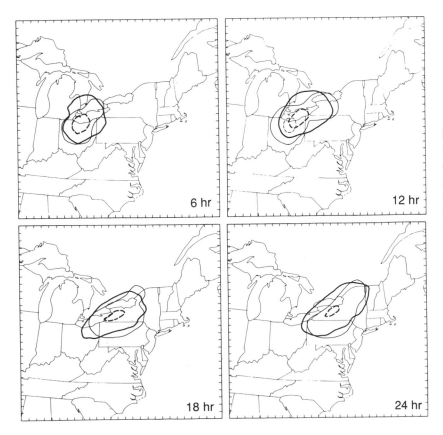

Figure 15.13 Concentrations of the inert tracer perfluoromethyl-cyclohexane, released over a 3-hr period at a single site and observed at 86 stations throughout the northeastern United States. The light line shows where measured volume mixing ratios were 1×10^{-14} molecules of tracer per molecule of air. The heavy line shows the prediction of the model. The four diagrams indicate measured and predicted results for the first four 6-hr observational periods following release. (R. A. Brost, P. L. Haagenson, and Y.-H. Kuo, The effect of diffusion on tracer puffs simulated by a regional scale Eulerian model, *Journal of Geophysical Research*, 93, 2389–2404, 1988.)

model, the planet is divided horizontally into 10° longitude by 10° latitude segments and vertically into 100-mbar grid boxes. Mean wind velocities in the east–west, north–south and vertical directions over 12 one-month periods were taken from meteorological observations. Over such long periods, the mean wind vectors will not necessarily describe the transport of chemical species correctly, because the deviations from the mean wind vectors will be substantial. This complication is taken into account by introducing eddy diffusion coefficients in the three coordinate directions, chosen to be proportional to the monthly variances of the corresponding wind components multiplied by time scales that represent typical durations of the synoptic disturbances. The meteorological suitability of the model can be determined by comparison between observed and calculated distributions of chemical tracers that have well-known sources at Earth's surface, and either well-known or nonexistent sinks. Examples of such molecules are the chlorofluorocarbons ($CFCl_3$ and CF_2Cl_2) and the radioactive ^{85}Kr produced by nuclear power plants. A good test of the behavior of the models is to compute HO• the distribution and compare the computed and measured concentrations of species with well-known industrial sources and a sink solely or largely determined by reaction with HO•, such as the partially hydrogenated CFCs (HCFCs).

The advantage of a decoupled 3D model is that a greater opportunity is provided to treat photochemical processes in some detail. Because the model can be run very efficiently, it can be used for scenario calculation. As an example, Figure 15.14 in the color insert shows simulated surface ozone volume mixing ratios for the preindustrial era and for the present. The figure shows that surface ozone concentrations have more than doubled over the past century in the northern hemisphere, mostly as a result of growing industrial emissions of NO that serves as a catalyst in the production of tropospheric ozone. A comparison of these computer model results with the observations of Figure 13.10 suggests that the chemistry of tropospheric ozone generation is now understood quite satisfactorily.

Simplified Time Integrations of Chemical Compounds

Typical models of stratospheric and tropospheric chemistry now contain several tens to hundreds of chemical reactions that must be integrated in time to simulate the chemical composition of the atmosphere, or parts thereof. As already mentioned, this is a formidable task, which in many applications cannot be handled without approximations. As the average atmospheric chemical lifetimes of several species are of the order of microseconds or less, it is clearly prohibitive to integrate the system of chemical equations with comparable short time steps. To overcome this difficulty, it is common in atmospheric chemistry modeling to assume the concentrations of the shortest lived compounds to be in steady state, i.e.,

$$\frac{\partial [i]}{\partial t} = 0 . \qquad (15.20)$$

As an example of such a treatment, let us consider the photochemical system, that describes some of the main features of the ozone and atomic oxygen concentration distributions in the daytime stratosphere, as first discussed by the British scientist Sidney Chapman in 1930 and given in Table 15.1. Diatomic oxygen is by far the most abundant form, with O_3 ranking next below 65 km and $O = [O(^1D)] + [O(^3P)]$ ranking next at higher altitudes. (Recall that the notation ^1D and ^3P indicates different electron excitation states.)

TABLE 15.1. The Vertical Distribution of Oxygen Compounds in the Middle Atmosphere[a]

ALTITUDE (KM)	CONCENTRATION (molec cm^{-3})			
	$[O_2]$	$[O_3]$	$[O(^3P)]$	$[O(^1D)]$
10	1.7(18)	1.0(12)	1.3(4)	—
15	8.1(17)	1.1(12)	5.5(4)	—
20	3.6(17)	2.9(12)	9.4(5)	9.0(−1)
25	1.6(17)	3.2(12)	6.7(6)	5.0(0)
35	3.5(16)	2.0(12)	2.4(8)	1.0(2)
40	1.7(16)	1.0(12)	1.2(9)	3.3(2)
45	8.9(15)	3.2(11)	3.7(9)	6.0(2)
50	4.8(15)	1.0(11)	6.5(9)	6.1(2)
55	2.6(15)	3.2(10)	8.4(9)	4.4(2)
60	1.5(15)	1.0(10)	6.5(9)	2.6(2)
65	8.2(14)	3.2(9)	5.0(9)	1.5(2)
70	4.2(14)	1.0(9)	4.0(9)	9.6(1)
75	2.0(14)	3.2(8)	3.8(9)	6.7(1)
80	9.0(13)	1.4(8)	1.4(10)	7.0(1)
85	3.7(13)	1.0(8)	3.0(10)	1.2(2)
90	1.3(13)	1.1(8)	3.0(11)	4.2(2)
95	4.7(12)	1.3(7)	3.3(11)	4.9(2)
100	1.9(12)	1.7(6)	3.2(11)	1.2(3)

a. From G. Brasseur and S. Solomon, *Aeronomy of the Middle Atmosphere*, 2nd Ed., D. Reidel, Dordrecht, 1986.

Chapman's proposal consisted of four reactions:

$$O_2 + h\nu \,(\lambda \leq 240 \text{ nm}) \;\rightarrow\; 2O \qquad (15.21)$$

$$O + O_2 + M \rightarrow O_3 + M \qquad (15.22)$$

$$O_3 + h\nu \,(\lambda \leq 1160 \text{ nm}) \rightarrow O + O_2 \qquad (15.23)$$

$$O + O_3 \rightarrow 2O_2. \qquad (15.24)$$

The scheme basically contains two kinds of reactions: (15.21) and (15.24), which form and destroy odd oxygen (the sum of O and O_3), and (15.22) and (15.23), which recycle O and O_3 but do not affect the sum of their concentrations. As it turns out, the chemical adjustment times to equilibrium for O and O_3 individually are both very short. That of atomic oxygen, τ_0, can be derived from the time-dependent equation

$$\frac{\partial [O]}{\partial t} = 2J_{15.21}[O_2] - k_{15.22}[O][O_2][M]$$
$$+ J_{15.23}[O_3] - k_{15.24}[O][O_3] \,, \qquad (15.25)$$

where M denotes an air molecule that removes excess energy from the reaction products. (Refer to Chapter 7 for a discussion of the formation of the rate terms.) Then

$$\tau_0 \leq \frac{1}{k_{15.22}[O_2][M]} \,, \qquad (15.26)$$

where we have neglected the relatively unimportant influence of reaction 15.24. Similarly, the lifetime of O_3 equals

$$\tau_0 \leq \frac{1}{J_{15.23}} \,. \qquad (15.27)$$

The above lifetimes can be easily estimated. τ_O ranges from 15 min at 65 km altitude to 10^{-5} s at Earth's surface; τ_{O_3} ranges from about 2 min above 40 km to less than 1 hr at Earth's surface. The individual chemical adjustment times of atomic oxygen and ozone toward an equilibrium state mostly defined by reactions 15.22 and 15.24 are, therefore, very short, so we may write

$$[O] = \frac{J_{15.23}[O_3]}{k_{15.22}[O_2][M]} = \phi[O_3] \,. \qquad (15.28)$$

This equation should be interpreted to mean that either the concentration of O atoms is given by that of O_3, or vice versa. What the equations do not give any information about, however, is the concentration of the sum of O and O_3, termed odd oxygen . Instead of integrating the concentrations of ozone in time, one can do this for odd oxygen instead, so

$$\frac{\partial ([O] + [O_3])}{\partial t} = 2J_{15.21}[O_2] - k_{15.24}[O][O_3] \qquad (15.29)$$

Making the above substitution for [O], we can transform Equation (15.29) into a time-dependent equation for ozone:

$$\frac{\partial ([O] + [O_3])}{\partial t} = 2J_{15.21}[O_2] - k_{15.24}\phi[O_3] \qquad (15.30)$$

The equilibrium concentration is thus given by the expression

$$[O_3]_e = \left(\frac{J_{15.21}k_{15.22}[M][O_2]^2}{k_{15.24}\,J_{15.23}} \right)^{1/2} \,. \qquad (15.31)$$

Because [M] and $[O_2]$ scale with air density, Equation (15.31) implies a 3/2 power dependence of $[O_3]_e$ on density provided the other terms in the equation do not change. However, $J_{15.21}$ increases with altitude and $J_{15.23}$ decreases with altitude. The result turns out to be a maximum ozone concentration at about 25 km.

The photochemical restoration time of the sum of O and O_3 is given by

$$\tau_{(O+O_3)} = \frac{(1 + \phi)}{4k_{15.22}\phi [O_3]_e} . \qquad (15.32)$$

The lifetime is much longer than that for each compound individually, of the order of days above 30 km and months to years at lower altitudes. This formulation makes the solution of the time-dependent equation much more practical, allowing us to integrate the sum of the concentrations of O and O_3 in time while also considering the effect of transport processes. The subdivision of odd oxygen into O and O_3 is accomplished by Equation (15.28). The longer photochemical restoration time for odd oxygen also means that as altitude decreases transport processes become increasingly important in order to describe the ozone distribution.

The foregoing treatment is an example of a general principle. Often two species (O and O_3 in this case) undergo rapid conversion to each other, but much slower conversion to a third species [in this case, to O_2 through Reaction (15.24)]. Forming the sum of the two (here O plus O_3) causes the rapid and numerically dominating terms to cancel, leaving only reactions with longer time scales. With no loss of information, inclusion of the sum rather than the individual components then permits a more efficient numerical integration of the chemical equations.

OCEAN MODELS

Oceanic model building is a specialty in itself, and one that is not of interest here per se. The dynamics and chemistry of the oceans are often crucial to atmospheric change, however. Examples are easy to list: (1) ocean mixing determines the sequestering of atmospheric carbon in sediments, thus strongly influencing the global carbon cycle (see Figure 14.7); (2) ocean currents determine much of the transport of heat from low to high latitudes (see Chapter 4), thus playing a major role in climate; (3) the importance of atmospheric nutrients to oceanic productivity (see Chapter 5) is dependent on sea surface temperatures and chemistry; (4) the emissions of chlorinated and sulfurated gases from the sea surface play major roles in stratospheric ozone chemistry (see Chapter 8) and the rate of cloud formation (see Chapter 5). Because climate and environmental chemistry cannot adequately be considered from only an atmospheric perspective, therefore, we present some information on current ocean models and then on efforts to link these models with those of the atmosphere.

Oceans and atmospheres are both fluid systems in which changing concentrations of trace species are of interest. These commonalities carry over to many aspects of model formulation. As with their atmospheric analogues, ocean models begin with the equations of motion and of continuity for chemical and physical properties such as momentum, heat, and salinity. One might picture an ocean model to some extent as an inverted atmospheric model (see Figure 15.6) with high resolution (i.e., many thin layers) near the surface and low resolution (i.e., a few thick layers) at great depths. Just as the atmospheric models must consider motion past such obstacles as mountain ranges, ocean models must deal with the topography of the ocean floor and the continents. The boundary conditions include input from rivers, loss to evaporation, response to surface winds, and a host of other factors.

As an example of the performance of current models, we show in Figure 15.15 in the color insert samples of the results of a 3D ocean model by Albert Semptner of the Naval Postgraduate School and Robert Chervin of the National Center for Atmospheric Research, both in the United States. Their model has 20 vertical levels and a horizontal grid size of 1/2° by 1/2°. It has been optimized for the purpose of simulating ocean circulation on modern supercomputers. Just as atmospheric model calculations predict jet streams and other flow features, so the ocean model results predict large and small flow features that can be compared with observations.

Figure 15.15 in the color insert displays computer model results at a depth of 160 m for all the world's oceans and can be compared with the very simplified conceptual ocean transport picture of Figure 6.5. Both the conceptual flow and the flow calculated by the computer model have a warm, shallow component that moves from east to west across the central Pacific Ocean. The detailed model does not predict this flow to be laminar, however, but to occur as a train of warm-core eddies. Many of the details of this eddy structure are seen in the limited observations of undersea ocean currents. Similar encouraging results are achieved for more localized flow structures around continental boundaries and undersea mountain barriers.

Like atmospheric models, ocean models are undergoing vigorous development and improvement. At their present state of readiness, however, they clearly reproduce most of the major observable features of oceanic currents and heat transfer. They are thus conceptually capable of serving as valuable partners to atmospheric models in analyzing the behavior of the planet in its present condition and under a variety of perturbations.

COUPLING THE OCEAN AND ATMOSPHERE

If both ocean models and atmosphere models are important to problems of global change, and examples of both types of models exist, why do we not have models that treat both the ocean and atmosphere together? The answer is related to the great differences in approach in the two types. The most detailed 3D ocean models currently have grid cells of about 1/2° by 1/2° and 10–20 layers. In contrast, the most detailed atmospheric models have the same number of vertical layers but grid cells of about 4° by 5°, a difference of about a factor of 80 ×

20 in the number of cells for which computations must be made. A second difference is in the processes that are treated in most detail. At present, ocean models devote great effort and computer resources to accurate simulation of flow at various depths; little effort is expended on trace species chemistry. For many atmospheric models, on the other hand, the air motions are secondary to such computer-hungry computations as deducing changing ozone concentrations as a function of chlorofluorocarbon inputs.

Given this disparity of approach, does coupling of ocean and atmosphere models occur at all? The answer is yes, to some extent. As an example, we reproduce in Figure 15.16 the results of an assessment of the effects of changing ocean heat transport on climate. In this study a standard 3D atmospheric model was used, but its boundary conditions for the present epoch were determined by comparison with ocean model results and ocean observations. Figure 15.16 compares the results of a computation for present conditions with one in which 15% less sea ice was present, as might have occurred in the past and might again in the future. Strong surface warming is seen to occur near the poles in regions formerly covered by sea ice. This warming is a consequence of the fact that open oceans absorb more solar radiation than does sea ice and can then transfer some of that radiant energy to the lower atmosphere.

Figure 15.16 The annual surface air temperature change (°C) between a calculation representing preindustrial ocean-atmosphere interactions and a calculation with 15% less sea ice. (D. Rind and M. Chandler, Increased ocean heat transports and warmer climate, *Journal of Geophysical Research, 96,* 7437–7461, 1991.)

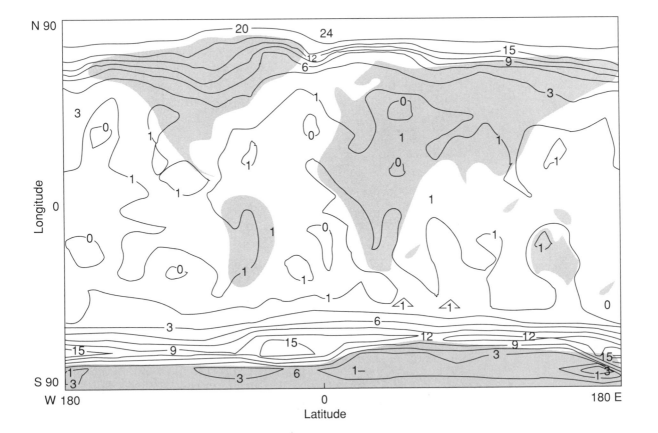

The example above is a reflection of the sophistication of current model coupling, in which one model or the other is used to provide boundary conditions to the other, perhaps at several different times during the computation, but true simultaneous computation is not achieved. The principle limitation to such interaction is the enormous amount of computer resources that would be required. As more powerful computers and more clever programming techniques are developed over the next decade or so, model coupling will become better and better. Full coupling is not likely to occur, however, until well into the twenty-first century, and this will continue to restrict the ability of scientists to predict the behavior of the Earth system under changing stresses.

SUMMARY: THE UTILITY OF EARTH SYSTEM MODELS

The discussion of chemical and physical models in this chapter demonstrates two things clearly: that much has been accomplished through the use of several varieties of models and that all of the models have their individual limitations. We conclude the discussion by summarizing which models are used for which applications.

Because of the time-consuming job of model building and the expense of computer use, the tendency is to use the simplest model that will give useful results for a particular problem. When one knows very little about a problem or when few data are available to guide model construction and verify model performance, a box or 1D model is most practical. When a horizontal dimension is important, as it is in studying the effects of emission changes at different latitudes, a 2D model is appropriate. The most complete answers are, in principal, provided by 3D models, but computing resources and model building constraints still severely limit their use.

From a scientific point of view, models are most useful not when they provide results that can be thought of as certain, but when they pose new scientific questions or indicate possibilities that can be confirmed by measurement. Measurements, in turn, often stimulate the development of new models or the modification of existing models so that specific segments of the model can be improved or its results investigated from new perspectives. This interplay of theory and experiment is the chief way in which environmental chemists forge new understanding, a process typical of many fields in science. Modeling is an especially important tool of Earth system chemistry because of the many interactions that exist. These interactions not only are of importance for chemistry per se, but also play a role in Earth's climate. Conversely, climate variability influences atmospheric and surface chemistry through changing emissions of biogenic trace gases at Earth's surface. The interaction between biosphere, atmosphere, hydrosphere, and human activities will require the development of increasingly complex global system models, and hence the use of more extensive computational facilities.

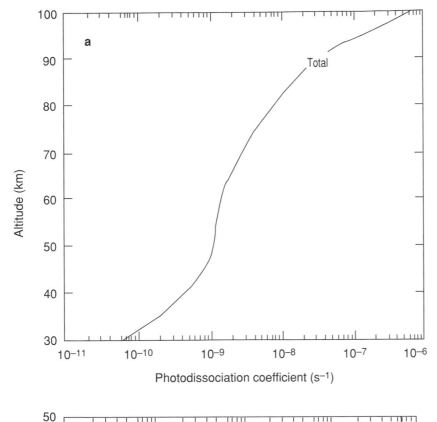

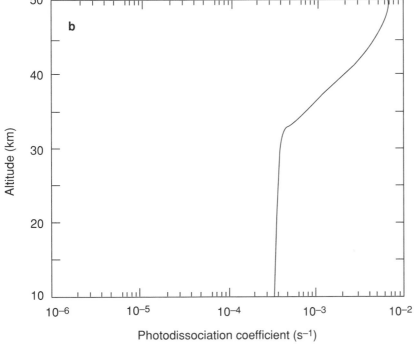

Figure 15.17 (a) The total photodissociation coefficient of molecular oxygen as a function of altitude. (b) The total photodissociation coefficient of ozone as a function of altitude. These curves are the sums of individual absorption band systems. (Reproduced with permission from G. Brasseur and S. Solomon, *Aeronomy of the Middle Atmosphere*, 2nd Ed. Copyright 1986 by Kluwer Academic Publishers.)

EXERCISES

15.1 What sort of processes are represented by each of the four terms of Equation (15.1)?

15.2 Imagine that you wish to use a box model to describe the relationships among ozone and the oxides of nitrogen in an urban atmosphere. You have decided to utilize the following set of chemical equations and rate constants. Write the equations that will be used to compute the changes in concentration of each of the species *as a consequence only of chemical reactions*, i.e., in the general form of Equation (15.2).

$$NO_2 + h\nu\ (\lambda \leqslant 420\ nm) \rightarrow NO + O \qquad (P1)$$

$$O + O_2 + M \rightarrow O_3 + M \qquad (P2)$$

$$O_3 + h\nu\ (\lambda \leqslant 360\ nm) \rightarrow O_2 + O \qquad (P3)$$

$$NO + O_3 \rightarrow O_2 + NO_2 \qquad (P4)$$

$$NO_2 + O_3 \rightarrow NO_3 + O_2 \qquad (P5)$$

$J_{P1} = 9.9 \times 10^{-3}\ s^{-1}$;

$k_{P2} = 6.0 \times 10^{-34}\ cm^6\ molecule^{-2}\ s^{-1}$;

$J_{P3} = 2.6 \times 10^{-5}\ s^{-1}$;

$k_{P4} = 1.0 \times 10^{-14}\ cm^3\ molecule^{-1}\ s^{-1}$;

$k_{P5} = 3.2 \times 10^{-17}\ cm^3\ molecule^{-1}\ s^{-1}$.

15.3 Using the answer to Exercise 15.2, write the differential equations that describe the overall change in concentration of the chemical species in the box model. (The equations will take the form of Equation 15.1.)

15.4 Construct a computational flow diagram describing the steps in the calculation of the variation in ozone concentrations in the box model as a function of time of day.

15.5 Derive Equations (15.26), (15.27), and (15.32).

15.6 Given the O_2 concentrations as a function of altitude of Table 15.1 and the O_2 and O_3 total photolysis rate coefficients of Figure 15.17, calculate photochemical steady state concentrations of O and O_3 from Equations (15.28) and (15.31) at altitudes of 100, 85, 70, 55, 40, 30, 25, 20, and 15 km. The rate constants $k_{15.22}$ and $k_{15.24}$ may be taken to be independent of temperature and equal to $5 \times 10^{-15}\ cm^3\ molecule^{-1}\ s^{-1}$ and $1 \times 10^{-33}\ cm^6\ molecule^{-2}\ s^{-1}$, respectively. How do you explain that the calculated O_3 concentrations are substantially larger than those observed?

FURTHER READING

G. Brasseur and S. Solomon, *Aeronomy of the Middle Atmosphere*, 2nd Ed., D. Reidel, Dordrecht, 1986.

R. L. Dennis, W. R. Barchet, T. L. Clark, S. K. Sellkop, and P. M. Roth, *Evaluation of Regional Acidic Deposition Models*, State of Science and Technology Report 5, National Acid Precipitation Assessment Program, Washington, D.C., 1990.

J.S. Chang, R. A. Brost, I. S. A. Isaksen, S. Madronich, P. Middleton, W. R. Stockwell, and C. J. Walcek, A three-dimensional Eulerian acid deposition model: Physical concepts and formulation, *Journal of Geophysical Research*, 92, 14681–14700, 1987.

L. T. Gidel, P. J. Crutzen, and J. Fishman, A two-dimensional photochemical model of the atmosphere. 1: Chlorocarbon emissions and their effect on stratospheric ozone, *Journal of Geophysical Research*, 88, 6622–6640, 1983, and P. J. Crutzen and L. T. Gidel, A two-dimensional photochemical model of the atmosphere. 2: The tropospheric budgets of the anthropogenic chlorocarbons CO, CH_4, CH_3Cl and the effect of various NO_x sources on tropospheric ozone, *Journal of Geophysical Research*, 88, 6641–6661, 1983.

J.D. Mahlman, H. Levy III, and W. J. Moxim, Three-dimensional simulations of stratospheric N_2O: Predictions for other trace constituents, *Journal of Geophysical Research*, 91, 2687–2707, 1986.

U. Mikolajewicz, B. D. Santer, and E. Maier-Reimer, Ocean response to greenhouse warming, *Nature, 345*, 589–593, 1990.

CHAPTER **16**

Regional Futures

The history of the atmosphere encompasses and demonstrates on a vast scale the workings of the planet and of the interactions of living organisms with that natural system. The perspectives gained from the atmosphere's history can be and have been used in a predictive way as well: to anticipate the chemical and physical characteristics of the atmospheres of the future. These predictions are of as much intellectual interest as are the historical data. They are useful as well for forecasting and planning the future course of life on the planet, as its peoples contend with an environment that seems almost certain to differ substantially from those of the past.

    Any attempt to predict the environments of the future will naturally look to the past for guidance on the extent and role of human

339

activities and policies, and for the impacts that those activities and policies have had on characteristics that can be measured. That information is then used to propose corrective measures and alternatives for planning and implementation. It is at this point that the physical science approach utilized in measuring and understanding past and present conditions must interact with economics, technology, and the social sciences so that the assumptions on which predictions of the future are based are tenable.

A set of assumptions regarding future activities and actions is termed a *scenario* and, in the best case, is constructed by competent professionals from reliable information. Notwithstanding that expertise, scenarios can never be more than frameworks for thought; in other words, scenarios are not predictions, because they seldom or never anticipate the unexpected twists and turns produced by major historical or ecological events.

This chapter presents several examples of predictions of air quality on different local and regional scales. The intent is not to be comprehensive but to provide examples of the capabilities and limitations of predictive models on distance scales similar to those of cities, states, and nations.

URBAN AIR QUALITY

Urban air quality assessments are generally designed to determine the actions needed to bring atmospheric constituent concentrations within levels established to ensure the well-being of the population. In cases where atmospheric chemistry does not come strongly into play, such as with total particulate matter or with the relatively unreactive carbon monoxide gas, the assessment techniques can be quite simple and progress can be monitored by regular analytical studies of air quality. The situation is more complex if the species being studied is reactive. Ground-level ozone presents the most difficult situation of all, because it is a species formed entirely by chemical reactions and has no direct emissions sources. Discussing ozone assessment and control therefore illustrates most of the characteristics of urban air quality projections.

Recall from Chapter 8 that ozone is produced by photolysis of nitrogen dioxide:

$$NO_2 + h\nu\,(\lambda \leq 410\,\text{nm}) \rightarrow NO + O\,, \qquad (16.1)$$

followed by the three-body recombination of the oxygen atom with molecular oxygen:

$$O_2 + O + M \rightarrow O_3 + M\,. \qquad (16.2)$$

Because this pair of reactions is often followed by

$$O_3 + NO \rightarrow NO_2 + O_2\,, \qquad (16.3)$$

the reaction sequence can generate ozone in abundance only if some of the NO is oxidized to NO_2 without removing ozone. An efficient method for doing so in the urban atmosphere is

$$RO_2\cdot + NO \rightarrow NO_2 + RO\cdot ,\qquad (16.4)$$

the $RO_2\cdot$ radicals being derived from many hydrocarbons emitted into the air:

$$RH + HO\cdot \rightarrow R\cdot + H_2O \qquad (16.5)$$

$$R\cdot + O_2 + M \rightarrow RO_2\cdot + M ,\qquad (16.6)$$

where RH denotes a hydrocarbon compound. It is important to note that reactions (16.4), (16.1), and (16.2) can use hydrocarbons to cycle molecular oxygen to ozone, with NO_x (NO and NO_2) functioning as a catalyst, as can be seen by the sum of the three reactions:

$$RO_2\cdot + O_2 \rightarrow RO\cdot + O_3 .\qquad (16.7)$$

The consequence of this cyclic chemistry is that control of ground-level ozone concentrations can, in principle, be accomplished by regulation of either reactive nonmethane hydrocarbon (NMHC) or oxides of nitrogen emissions or both. In practice, the degree to which emission controls are available and practical often constrains the emission control approaches.

Calculations for ozone concentrations in a number of different urban areas have been performed with models of different dimensions and complexities. An approach that has seen wide use has been to use a 1D model to make numerous calculations for a variety of emissions scenarios and to plot selected portions of the results for ready analysis. Figure 16.1, by Robert Bilger of the University of Sydney, Australia illustrates such an effort. The diagram refers to conditions in the city and suburbs of Sydney and is used in the following way. Morning rush-hour concentrations of NO_x and NMHC are determined, often in the form of monthly, seasonal, or annual averages, to reflect the characteristic natural meteorological variations that influence concentrations. These NO_x and NMHC concentrations locate a point on Figure 16.1. The curves on the figure are isoquots (lines of constant concentration) of the maximum ozone levels that the model predicts will be reached during a day in which the morning concentrations of the precursors are as specified.

It is instructive to study Figure 16.1 for a situation where the NMHC and NO_x levels are changing. Consider the situation where average precursor concentrations are at point 1 on the figure. If emissions of both NO_x and NMHC increase and their concentrations move to point 2, the maximum ozone concentration will increase from about 100 ppbv to about 140 ppbv. (We say "about" here because the models are generally not considered accurate enough to give more than

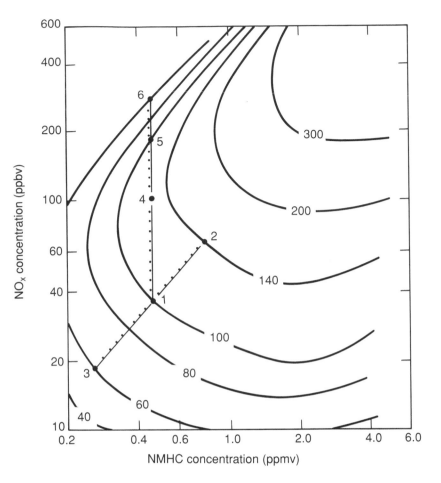

Figure 16.1 Isopleth diagram for maximum ambient daily ozone concentrations (in ppbv) derived for Sydney, Australia, given average morning rush-hour concentrations of NO_x and nonmethane hydrocarbons (NMHC). The numbers 1–6 on the figure are described in the text. (Reproduced with permission from R. W. Bilger, Optimum control strategy for photochemical oxidants, *Environmental Science and Technology, 12*, 937–940. Copyright 1978 by American Chemical Society.)

approximate results.) Conversely, if emissions of both precursors are reduced enough so that the concentrations fall on point 3, the maximum ozone level will drop to about 60 ppbv. Now consider what happens if NMHC emissions are held constant while those of NO_x increase. If the concentration point goes from 1 to 4, the ozone maximum will increase. If NO_x emissions increase further, to point 5, the ozone maximum will retreat to approximately its former level. If NO_x emissions increase still further, to point 6, the ozone maximum will retreat still more, to below the level of point 1. The reason for this complex behavior is that, with increasing emissions of the primary pollutant NO, the formation of ozone is both favored, because NO serves as a catalyst (see above), and disfavored, because O_3 is lost in the oxidation reaction from NO to NO_2 and as NO_2 removes HO• from the chemical system via

$$NO_2 + HO• + M \rightarrow HNO_3 + M , \qquad (16.8)$$

thus counteracting the formation of radicals. The exact balance struck by those opposing forces is a nonlinear function of the species concentrations. It should also be noted that Figure 16.1 says nothing about ozone levels far downwind of the emissions area. In those locales, diluted NO_x can react with additional emitted NMHC, including those from vegetation, to produce enhanced ozone, as has been observed downwind of large urban centers such as London, Los Angeles, and New York.

Given the above perspective, it is interesting to reflect upon the increasing ozone concentration trend in Europe, illustrated in Figure 13.8. This trend would suggest that either NO_x or NMHC concentrations or both are increasing in Europe. The data to study this issue are incomplete, but trends in NO_x emissions in several different countries are shown in Figure 16.2; in most cases, including West Germany and France, the emissions are stable or decreasing slightly. Figure 16.1 may thus imply that NMHC concentrations are increasing, because in most areas of the diagram a stable NO_x concentration and increasing NMHC concentrations produce increased ozone.

It is generally the case that engineering or economic considerations limit the type and amount of emission controls that may be used, so movement in certain directions on Figure 16.1 is constrained or not possible. For example, NMHC emitted from vegetation can be important in generating ozone in certain locations and seasons, as shown for Atlanta, Georgia by William Chameides and colleagues at the Georgia Institute of Technology. In addition, Figure 16.1 is only an approximation of a complex chemical reaction set combined with a full spectrum of meteorological variables and local conditions. It is not surprising that control of ground-

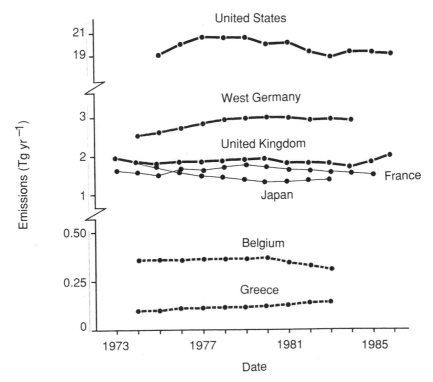

Figure 16.2 Trends in emission of oxides of nitrogen in different countries, 1973–1986. (After Global Environment Monitoring System, *Assessment of Urban Air Quality*, United Nations Environment Program, Geneva, 1988.)

level ozone concentrations has been so difficult and that some urban areas, despite major expenditures and years of effort, have made disappointingly little progress toward reducing ozone concentrations.

OZONE AND PRECIPITATION ACIDITY IN NORTHEASTERN NORTH AMERICA

Many predictive models are designed to study regions significantly larger than an urban area. An example is the Regional Acid Deposition Model (RADM), already described briefly in Chapter 15. The goal of this 3D Eulerian model is to predict the chemistry of precipitation in the northeastern United States and southeastern Canada, given a scenario of emissions and meteorological conditions. An overview of the model components is shown in Figure 16.3. The model divides the atmosphere into six vertical levels, each level having 30 by 30 horizontal grid cells and each cell being approximately 80 x 80 km^2 in size. In each grid cell, hourly meteorology and trace gas emission data are specified for a period of several days. The model uses this information to compute time-varying 3D distributions of trace gases, particles, and droplets, as well as temporal and spatial distributions of the deposition of trace species to surfaces in the region.

RADM needs to deal with the gas-phase chemistry of the atmosphere, the transport of species between gas, aqueous, and solid phases, and the chemistry within liquid droplets. In addition, its emphasis on precipitation and air transport requires that it have very sophisticated treatment of meteorology and cloud physics. All of these topics have been discussed in earlier chapters of this book. RADM is highly advanced, and the complete model is large, intricate, and expensive.

Figure 16.3 An overview of the processes and computational flow of the Regional Acid Deposition Model (RADM). (J. S. Chang, et al., A three-dimensional Eulerian acid deposition model: Physical concepts and formulation, *Journal of Geophysical Research, 92,* 14681–14700, 1987.)

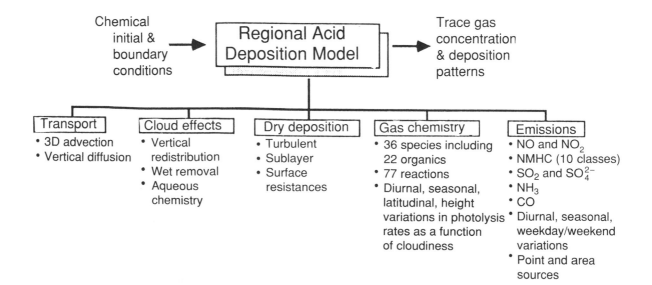

As an example of the output of the RADM study, Figure 16.4 shows the computed dry deposition (uptake by vegetation on Earth's surface under nonprecipitating conditions) of nitric acid to two different types of ground surfaces. The deposition flux in these two cases is quite different, because the higher surface roughness and large leaf surface area of forests causes them to absorb HNO_3 at a significantly faster rate than is the case for other surface types. The HNO_3 depositions are much more broadly dispersed than are those of sulfate (also computed by RADM), because the SO_2 precursor emissions tend to be concentrated at particular point sources, whereas emissions of NO_x, which are just about evenly divided between motor vehicles and power plants, are spatially more uniformly distributed.

A number of calculations have been made by RADM in which emissions in the northeastern United States were changed in various ways. Large spatial differences were noted. It is particularly interesting to note that decreases in the emissions of gaseous SO_2 are not reflected everywhere in proportionally diminished sulfur deposition to surfaces. Any sulfur that is not deposited in the United States is carried by winds across the border to Canada or to the east over the Atlantic Ocean. Concomitantly, depending on the air flow, some of Canada's emissions are deposited on the United States. These findings, suggested earlier by others and confirmed more recently by RADM, provided much of the impetus for research in the U.S. National Acid Precipitation Assessment Program (NAPAP) and subsequently for legislation in the United States and Canada to limit the emissions of sulfur compounds.

Figure 16.4 The spatial distribution of dry HNO_3 deposition flux (kg per hectare) to surfaces of different types during a RADM model calculation simulating the period 22–24 April 1981. H, locations of local deposition flux maxima. (a) Agricultural regions; (b) forested regions. (Reproduced with permission from C. Walcek and J. Chang, A theoretical assessment of pollutant deposition to individual land types during a regional-scale acid deposition episode, *Atmospheric Environment, 21*, 1107–1113. Copyright 1987 by Pergamon Press, plc.)

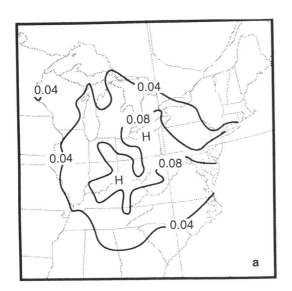

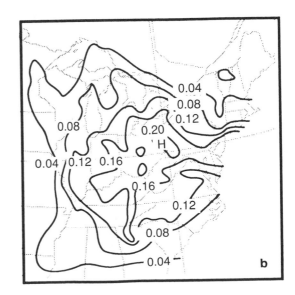

SULFUR DEPOSITION AND ITS EFFECTS IN EUROPE

Overview and Deposition Calculation

A model that makes an interesting contrast with RADM is the RAINS (Regional Acidification INformation and Simulation) systems analysis model developed at the International Institute for Applied Systems Analysis (IIASA) in Austria. The emphasis of the model is on spatial and temporal overviews of transborder air pollution transport in Europe. Its time horizon extends from 1960 to 2040 to accommodate comparison with historical data and to permit long term examination of the effects of policy options.

In its most recent form, the RAINS model deals with sulfur and nitrogen, while retaining its emphasis on sulfur deposition in precipitation. This constraint prevents the model from calculating the rates of chemical processes by first principles; for example, no set of gas-phase chemical reactions is solved. In lieu of this, RAINS utilizes transfer coefficients derived from models formulated by other institutes or agencies in Europe. Working within this approach, RAINS formulates a rather complete mathematical treatment of emission fluxes and atmospheric depositions and impacts, the so-called *source-receptor relationships*, as shown in Figure 16.5. The sum of the wet and dry deposition of the sulfur and nitrogen species constitutes an assessment of acid deposition. Note that the RAINS model addresses a number of impacts of acidic precipitation directly and extensively rather than concentrating on the chemical interactions that are responsible for the impacts.

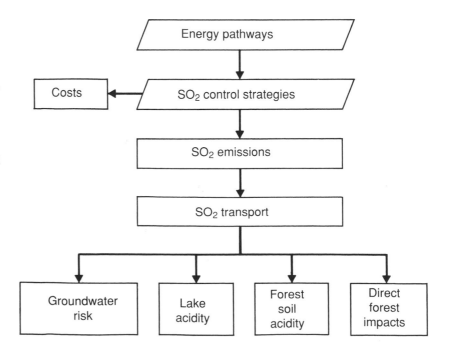

Figure 16.5 A schematic overview of the major components of the Regional Acidification INformation and Simulation (RAINS) model. (Reproduced with permission from J. Alcamo, et al., Acidification in Europe: A simulation model for evaluating control strategies, *Ambio, 16,* 232–245. Copyright 1987 by The Royal Swedish Academy of Sciences,)

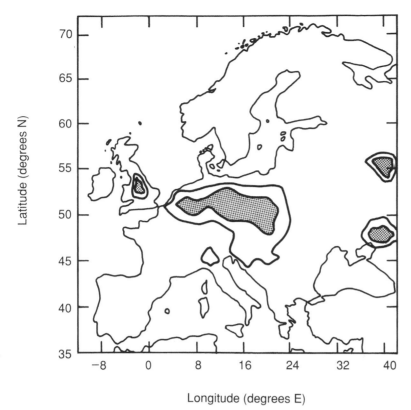

Figure 16.6 Total European sulfur deposition in 1980 (the heavy line is the isopleth for deposition of ≥ 5 g S m^{-2}) compared with the same isopleth predicted by the RAINS model for a 30% uniform reduction of sulfur emissions in all countries (the predicted isopleth encloses the shaded areas). (Reproduced with permission from L. Hordijk, Towards a targetted emission reduction in Europe, *Atmospheric Environment, 20*, 2053–2058. Copyright 1986 by Pergamon Press, plc.)

An example of the output of the RAINS model is shown in Figure 16.6. Here the sulfur deposition in 1980 is compared with that predicted by RAINS should a uniform 30% reduction in sulfur emissions occur throughout Europe. The sizes of the areas within which high rates of sulfur deposition occur are seen to shrink substantially.

Computing Emissions of Sulfur Gases

As an example of the level of detail required, consider the RAINS computation of emissions of sulfur. A separate computation is performed for each sulfur-containing fuel, using the relationship

$$S_{i,j,k}(t) = \sum_l E_{i,j,k,l}(t) \frac{sc_{i,j,k}}{Hv_{i,j}} (1 - sr_{j,k})(1 - x_{i,k,l}) \qquad (16.9)$$

where the symbols have the following meanings:

E	Energy use
Hv	Heat value of fuel
sc	Sulfur content
sr	Fraction of emissions retained in ash
x	Fraction of emissions removed by control devices
S	Sulfur emissions
i	Country

i   Fuel type
k   Economic sector
l   Abatement technology
t   Time

Following these calculations, the total emissions of sulfur per country are given by

$$S_i(t) = \sum_i \sum_K S_{i,j,k}(t) + S_i^p(t) . \qquad (16.10)$$

where S_i^p indicates the sulfur emission rates from noncombustion processes. Similar computational complexities apply to most of the processes indicated in Figure 16.5.

Figure 16.7 The distribution in the year 2040 of central European forest soils by pH classes for scenarios of 30% (solid bars) and 58% (open bars) reductions in emitted sulfur dioxide by the year 2000, as computed by the RAINS model. A, Austria; CH, Switzerland; CS, Czechoslovakia; D(E), Germany (East); D(W), Germany (West); H, Hungary; NL, Netherlands; P, Poland. The bar chart at the right gives the aggregate distribution for all countries shown; the vertical scale is percentage. (Reproduced with permission from J. Alcamo, et al., Acidification in Europe: A simulation model for evaluating control strategies, *Ambio, 16,* 232–245. Copyright 1987 by The Royal Swedish Academy of Sciences.)

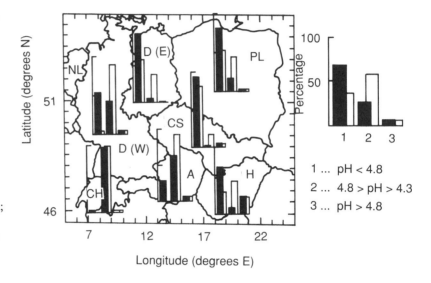

Soil Acidification

Because the RAINS model includes modules for computing the interactions between deposition and impacts, typical impact results are produced as part of the calculation. Soil acidification is one such impact, important because the capacity of soil to buffer acidic deposition is a key factor in regulating the chemistry of lake water and groundwater. Soil acidification has also been linked, with less certainty, to forest damage.

Among the data in the model are chemical information on European soil types and soil thicknesses. To compute soil acidity, the model compares the cumulative load of deposited sulfate to the individual soil buffer capacities. A selection of soil acidity results for the year 2040 is given in Figure 16.7, for two different computations. In one of these, sulfur emissions throughout Europe were reduced by 30%; in the other, by 58%. The resulting percentages of soil in three different acidity ranges in several countries are plotted in the figure. Examine first the results for the western part of Germany (left center). For the 30% reduction scenario, about half of the surficial soil is predicted to be highly acidified (pH < 4.0). For 58% reductions, most of the soil area is predicted to be of intermediate acidity. In Switzerland (lower left), the mountainous soils respond little to the changes in emissions. The soils of Poland (upper right) respond qualitatively as do those of the western part of Germany, but the contrast between the two emissions scenarios is less marked in Poland because of differences in soil chemistry, source locations, and atmospheric transport.

Lake Acidification

A second impact treated explicitly by the RAINS model is the acidification of lakes. This impact is related to the acidification of soils, because runoff and groundwater that are inadequately buffered by soil in the lake's catchment area can produce enhanced acidification of the lake itself. Lake acidification also depends on the amount of snowmelt, the flow paths of runoff, the chemistry of the lakes, and several other factors.

When these conditions are combined with the deposition and transport components of the overall calculation, lake acidification at designated times and places can be computed. Figure 16.8 shows an example, giving the distribution of pH classes in the lakes of Finland, Norway, and Sweden for the year 2000, for sulfur dioxide emission reductions of 30% and 58%. The calculations indicate that regions that are now heavily influenced by sulfur emissions, especially south-

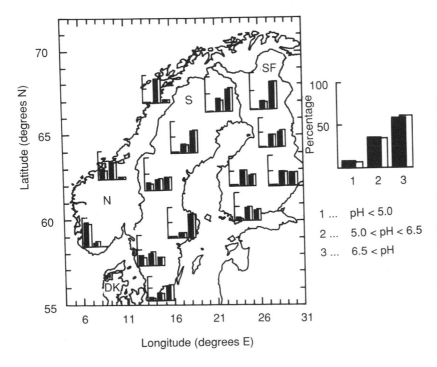

Figure 16.8 The distribution in the year 2000 of pH classes in Scandinavian lakes, as computed by the RAINS model. The sulfur emission reduction scenarios are those of Figure 16.7. DK, Denmark; N, Norway; S, Sweden; SF, Finland. The bar chart at the right gives the aggregate distribution for all countries shown; the vertical scale is percentage. (Reproduced with permission from J. Alcamo, et al., Acidification in Europe: A simulation model for evaluating control strategies, *Ambio, 16*, 232–245. Copyright 1987 by The Royal Swedish Academy of Sciences.)

ern areas of Sweden and Finland, can show modest improvements in the most severe lake acidification class if major sulfur controls are adopted. In most other regions of Scandinavia, little difference is present in the results from the two emission scenarios.

Other Impacts of Sulfur Deposition

RAINS also is capable of computing groundwater acidity as a function of space and time for different emission scenarios and of doing the same for direct forest damage impact. In the former case, variables such as soil type, aquifer size, mineral composition, and water available for recharge must be considered. In the latter case, direct foliar damage due to oxidants as well as SO_2, nitrogen overfertilization as a result of pollutant nitrogen deposition, and climatic factors in combination with natural stresses must be taken into account. Like the maps of soil and lake acidity, maps of impacts on these other environmental characteristics can be produced for different emission scenarios.

Summary

RAINS is of interest not only because it can be useful in guiding policy decisions on specific impacts of trace contaminants in the air, but also because it attempts to examine several environmental impacts in the same analysis. Such an approach emphasizes that many impacts are linked together and that few impacts have a single cause. Nonetheless, the present model deals in detail only with the effects of sulfur emissions (although deposition calculations for nitrogen compounds are also performed) and presents the results individually, without a convenient and efficient way to communicate the overall results of the analysis. That difficult but important task, and a method for accomplishing at least some of it, make up the remainder of this chapter.

AN ANALYTIC FRAMEWORK FOR ENSEMBLE ASSESSMENT

Air quality and its impacts are often discussed in fragmentary fashion, as if the sources for one atmospheric species bore no relationship to the sources for another and as if the air quality problems on a local or regional scale bore little relationship to problems on a global scale. The contrasting approach of ensemble assessment attempts to view impacts from a common framework so that they may be compared with each other and so that a perspective may be reached concerning the multiplicity of impacts. The design for such a framework seeks to establish the causal relationships between "critical environmental properties" and potential sources of environmental change. Critical environmental properties, in the sense used here, are those few attributes of the environment that parties involved in the assessment agree to be crucially important. Which environmental properties are

TABLE 16.1. Definitions of Critical Atmospheric Properties

CRITICAL PROPERTIES	DEFINITION
Ultraviolet radiation absorption	This property reflects the ability, especially of stratospheric ozone, to absorb ultraviolet solar radiation, thus shielding Earth's surface from its deleterious effects.
Greenhouse warming	This property reflects the balance among processes through which the atmosphere transmits much of the energy arriving from the sun at visible wavelengths while absorbing much of the energy radiated from Earth and its atmosphere at infrared wavelengths.
Photochemical smog (regional oxidants)	This property reflects the production of a variety of highly reactive and oxidizing gases, especially tropospheric ozone.
Precipitation acidification	This property reflects the acid–base balance of the atmosphere, surface water, and soils, as reflected in the chemical composition of rain, snow, and fog.
Visibility degradation	Visibility is reduced when light of visible wavelengths is absorbed or scattered by gases or particles in the atmosphere.
Oxidation (cleansing) efficiency of the atmosphere	This property reflects the efficiency with which gases emitted into the atmosphere are broken down by reactions with the hydroxyl radical or other oxidizers.
Materials corrosion	This property reflects the ability of the atmosphere to corrode materials exposed to it, often through the chloridation or sulfurization of marble, masonry, iron, aluminum, copper, and other materials.

valued in a particular case will depend upon specific social, political, and environmental circumstances. A clear lesson from the past two decades is that any analysis is unlikely to be useful for comprehensive policy-making unless some definite—and preferably short—list of critical environmental properties is specified as a focus for assessment.

For purposes of the present discussion, we have selected seven critical atmospheric properties (see Table 16.1). Six of these have been discussed earlier. The seventh, materials corrosion, is included here because it is a societal problem of enormous proportions, both financially and culturally. In the United States alone, it has been estimated that the annual cost of corrosion, much of which is related to atmospheric exposure, is 30 billion dollars annually. Examples of these processes include practical issues such as rusted bridge supports and esthetic issues such as deteriorating statuary. The principal atmospheric corrodents are sulfur- and chlorine-containing species, and the most common susceptible materials are stone (marble, limestone, masonry) and the "engineering metals": steel, copper, zinc, aluminum, lead, and tin. Materials such as polymers, fabrics, photographic film, and paper are often susceptible as well, particularly as they can be exposed to corrosive environments even when indoors.

The goal of the framework we wish to develop is to understand the relationships between the seven critical properties of the atmosphere and the natural fluctuations and human activities that might be sources of significant change in them. The present knowledge regarding the critical atmospheric properties affected by changes in specific atmospheric chemicals is given qualitative expression in Figure 16.9, where both direct and indirect chemical effects are indicated. Thus, changes in ozone concentrations are shown to directly affect the critical atmospheric component "ultraviolet energy absorption" because the ozone molecules themselves have the ultimate impact. Halocarbons and nitrous oxide, although surely relevant to ultraviolet radiation absorption, are shown to affect this critical atmospheric component

Figure 16.9 Major impacts of atmospheric chemistry on critical atmospheric properties. The squares indicate that the listed chemical is expected to have a significant impact on the listed property of the atmosphere. Solid squares indicate that the impact is direct, hatched squares that it is indirect. Definitions of the atmospheric properties are given in Table 16.1.

Chemical constituents	Ultraviolet radiation absorption	Greenhouse warming	Photochemical smog	Precipitation acidity	Visibility degradation	Materials corrosion	Oxidation efficiency
C (soot)		■			■		
CO_2		■					
CO			▨				■
CH_4		■	▨				■
C_xH_y			■		▨		
NO_x			■	■	■		■
N_2O	▨	■					
NH_3/NH_4^+				■	▨		
SO_2		▨		■	■	■	
H_2S						■	
COS		▨				■	
Organic S						▨	
Halocarbons	▨	■					
HCl						■	
O_3	■	■	■				■

indirectly, because they act by changing the concentration of ozone. It is important to note from Figure 16.9 that a significant number of chemicals have multiple impacts. Sources of disturbance or international policies that affect these chemicals must therefore be assessed in terms of these multiple impacts on the atmosphere.

The principal sources of emissions to the atmosphere have been discussed in previous chapters. Present knowledge regarding the emission of specific chemicals is expressed qualitatively in Figure 16.10. The figure shows that the levels of a number of atmospheric chemicals are affected by emissions from many different types of sources.

Figures 16.9 and 16.10 can be combined to provide an overall analytic framework for atmospheric assessment. One begins with a critical atmospheric component like precipitation acidification and its direct and indirect chemical causes (Figure 16.9), and identifies the sources responsible for initiating those interactions (Figure 16.10). The result is Figure 16.11, a matrix that shows the impact of each potential source of atmospheric change on each critical atmospheric component. The assessment is qualitative, reflecting to some extent the present state of knowledge. It also includes estimates of the reliability of that knowledge, an important component of such an assessment effort.

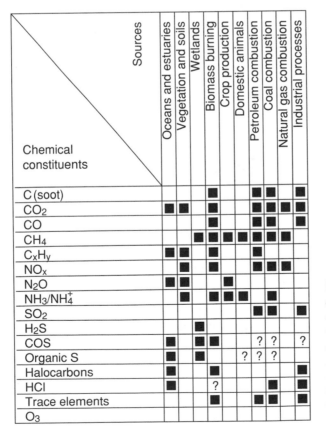

Chemical constituents	Oceans and estuaries	Vegetation and soils	Wetlands	Biomass burning	Crop production	Domestic animals	Petroleum combustion	Coal combustion	Natural gas combustion	Industrial processes
C (soot)				■			■	■		■
CO_2	■	■		■			■	■	■	■
CO				■			■	■		■
CH_4			■	■	■	■	■	■	■	
C_xH_y	■	■		■			■			
NO_x		■		■			■	■	■	
N_2O	■	■				■				
NH_3/NH_4^+		■		■	■	■		■		
SO_2							■	■		■
H_2S			■							
COS	■	■	■					?	?	?
Organic S	■	■					?	?	?	
Halocarbons	■			■						■
HCl	■				?			■		■
Trace elements				■			■	■		■
O_3										

Figure 16.10 Sources of major disturbances to atmospheric chemistry. The squares indicate that the listed source is expected to exert a significant effect on the listed chemical. A question mark indicates that emission of a significant amount of a certain compound from a certain source in uncertain.

Figure 16.11 An initial ensemble assessment of impacts on the global atmosphere. The critical atmospheric properties defined in Table 16.1 are listed as the column headings of the matrix. The sources of disturbances to these properties are listed as row headings. Cell entries assess the relative impact of each source on each component and the relative scientific certainty of the assessment. The significance of the letters in the cells is described in the text.

The reasons for each of the impact assessments pictured in Figure 16.11 can be summarized as follows:

1A Release of CH_3Cl and N_2O and thus the influence on stratospheric ozone chemistry via NO_x and ClO_x catalytic cycles.

1F Release of corrosive halogen and sulfur gases.

2A Release of N_2O (see 1A).

2B Release of N_2O; serves as greenhouse gas.

2C Release of nonmethane hydrocarbons (NMHC) in NO-rich environments.

2D Release of NMHC, followed by oxidation to organic acids.

3A   Release of CH_4, which plays a chemical role in stratospheric ozone chemistry via effects on ClO_x and HO_x catalytic cycles.

3B   Release of CH_4; serves as greenhouse gas.

3F   Release of corrosive sulfur gases.

3G   Release of CH_4 and its influence on tropospheric HO•.

4A   Release of CH_4 (see 3A).

4B   Release of CO_2 (deforestation) and CH_4.

4C   Release of NMHC and NO_x and their enhancement of tropospheric ozone production.

4D   Release of NO_x followed by oxidation to nitric acid.

4E   Release of smoke.

4G   Release of CH_4 and CO and their influence on tropospheric HO•.

5A   Release of N_2O (see 1A).

5B   Release of N_2O and CH_4.

5G   Release of CH_4 and its influence on tropospheric HO•.

6A   Release of CH_4 (see 3A).

6B   Release of CH_4; serves as greenhouse gas.

6D   Release of NH_3, an alkaline substance.

6E   Release of NH_3 and incorporation in aerosol.

6G   Release of CH_4 and its influence on tropospheric HO•.

7A   Venting of CH_4 during petroleum production.

7B   Release of CO_2 and CH_4.

7C   Release of NMHC and NO_x (see 4C).

7D   Release of SO_2 and NO_x followed by oxidation to sulfuric and nitric acid.

7E   Release of SO_2 and soot.

7F   Release of SO_2 and soot.

7G   Release of CH_4, CO, and NO_x and their influence on tropospheric HO•.

8A   Release of CH_4 during coal mining.

8B   Release of CO_2 and CH_4.

8C   Release of NMHC and NO_x (see 4C).

8D Release of SO_2 and NO_x followed by oxidation to sulfuric and nitric acid.

8E Release of SO_2 and soot.

8F Release of SO_2 and soot.

8G Release of CH_4, CO, and NO_x and their influence on tropospheric $HO\cdot$.

9A Release of CH_4 during production and distribution (see 3A).

9B Release of CH_4 during production and distribution.

9G Release of CH_4 during production and distribution and its influence on tropospheric $HO\cdot$.

10A Production of CFC and halon gases and release of NO_x and H_2O from aircraft, and thus the destruction of ozone by ClO_x, BrO_x, NO_x, and HO_x catalytic cycles.

10B Release of CFCs and of NO_x and H_2O from aircraft.

10C Release of NMHC and NO_x (see 4C).

10D Release of SO_2 and NO_x followed by oxidation to sulfuric and nitric acid.

10E Release of SO_2 and soot.

10F Release of SO_2 and soot.

10G Release of CH_4, CO, and NO_x and their influence on tropospheric $HO\cdot$.

The simplest atmospheric impact assessments involve only a single cell of the matrix. A typical example is the study of the impacts of a single source, such as a new coal-fired power station, on a single critical atmospheric component, such as precipitation acidification (location *a* in Figure 16.11). More complex atmospheric assessments have addressed the question of aggregate impacts across different kinds of sources. A contemporary example is the study of the net impact on Earth's thermal radiation budget caused by chemical perturbations due to fossil fuel combustion, biomass burning, land-use changes, and industrialization (e.g., locations *b* in Figure 16.11). An alternative approach, especially useful for the purposes of policy and management, is to assess the impacts of a single source on several critical atmospheric properties. The coal combustion study noted above would fall into this category if the impacts were assessed not only on acidification but also on photochemical oxidant production, materials corrosion, visibility degradation, and so on (e.g., locations *c* in Figure 16.11). If desired, the columns could be summed in some way to give the net impact of the ensemble of sources on each critical property. Similarly, the rows could be summed in some way to give the net effect of each source on the ensemble of properties.

Figure 16.11 shows that the sources of most general concern, as indicated by their impact ratings, are almost wholly anthropogenic in nature: fossil fuel combustion (especially coal and petroleum), biomass combustion, and industrial processes. Emissions from crop production, especially CH_4 from rice paddies, also have significant effects on climate and atmospheric chemistry.

Among the most troublesome interactions between development and environment are those that involve cumulative impacts. In general, cumulative impacts become important when sources of perturbation to the environment are grouped sufficiently closely in space or time that they exceed the natural system's ability to remove or dissipate the resultant disturbance. The basic data required to structure such assessments are the characteristic time and space scales of the atmospheric constituents and development activities (see Figure 15.1). For example, perturbations to gases with very long lifetimes accumulate over decades to centuries around the world as a whole. Today's perturbations to those gases will still be affecting the atmosphere decades or centuries hence, and perturbations occurring anywhere in the world will affect the atmosphere everywhere in the world. Long-lived emittants tend to be radiatively active, thus giving the greenhouse syndrome its long-term, global-scale character. At the other extreme, heavy hydrocarbons and coarse particles, being short lived, drop out of the atmosphere in a matter of hours, normally traveling a few hundred kilometers or less from their sources. The atmospheric properties of visibility reduction and photochemical oxidant formation associated with these chemicals thus take on their acute, relatively local or regional character. Species with moderate atmospheric lifetimes include gases associated with the acidification of precipitation and fine particles, all with characteristic scales of a few days and a few thousand kilometers.

These concerns, together with the evidence that the concentrations of many chemical compounds in the biosphere are increasing, indicate that additional extensions to the conceptual framework are needed to provide it with spatial and temporal dimensions. The effort described earlier was focused on present-day impacts. One can envision, however, preparing separate versions of the framework shown in Figure 16.11 for global, regional, and local interactions, and for different epochs in time. As suggested in Figure 16.12, this chronology might consist of multiple versions of Figure 16.11 created to reflect "slices in time" through the evolving conditions at 25- or 50-yr intervals. The result, it is hoped, will help to put the changing character of interactions between human activities and the environment into a truly synoptic historical and geographical perspective.

Properties

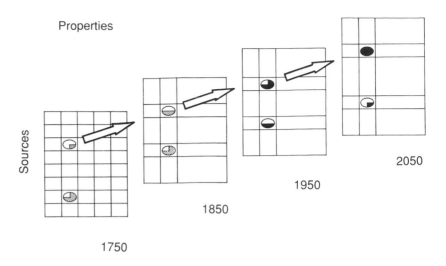

Sources

1750

1850

1950

2050

Figure 16.12 A history of disturbances to the atmosphere, expressed through a time series of source-impact matrices such as that shown in Figure 16.11. This display suggests how two of the matrix elements would be evaluated at each of the "time slices," perhaps becoming more or less significant with time, as indicated here. The full assessment would include such an evaluation for each individual matrix element at each time slice.

EXTENDING THE ASSESSMENT TO DIFFERENT REGIMES

It is important to recall at this point that the previously presented assessments have been limited to impacts of development on the atmosphere. In fact, other regimes in the biospheric system are important as well. For example, the concern about the impact of acidification through time is based not on the accumulation of relevant chemicals in the atmosphere but on their accumulation in other media, such as soil and water. Expanding synoptic assessment frameworks to contend with additional environmental and developmental dimensions is a major part of sustainable development research. Beginnings have been made for water systems and soil systems, but much is yet to be done.

A relatively straightforward addition to the atmospheric framework would be the inclusion of one or more critical environmental properties that reflect the role of atmospheric chemicals as direct fertilizers or toxins for plants. This modification would allow the integrated treatment of phenomena such as the stimulation of plant growth by carbon dioxide and its inhibition by sulfur oxides—both products of fossil fuel combustion. Somewhat more ambitiously, the approach could be expanded beyond its present chemical focus to include the appropriate physical and biological processes and the sources of disturbance to them. Ultimately, the need is for a qualitative framework that puts in perspective the impacts of human activities and natural fluctuations, not just on the atmospheric environment, but also on soils, water, and the biosphere as a whole.

Although a comprehensive treatment has yet to be made for other regimes, it is possible to envision one way in which such an assessment might be accomplished. The first step would be to select regimes of interest other than the atmosphere, such as oceans or soils. The second step would be to use the same analytic approach taken for the atmosphere to determine the matrix elements on a source → impact diagram similar to that of Figure 16.11. A possible example for soils is shown in Figure 16.13. The matrix elements are then summed in some appropriate fashion to give a total regime impact for each source; these appear in the right column of Figure 16.13. The same sequence of operations is repeated for each regime of interest, not only for the present and perhaps for the past, but for future times as well.

In the final ensemble display, the total impact columns from each regime display for each time slice are extracted and combined, producing a display of the type shown in Figure 16.14. Such a diagram could be used to illustrate progress or retrogression over time.

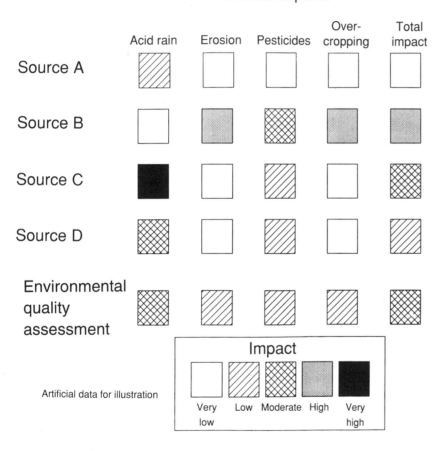

Figure 16.13 A concept for synoptic assessment of impacts on global soils. The data are artificial and appear for didactic purposes only. This figure for soil impacts is analogous to Figure 16.11 for atmospheric impacts.

EMISSION PROJECTIONS

The comprehensive assessment techniques presented in Figures 16.9–16.14 have not been utilized in any consistent and rigorous way thus far, not even in the case of the atmosphere. To indicate the potential usefulness of some of these approaches, however, we have completed a simpler operation: that of assessing for the present epoch, and projecting, under a specified set of development assumptions, the emissions (rather than the impacts) of a number of trace gases as a consequence of electrical energy generation. We selected seven countries, spanning the range of highly to slightly developed, searched out their present populations and electrical generating capacities, and made the perhaps not unrealistic assumption that by the year 2025 every citizen in all countries would be provided with electrical energy at half the rate as today's average Italian, except that no country will undergo a decrease in per capita energy in order to achieve that level. Some of the pertinent figures involved in making this assessment are given in Tables 16.2 and 16.3. Two items are of special note. One is the great disparity in energy use per capita, a multiplier of about 100 separating the high and low extremes. The second is the truly awesome population increases anticipated in several of the countries, populations two or three times those of today being forecast for India, Kenya, and Nigeria only 35 yr from the present.

Figure 16.14 A concept for ensemble assessment of the impacts over time of sources related to biospheric development on several different environmental regimes. The data are artificial and appear for didactic purposes only. This figure is an extension across regimes of Figure 16.12.

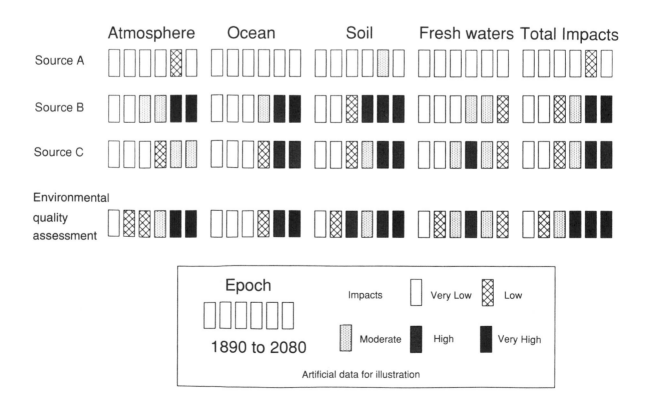

TABLE 16.2. Selected Country Statistics—1984[a]			
COUNTRY	ENERGY USE (Billion kWh)	POPULATION (millions)	ENERGY USE (MWh/person)
USA	2413	237.2	10.17
China	327	1034.9	0.32
Italy	161	57.0	2.82
Brazil	152	134.4	1.13
India	138	746.4	0.18
Nigeria	7.5	88.1	0.09
Kenya	1.7	19.4	0.09

a. Energy use data and 1984 populations are from *The World Almanac and Book of Facts*, Newspaper Enterprise Assn., New York, 1986.

TABLE 16.3. Prospective Country Statistics—2025[a]		
COUNTRY	2025 POPULATION (millions)	POP. INCREASE, 1984–2025 (%)
USA	296	25
China	1493	44
Italy	69	21
Brazil	235	75
India	1445	94
Nigeria	303	244
Kenya	80	312

a. Population projections are from N. Keyfitz, The growing human population, *Scientific American, 261* (3), 119–126, 1989.

The computation of emissions intensities as a consequence of electrical energy generation in each of the countries is given for each species i by

$$E_i = \frac{\left(\sum_j [M_j \cdot \epsilon_{i,j}] \right)}{A} , \qquad (16.11)$$

where M_j is the annual use of fuel j in the country, ϵ_{ij} is the overall average emission factor (grams of emittant i per gram of fuel j), and A is the country area. Electricity is generated almost entirely by a mix of five processes: coal combustion, petroleum combustion, natural gas combustion, nuclear power, and hydropower. The latter two processes emit no species to the atmosphere, with the possible exception of wetland methane emissions should new hydropower water reservoirs be created in the tropics. The year 2025 assessment requires adopting projections of population growth, per capita energy consumption, and the fractions of the needed electrical energy supplied by the different energy sources. We assume that hydropower will provide as much electricity as it is now providing, but no more, that nuclear power will provide for each country 150% of its current capacity by 2025, and that supplies of petroleum and natural gas will begin to diminish and to increase in cost by 2025.

To display the results of this exercise, we reflect the emissions intensities for the countries on a world map with the intensities hatched according to a logarithmic scale. The display for carbon dioxide, a product of all combustion of fossil fuel, is given in Figure 16.15. Carbon dioxide is of particular concern, of course, as a major component of the greenhouse effect. The 1980 diagram indicates that Italy has the highest CO_2 emissions intensity among the seven countries, even though the United States consumes significantly more energy per capita. India and China have intermediate emissions intensities (but note again that the scale is logarithmic, so the actual values are rather low), and the emissions intensities of the African and South American countries are low indeed.

The same analysis for the year 2025 gives quite different results. Italy retains its high ranking but shares it with India and Nigeria. Kenya and China are projected to have emissions intensities nearly as high. The United States and Brazil are high as well, but the intensity is restrained in those countries by the relatively large land areas per capita. It is clear that if actual emissions intensities prove to be anything like those anticipated by the scenario used here, the populous countries of Asia and Africa will have high emissions intensities by the end of the twenty-first century's first quarter and will be producing a large portion of the planet's total CO_2 emissions.

The maps for NO_x, largely a product of high temperature petroleum combustion, are shown in Figure 16.16. Nitrogen oxides, as we have seen, have impacts on a local scale, especially the generation of photochemical smog. The 1980 analysis shows Italy and the United States at the high end of the NO_x emissions intensities, with the other countries being moderately or very low. The 2025 picture is quite different, with increased energy demand pushing NO_x emissions in India, Nigeria, Kenya, and China to relatively high levels. In those countries, photochemical smog during sunlit summer months may become a severe problem.

Finally, Figure 16.17 depicts emissions intensities in the seven countries for sulfur dioxide. This pollutant is implicated in acidic precipitation and reduced visibility, among other effects, and is generated by the combustion of sulfur-containing fossil fuel, both coal and petroleum. The 1980 assessment shows Italy with high emissions intensity, most of the other countries at modest levels, and the African countries at low levels. In 2025, with much energy being supplied by readily available coal, the prediction is for very high SO_2 emissions intensities in Italy, Nigeria, and India and for high intensities in China

Figure 16.15 Carbon dioxide emission intensities as a result of electrical energy generation for selected countries for 1984 and 2025, assuming that the supply of per capita energy in 2025 has remained constant for the United States and Italy and has grown elsewhere to be equal to half that of Italy in 1984.

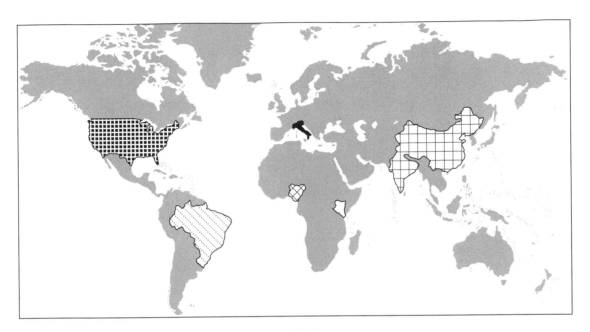

a 1980

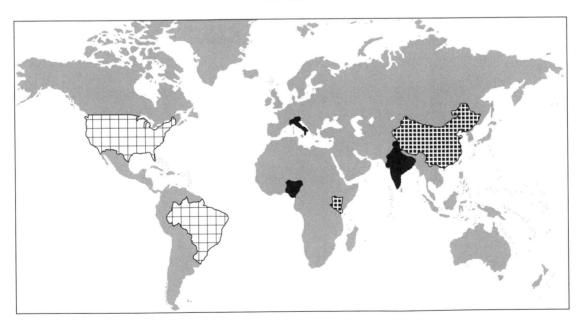

b 2025

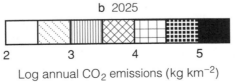

Log annual CO_2 emissions (kg km^{-2})

Figure 16.16 Nitrogen dioxide emissions intensities as a result of electrical energy generation for selected countries for 1984 and 2025, assuming that the supply of per capita energy in 2025 has remained constant for the United States and Italy and has grown elsewhere to be equal to half that of Italy in 1984.

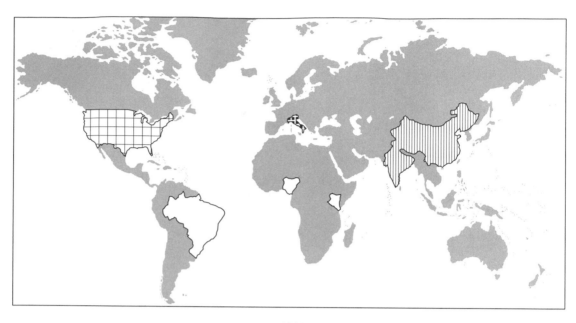

a 1980

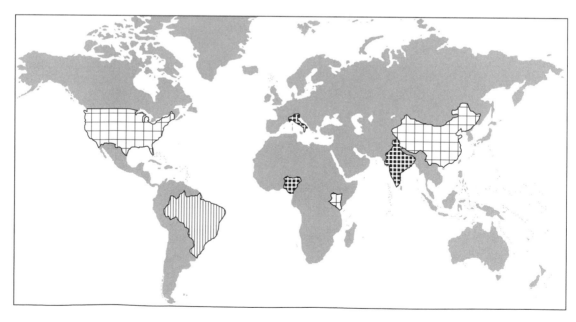

b 2025

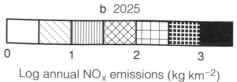

Log annual NO_x emissions ($kg\ km^{-2}$)

Figure 16.17 Sulfur dioxide emissions intensities as a result of electrical energy generation for selected countries for 1984 and 2025, assuming that the supply of per capita energy in 2025 has remained constant for the United States and Italy and has grown elsewhere to be equal to half that of Italy in 1984.

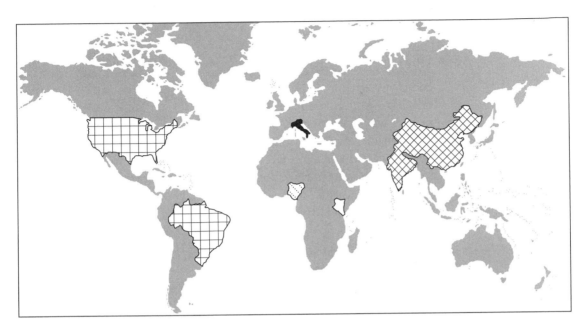

a 1980

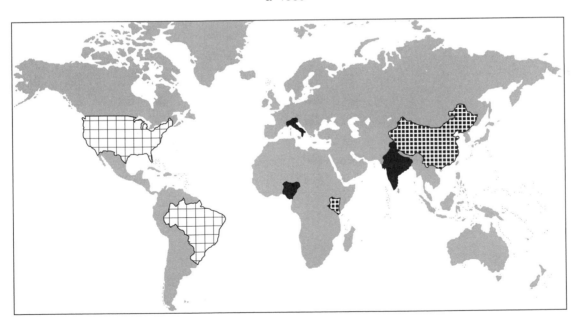

b 2025

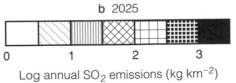

0   1   2   3

Log annual SO_2 emissions (kg km^{-2})

and Kenya. Brazil and the United States will be somewhat lower, aided by their large land areas, although the sources, of course, are not uniformly distributed. There is definite potential in much of Asia and Africa, particularly, for increases in corrosion and degradation of monuments, sculptures, and buildings, for acidification of precipitation and hence of lakes, and for decreases in visibility.

An alternative display for the information in Figures 16.15–16.17 is given in Figure 16.18, which is a variation of Figure 16.14 with impacts being replaced by emissions intensities and the seven countries serving as the sources. On this display, the transition of the developing countries from modest to major emission sources over the next 35 yr is clearly evident for emittants having a variety of impacts and spatial scales. It is inappropriate to place much reliance on the details of this simple reference case analysis, which was selected as much for its simplicity as for its potential accuracy. Nonetheless, its principal message, and that of more complex analyses, is that increased energy use, especially if it involves the combustion of large amounts of coal, seems likely to produce global impacts from CO_2 emissions and local to regional impacts from SO_2, NO_x, and other emissions. Although the impacts will be most severe in the countries with the highest emissions intensities, some impacts, such as those caused by CO_2, will be detectable all across the globe.

Figure 16.18 Ensemble assessment of emissions intensities as a result of electrical energy generation for selected countries in 1984 and 2025.

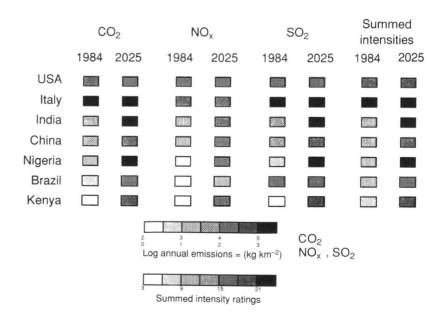

Energy generation involving the combustion of fossil fuels results in the generation and emission of a number of trace species to the atmosphere. These species are of various reactivities (hence lifetimes) and manifest their impacts on all spatial scales from local to global. Such an assessment does not mean that development should not occur; rather, it means that its impacts on planetary processes should be studied and minimized.

The spectrum of impacts indicates that neither individual trace species nor individual impacts should be treated in isolation but that the ensemble of emittants and impacts should be analyzed in a single concerted approach. We have indicated in this chapter one possible technique for such an analysis. Many variations of this approach are possible, but the importance of recognizing the concept seems paramount. Only by such approaches can the development that will inevitably take place be made to do so in a sustainable fashion.

EXERCISES

16.1 The morning rush-hour concentrations of NO_x and NMHC on an average day in Sydney, Australia are 40 ppbv and 1.0 ppmv, respectively. Using Figure 16.1, determine what concentration of ozone would be expected. On a day with little air movement, the same species have concentrations of 100 ppbv and 2.0 ppmv. Under those conditions, what ozone concentration is expected? If emissions evolve so that average morning rush-hour concentrations are 100 ppbv NO_x and 0.6 ppmv NMHC, will ozone concentrations be higher or lower than in the first case above? Why? What if average concentrations become 200 ppbv NO_x and 0.6 ppmv NMHC? Why?

16.2 Within a city square, traffic flow is as follows:

TIME	CARS/HR	TRUCKS/HR
6 AM	1,600	200
7 AM	2,500	350
8 AM	5,600	730
9 AM	8,200	970
10 AM	12,100	1,400
11 AM	15,500	2,300
12 Noon	18,000	3,000

Thirty percent of the vehicles have older emission controls, 70% newer. The square is 0.4 km across, and the average vehicle traverses two sides of the square before leaving it. The average automotive emission factors are CO (old) = 9.4 g km⁻¹; CO (new) = 2.1 g km⁻¹; NO_x (old) = 1.3 g km⁻¹; NO_x (new) = 0.6 g km⁻¹. For trucks, the factors are: CO (old) = 12.5 g km⁻¹; CO (new) = 6.3 g km⁻¹; NO_x (old) = 2.0 g km⁻¹; NO_x (new) = 1.1 g km⁻¹. Compute the emission fluxes of CO and NO_x to the square for each hour between 6 AM and noon.

16.3 Assume that the emission fluxes of CO and NO_x from Exercise 16.2 are dispersed into the atmosphere over the city square. Ninety percent are lost, but 10% remain. Each hour the buildup of pollutants is advected from the square and the slow buildup begins again. Because mixing is inhibited, the boundary layer is 1.0 km in depth. Compute the concentrations in ppmv of CO and NO_x each hour, assuming there is no loss to surfaces or to chemical reactions and that the concentrations within the volume of air over the square are uniform. Compare these numbers with the health standards of the World Meteorological Organization for 1-hr average concentrations: CO, 24 ppmv; NO_x, 0.25 ppmv.

16.4 Although damage to stone monuments by environmental factors almost certainly involves a number of different chemical species, there is general agreement that the amount of rain and the sulfur dioxide concentrations are highly significant. A study in the British Isles reported the following relationship between rain amounts, SO_2 concentrations, and loss of surface from a particular type of marble in urban locations:

$$L = 25.7 + 0.102 \, [SO_2]$$

where L is the loss rate in micrometers per meter of rain and SO_2 concentrations are in micrograms per cubic meter. A new statue, having surface features carved to an average depth of 3 mm, is to be placed in a Copenhagen park. Estimate how long it will be before the surface features are completely lost if the statue is placed in the park in 1980, 2030, and 2080, using hypothetical but not unreasonable data from the table on the next page. Repeat the calculation for New York City and Calcutta.

Precipitation and Sulfur Dioxide in Three Cities

CITY	ANNUAL PRECIP. (cm yr^{-1})	SO$_2$ CONCENTRATION (ppbv)		
		1980	**2030**	**2080**
Copenhagen	60	14	21	25
New York City	145	17	20	18
Calcutta	88	16	54	83

16.5 You wish to design a generating station to provide 100 MW of power, and you can utilize either coal, petroleum, or natural gas. The energy contents of the fuels are 6×10^3 cal g^{-1} (coal), 10.5×10^3 cal g^{-1} (fuel oil), and 13.2×10^3 cal g^{-1} (natural gas). What mass flows of each fuel would be needed? Compute the emission fluxes of CO_2, CO, NO_x, CH_4, SO_2, and HCl for each alternative, using the following table of emission factors.

Emission Factors[a] for Fossil Fuel Use

FUEL	CO$_2$	CO	CH$_4$	NO$_x$	SO$_x$	HCl
Coal	2.71	0.001	0.005	0.010	0.031	0.002
Petroleum	2.84	0.07	0.009	0.011	0.002	
Nat. Gas	2.90		0.01	0.004		

a. Values are grams of emitted gas per gram of fuel consumed.

16.6 Choose a country in Africa, Asia, or South America. From sources of your choosing and documentation, estimate the present-day population and per capita energy use, and determine reasonable projected values of population and per capita energy use in the year 2025. (Use Tables 16.2 and 16.3 as examples.) How do you think the needed energy will be supplied? Using the fossil fuel emission factors above, compare the emissions into today's atmosphere with those in 2025. Discuss the limitations of your assessments, including which of the limitations could be resolved by using more detailed research. How will any solid waste that is generated as a consequence of energy generation be disposed of? If you were prime minister, responsible for maintaining both the economy of the country and its environment, what actions would you take? Why?

16.7 Total motor vehicle registrations in 1980 in millions for the seven countries indicated in Figures 16.15 to 16.18 were United States, 155.9; Italy, 19.1; India, 2.0; China, 1.0; Nigeria, 1.1; Kenya, 0.2, Brazil, 11.6. Assume that the populations of these countries in 2025 will be those given in Table 16.3, that all motor vehicles will still use petroleum combustion for motive power, and that the new vehicle emission factors of Exercise 16.2 will decrease by 50%. Make reasonable projections for the growth of the motor vehicle populations in the different countries, discuss your reasoning or source of information briefly, and construct figures for 1980 and 2025 for CO_2, CO, CH_4, NO_x, and SO_2 from motor vehicles similar to those of Figures 16.15 to 16.17.

FURTHER READING

J. Alcamo, et al., Acidification in Europe: A simulation model for evaluating control strategies, *Ambio, 16*, 232-245, 1987.

W. C. Clark and R. E. Munn, eds., *Sustainable Development of the Biosphere*, Cambridge University Press, Cambridge, UK, 491 pp., 1986.

J. L. Helm, ed., *Energy: Production, Consumption, and Consequences*, National Academy Press, Washington, D.C., 296 pp., 1989.

J. W. Tester, D. O. Wood, and N. A. Ferrari, eds., *Energy and the Environment in the 21st Century*, MIT Press, Cambridge, MA, 1991.

CHAPTER 17

Global Futures

Many atmospheric research teams are hard at work constructing and running global climate models (GCMs) of various degrees of complexity to make predictions concerning the future of the Earth's atmosphere and climate. These models are both the basis for and the result of concern about the future concentrations of carbon dioxide and several other greenhouse gases, stratospheric and tropospheric ozone, the effects of chlorofluorocarbons, and so forth, and the resulting impacts, including surface temperature changes, increases in ultraviolet solar radiation at ground level, and the like. Because the chemistry is so intertwined, it is, in principal, possible for a single model to make predictions concerning many of the potential problems. In fact, without including in the model at least an approximation of the essential chemistry, a model is unlikely to be valid for looking at any single problem. However, model construction and analysis activities have tended to emphasize only a single attribute of the atmosphere rather than to provide comprehensive assessments. From a policy point of view, it is obviously desirable for scientists to recommend actions that will achieve the optimum balance among all the environment's problems while at the same time allowing for sustainable development of the planet and its resources. It is, however, difficult or impossible for such recommendations to be made in a convincing manner with the sophisticated but limited tools available. We will return to this point in the chapter summary.

In the next several pages, we present representative results of computer models that address the future of specific air quality parameters on a global basis. In one or two instances, the impacts of such changes are also presented. These examples have been chosen to illustrate the sorts of results that are possible with global models, not because they are the "best" or the "most fully certified"; one can only assert that the models and results that we present are among those that are currently well regarded and are being used to provide input to policy decisions.

STRATOSPHERIC OZONE REDUCTIONS FROM INCREASED CFCs

The potential depletion of stratospheric ozone has the potential to change the biological makeup and perhaps the climate of the planet. As a consequence, it has been among the most extensively investigated impacts and its possible evolution with time has been studied by computer model techniques. Both 1D and 2D models have been used, and the results, from the standpoint of usefulness to policy makers, are consistent. We present here the results of one such model with which we are familiar, that of Brühl and Crutzen of the Max-Planck-Institute for Chemistry in Germany, who consider in their study the consequences for stratospheric ozone of a number of different emission scenarios. The computer model scenarios assume specified increases with time of the following species: carbon dioxide, methane, nitrous oxide, carbon monoxide, and NO_X ($NO + NO_2$). Chlorofluorocarbon (CFC) emissions are treated separately in several scenarios. Scenario I assumes continued CFC production at 1974 levels. Scenario II assumes the CFC production restrictions specified in the Montreal Protocol of 1987: a 50% reduction for CFC-11 and CFC-12 by 1998. Scenario III assumes the CFC production restrictions specified in the 1990 London amendments to the Montreal Protocol: For CFC-11 and CFC-12, the maximum allowed percentages of 1986 production are 80 (1993), 50 (1995), 15 (1997) and 0 (2000). In addition, production of methyl chloroform (CH_3CCl_3) and carbon tetrachloride (CCl_4) is to cease by 2005.

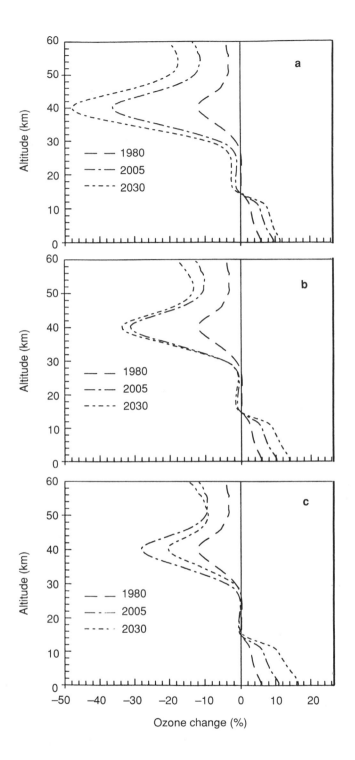

Figure 17.1 Percentage ozone changes in the southern hemisphere (referenced to the year 1965), computed with a $1^1/_2$ D model for three different chlorofluorocarbon emission scenarios. (a) Scenario I, assuming constant CFC emissions at 1974 levels. (b) Scenario II, assuming CFC emission reductions as specified in the Montreal Protocol of 1987. (c) Scenario III, assuming CFC emission reductions as specified in the 1990 London revisions to the Montreal Protocol. (C. Brühl and P. J. Crutzen, Scenarios for possible changes in atmospheric temperatures and ozone concentrations due to man's activities as estimated with a one-dimensional coupled photochemical climate model, *Climate Dynamics, 2*, 173–203, 1988.)

The ozone depletion percentages calculated for the three scenarios are shown in Figure 17.1. Because significant anthropogenic perturbations of the global atmosphere are recent phenomena, the results for years prior to 1965 are essentially those of the natural atmosphere and represent a balance between stratospheric ozone and natural emissions of nitrous oxide, methyl chloride, and other species involved in stratospheric chemistry.

The Scenario I results demonstrate that, if emissions of CFCs were to continue indefinitely at 1974 rates, maximum local ozone reductions at 40 km altitude would be as much as 30% by the year 2005 and nearly 50% by the year 2030. If the Montreal Protocol were fully implemented (Scenario II), it would have little effect on the situation in the year 2005, because the reductions would just have begun; however, little additional degradation would be anticipated between the years 2005 and 2030. For the London revision scenario (III), little effect would be seen at 2005, but by the year 2030 significant improvement would be evident. Because model results using different scenarios and different computational approaches can produce rather different results, the crucial aspect of the results for the present work is not the detailed quantitative agreement among models, but the fact that the predicted column ozone reductions over the next century suggest very substantial negative impacts toward the end of that period if significant CFC emission reductions do not occur as scheduled.

Recent studies on ozone chemistry in Antarctica (discussed in Chapter 8) have added considerable complexity to the picture of atmospheric ozone generation. The interactions of ice crystals with the ozone chemistry of the lower stratosphere have not yet been included in the model described here, so the results given in Figure 17.1 cannot be regarded as numerically of great accuracy. Nevertheless, model calculations of this kind have played a major role in convincing governments and industrial leaders that worldwide emissions of fully halogenated CFCs should be halted within the decade, and the results of future improved models will doubtless have similar impact.

REDUCTIONS IN THE SELF-CLEANSING POWER OF THE ATMOSPHERE

We have already noted several times (e.g., Chapter 8) that the hydroxyl radical is the crucial cleansing agent in the atmosphere and that a decrease in the concentration of HO$\cdot$ will therefore result in an increase in the concentrations of most trace gases emitted into the atmosphere. It is therefore worth looking more carefully into the sources and sinks of HO$\cdot$.

Reductions in the Self-
Cleansing Power
of the Atmosphere

375

The primary formation mechanism for is HO• the high-energy photolysis of ozone:

$$O_3 + h\nu \, (\lambda < 310 \text{ nm}) \rightarrow O(^1D) + O_2 \quad (17.1)$$

$$O(^1D) + H_2O \rightarrow 2HO• \, . \quad (17.2)$$

Once created, the radicals participate in many reaction chains. The atmospheric removal of HO• varies markedly from location to location and season to season, but the most significant processes on a global basis are reactions with CO (about 70% of the global total) and methane (about 30% of the global total):

$$CO + HO• \rightarrow CO_2 + H• \quad (17.3)$$

$$CH_4 + HO• \rightarrow CH_3• + H_2O \, . \quad (17.4)$$

These reactions are also the main ones removing CO and CH_4 from the atmosphere. It is therefore highly significant that the concentrations of both methane and carbon monoxide are increasing with time, as shown in Chapter 14. This could lower global HO• concentrations, which in turn would enhance the CO and CH_4 concentrations, and thus provide a positive feedback system. With this important perspective on the atmosphere's self-cleansing ability, Anne Thompson of NASA and Ralph Cicerone, then of the National Center for Atmospheric Research, have performed calculations with a 1D model to investigate the results of long-term changes in CH_4 and CO emission rates on the concentration of HO•. Their computations include several different choices for rates of emission, because one cannot, of course, be sure that any given prediction is the correct one. An example of their results is shown in Figure 17.2, which gives the vertical concentration profiles for HO• below 2 km for three different years: 1970, 2005, and 2035, and for two different assumptions about NO_x concentrations. In the first case, the relatively low NO_x concentrations have little effect on the overall chemistry, and the added CH_4 and CO cause a reduction in HO• concentrations with time. In the second case, with higher NO_x concentrations, the picture becomes quite complex. It shows concentrations of HO• increasing with time at low altitudes and decreasing at higher altitudes. The reason for the high altitude decrease is again the methane and carbon monoxide reactions with the hydroxyl radical.

At the lower altitudes, however, NO_x-catalyzed O_3 and HO• production comes into play, with the result that ground level HO• is higher than at earlier times. The total column abundance of HO•, however, has still decreased substantially. As noted by Thompson and Cicerone, the very different chemical lifetimes of NO_x (days), CO (weeks to months), and CH_4 (about 10 yr), and the high spatial heterogeneity of emissions around the world, would require 2D or 3D model calculations to achieve a more comprehensive treatment.

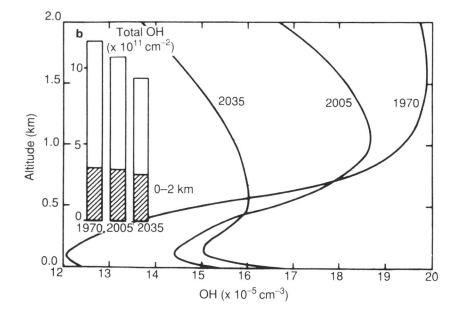

Figure 17.2 Vertical concentration profiles of the hydroxyl radical (HO•) within the lower 2 km of the atmosphere for (a) low and (b) high NO_x concentrations. The computer model assumes increased concentrations of methane and carbon monoxide with time. The bar graphs show the total column abundance, that is, the total abundance from the ground to the top of the atmosphere; the shaded portion of the bars show the portion of the column abundance contained in the lowest 2 km of the atmosphere. (A. M. Thompson and R. J. Cicerone, Possible perturbations to atmospheric CO, CH_4, and OH, *Journal of Geophysical Research, 91*, 10853–10864, 1986.)

ANTHROPOGENIC ENHANCEMENT OF THE GREENHOUSE EFFECT

The basic source of energy for most terrestrial processes, and certainly for climate, is the absorption of solar radiation. If the climate system is to be in equilibrium, the solar radiation that is absorbed by Earth must be balanced by outgoing thermal radiation from Earth.

Much of the current concern over the impact of human activities on climate arises from the ability of human development of Earth to alter the amount of absorbed or emitted radiation, or to change the hydrologic cycle of the planet. In assessing these impacts, one must place them in perspective with natural variations in climate and driving forces. We have mentioned earlier that many different trace gases have the potential to alter Earth's climate. The most important is water vapor, whose atmospheric concentrations cannot be directly influenced by human activity. Next in importance is carbon dioxide, whose effect is calculated to be large over the next century. Almost equally important is a group of other gases with long atmospheric lifetimes and absorption characteristics suitable for interaction with terrestrial infrared radiation. It is important to realize that the combined effect of these gases is enhanced by about a factor of 2 as a result of the increase in water vapor that can be expected to accompany any temperature rise at Earth's surface as a consequence of increased evaporation from the oceans.

An increase in the concentration of a greenhouse gas initially decreases the flux of long-wave terrestrial radiation to space as more of the radiation is trapped in the troposphere. The effect will be a temperature rise at Earth's surface, the magnitude depending in part on related process such as changes in water evaporation rates or cloudiness. When comparing the heating powers of different gases, it is therefore advantageous to estimate for each the reduction in infrared radiation leaving Earth per unit increase in atmospheric abundance. This quantity is called the *radiative forcing* of a specific gas. Such approaches reveal that the radiative forcing of methane is about 25 times larger than that of CO_2, on a molecule to molecule basis. (Radiative forcings are often given for comparison by weight; in that case the above number must be multiplied by 2.75, the ratio of the molecular weights of CH_4 and CO_2.)

Why is the radiative forcing by methane so much larger than that of carbon dioxide? The reason is that there is already so much CO_2 in the atmosphere that in many spectral regions the absorption by CO_2 is already nearly complete, and added CO_2 is left without a major role (see Figure 3.3). In contrast, several less abundant greenhouse gases, including methane, have absorption bands in regions where little absorption is now occurring, and each added molecule provides strong new absorption capabilities. Figure 17.3 compares the relative heating produced by several of the gases; it shows that a doubling of the atmospheric CO_2 concentration from 250 to 500 ppmv increases greenhouse warming by only about 3.3 W m^{-2}, whereas an increase from 1 to 4 ppmv of methane gives an additional heating of 1.3 W m^{-2}. The results are even more dramatic for CFC-11, CFC-12, and N_2O.

Several research groups have used computer models to examine the effects of increases in the concentrations of radiatively active trace gases. The results of one such study are pictured in Figure 17.4, which shows the relative contributions to radiative forcing of the different greenhouse gases during several epochs. The results show how the influence of gases other than CO_2 has grown over the years, from almost negligible prior to 1900 to nearly equal as we approach the end of the century.

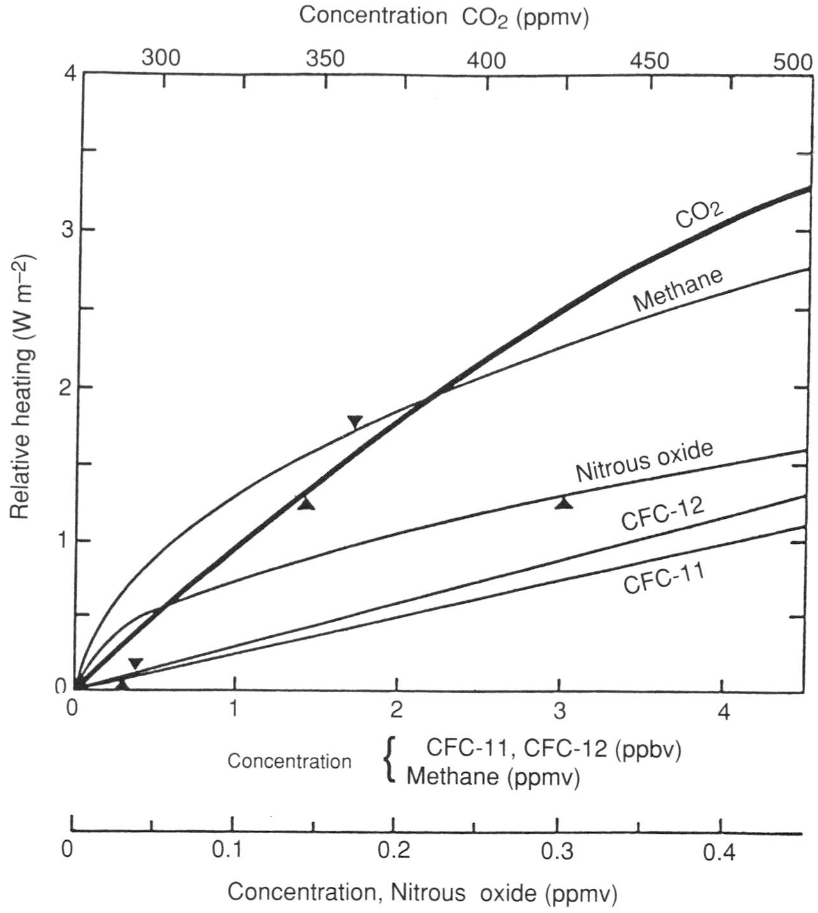

Figure 17.3 Greenhouse heating due to different concentrations of selected gases. The baseline for CO_2 is 275 ppm (approximately the preindustrial level); it is zero for the others. The triangles denote 1990 concentrations. (J .T. Houghton, G. J. Jenkins, and J. J. Ephraums, eds., *Climate Change: The IPCC Scientific Assessment*, Cambridge University Press, Cambridge, UK, 1990.)

Computing Global Warming Potentials

Global warming potentials express both atmospheric lifetime information and the ability of a molecule, once present, to absorb solar radiation. The computation is done by defining an integrated heating function $\Xi_i(T)$, which integrates the radiative forcing over a specific time period of interest:

$$\Xi_i(T) = \int_0^T RF_i \exp(-\tau/\tau_i)\, dt \quad , \qquad (17.5)$$

where RF_i is the radiative forcing of gas i and τ_i is its atmospheric lifetime, assumed here for simplicity to follow exponential decay. We then compute the integrated heating function for CO_2 and define the *global warming potential* $GWP_i(T)$ of a specific greenhouse gas with respect to carbon dioxide as

$$GWP_i(T) = \frac{\Xi_i(T)}{\Xi_{CO_2}(T)} \quad . \qquad (17.6)$$

In the future, the GWP may become a useful tool for international policy decisions, allowing nations to fulfill calculated and negotiated greenhouse warming reduction quotas.

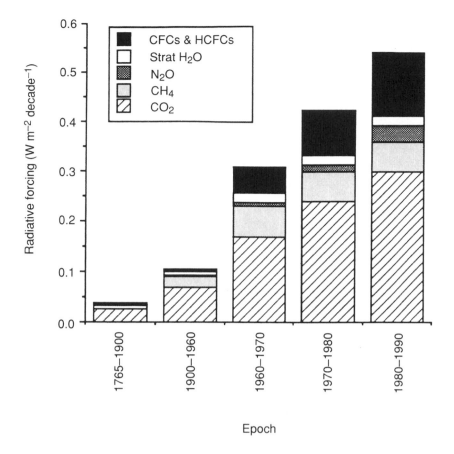

Figure 17.4 The degree of radiative forcing produced by selected greenhouse gases in five different epochs. Until about 1960, nearly all the forcing was due to CO_2; today the other greenhouse gases combined nearly equal the CO_2 forcing. (J. T. Houghton, G. J. Jenkins, and J. J. Ephraums, eds., *Climate Change: The IPCC Scientific Assessment*, Cambridge University Press, Cambridge, UK, 1990.)

The radiative forcing indicates the warming produced by the presence of an added quantity of a specific gas in the atmosphere but does not allow comparison of the true relative impacts of gaseous emissions, because different gases have different atmospheric lifetimes. To take that property into account, a greenhouse warming potential has been defined and computed for many different gases. Some of the results of these calculations are given in Table 17.1, where the second column lists the very great differences in impact of the different gases on a mass basis. The emission rates tend to be skewed in the opposite direction, however (column 3), with the result (column 4) that carbon dioxide, methane, and the CFCs are expected to contribute the most to global warming over the next centruy. To stabilize the concentration of these molecules at their present levels, quite substantial decreases in the anthropogenic emission rates are required (column 5).

TABLE 17.1. Global Warming Potentials and Other Relevant Data for Radiatively Important Gases[a]

GAS	GWP[b]	1990 EMISSIONS (Tg)	RELATIVE CONTRIBUTION OVER 100 yr (%)	STABILIZATION REDUCTION (%)[c]
CO_2	1	26,000	61	~60
CH_4	21	300	15	15–20
N_2O	290	6	4	70–80
CFCs	3000-8000	0.9	11	70–85
HCFC-22[d]	1500	0.1	0.5	40–50

a. Source: J. T. Houghton, G. J. Jenkins, and J. J. Ephraums, eds. *Climate Change: The IPCC Scientific Assessment*, Cambridge University Press, Cambridge, UK, 1990.
b. Global warming potential per 1 kg of gas, referenced to CO_2.
c. These are the percentage reductions in the rates of anthropogenic emissions that would be required to stabilize concentrations at present-day levels.
d. HCFC is an abbreviation for a partially hydrogenated CFC. HCFC-22 is a designation for $CHClF_2$, chlorodifluoromethane.

When the radiative properties of greenhouse gases are combined with emission scenarios, computer models may be used to indicate likely temperatures for Earth's surface at different times in the near and intermediate future. For example, Figure 17.5 illustrates how several calculations involving projections of future atmospheric emissions have been grouped to show possible low, intermediate, and high temperatures throughout the next half-century. The ranges reflect current uncertainties in the degree to which climate is sensitive to changes in the greenhouse gas concentrations, the degree and rate with which the oceans can absorb increased heat from the air, and other factors. It seems very likely that the actual climate future for the next half-century will be somewhere within the ranges indicated should no significant restraint be placed on the emissions of greenhouse gases. Even under severe limitations on emissions, other calculations indicated a minimum global average temperature increase of 2 °C by the year 2050.

An average global temperature change of one or two degrees does not seem very large, but that impression is altered when the change is put into perspective with past climate oscillations. As we have said earlier, the Little Ice Age of 1450–1890 was about 1 °C cooler than the present. Just that small amount of cooling forced significant changes in agricultural practice and habitation, so global temperature changes of that magnitude are definitely important. Should warming of one or two degrees occur, the planet would be at a temperature probably not seen for 120 kyr (see Figure 10.15). Computer models estimate that among the results of such warming would be a change

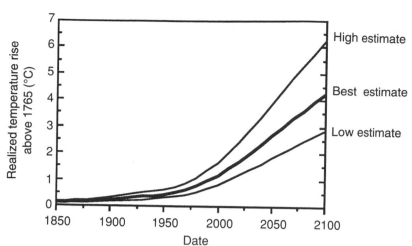

Figure 17.5 The annual mean surface air warming (in °C) for computations that assume "business as usual" rates of greenhouse gas emissions, but consider a variety of climate sensitivities, oceanic heat uptake rates, and other variables. (J. T. Houghton, G. J. Jenkins, and J. J. Ephraums, eds., *Climate Change: The IPCC Scientific Assessment*, Cambridge University Press, Cambridge, UK, 1990.)

in global mean sea level of perhaps 20 cm by the year 2030 and perhaps 45 cm by 2070, mostly due to thermal expansion of the oceans and the increased melting of mountain glaciers. Should such a change occur today, it would cause major disruption to the large fraction of the world's population living in coastal regions, especially developing nations in Southeast Asia. Finally, suppose that the high warming scenarios occur. Changes of three degrees or more are similar to the temperature change between the last ice age and the present, a transition that took place over several thousand years. In contrast, humankind would be producing the warmest world in millions of years, at least, and within a period of only a century. Such a rate of change in planetary tem-perature would be, as far as we know, entirely unprecedented.

In addition to predicting the effects of changes in the concentrations of present greenhouse gases, it is worth asking whether other gases will be of concern in the future. This inquiry leads us to consider the characteristics of an efficient greenhouse gas. Recall from Chapter 3 that we have identified the following as (currently) the most effective greenhouse gases: H_2O, CO_2, CH_4, O_3, N_2O, $CFCl_3$, and CF_2Cl_2. What common characteristics do they possess? First, they are unreactive toward HO• and O_3, the principal atmospheric reaction initiators. (Methane is an apparent exception, but its C–H bonds turn out to be the strongest of all C-H bonds as a consequence of methane's symmetry, so the CH_4 + HO• reaction rate is very slow.) The requirement for low or negligible atmospheric reactivity eliminates virtually all organic species from consideration as potential greenhouse gases, because their exposed hydrogen atoms are readily attacked by HO•. The same argument holds for reduced inorganic species such as NH_3 and H_2S. Also eliminated are inorganic acid precursor molecules such as NO and SO_2, which form addition compounds with HO•. A second requirement for greenhouse gases is that they are not subject to dissociation by solar photons in the visible and near ultraviolet portions of the spectrum. This requirement eliminates as potential greenhouse gases any molecules with O–O bonds, such as H_2O_2, and those with weak NO–O bonds such as NO_2 and NO_3. (Ozone is an apparent exception, but the oxygen atom liberated by photolysis reforms ozone with such a high probability that photolysis is not an important sink.) Third, greenhouse gas molecules are insoluble, or nearly so, in water. This requirement eliminates as potential greenhouse gases any organic molecule with an –OH group, such as HCOOH, and many inorganics such as HCl and HNO_3. Fourth, greenhouse gas molecules must have bonds that absorb radiation within the 8–13 μm window left by CO_2 and H_2O. The gases O_3, CH_4, N_2O, the CFCs, CF_4, and C_2F_6 meet this requirement. We are fortunate that this list is quite restrictive, and, except for those few atmospheric molecules that happen to be effective greenhouse gases, it seems likely that only carefully designed anthropogenic molecules have the potential to be added to the current greenhouse gas list.

Feedbacks in the Climate System

Even when the magnitude of climate forcing by changes in external factors such as solar radiation or absorbing gas concentration is known, it is a major task to translate such changes into an accurate assessment of the resulting impact on climate. This difficulty occurs primarily because the terrestrial climate system can respond to the imposed changes by reinforcing their magnitude (a *positive feedback*) or by damping their magnitude (a *negative feedback*). Without good information on the degree to which systems will engage in feedback, a scientist may to limited to a qualitative statement along the lines of "An increase in greenhouse gas concentrations will lead to a warmer global climate" rather than being able to make a much more useful statement concerning the magnitude of that warming. Without being comprehensive, we list and discuss here a few of the better known climate feedback processes.

Ice–albedo feedback is the process in which an increase in global surface temperatures may lead to accelerated melting of ice sheets and glaciers. When melting occurs, the planetary albedo decreases, thereby allowing more solar radiation to be absorbed at high latitudes and causing additional warming.

Water vapor–temperature rise feedback is the process in which an increase in global surface temperatures will lead to enhanced evaporation of water vapor from the oceans. Because water vapor is the most important of the greenhouse gases, an increase in its atmospheric concentration will amplify the climate warming.

Chemical feedbacks involve the effects of an individual chemical species on another species that is central to more than one chemical cycle. The best example is probably for methane, increases in which may lead to a global decrease in HO• concentrations. Because the primary CH_4 sink is reaction with HO•, a decrease in the latter further enhances the CH_4

increase. The hydroxyl radical is, as has been pointed out, the species that removes most reactive molecules from the atmosphere; thus, any decrease in the HO• concentration will permit the buildup of higher concentrations of many species. In addition, because CH_4 is also a greenhouse gas, it is involved in both chemical and climate feedbacks.

Cloud feedbacks are complicated and important. An increased rate of water vapor evaporation from Earth's surface provides an increased probability of cloud formation. The increased cloud cover leads to an increase in Earth's albedo, because clouds reflect radiation much more readily than does ocean or land. A cooling of the surface does not necessary follow, however, because clouds also play an important role in the emission and absorption of terrestrial infrared radiation. For low-lying clouds such as the horizontally extensive stratus, the infrared effect is not very large because their temperatures do not deviate very much from those of Earth's surface. Consequently, low clouds act to cool the surface. However, cirrus clouds in the middle and upper troposphere behave quite differently. These clouds are at lower temperatures and generally have lower albedos, so they transmit a substantial fraction of incoming solar radiation. At the same time, they are able to absorb larger quantities of upward-directed radiation emanating from the warmer lower troposphere and surface. Thus, an increase in high clouds will lead to a warming at Earth's surface. In addition to the difficulty of deducing the effects of changes in forcing on cloud formation rates, cloud feedbacks that also may exist involve liquid water content, droplet size distribution, cloud lifetimes, and the interplay between temperature and the presence of ice crystals or liquid droplets.

Perhaps the most poorly understood feedbacks are the *biotic feedbacks*. It is known that higher surface temperatures may lead to enhanced soil respiration rates, thereby causing release of additional CO_2 to the atmosphere and thus stronger greenhouse forcing. However, increased atmospheric CO_2 concentrations also stimulate plant growth, the so-called CO_2 *fertilization effect*. Plant growth will remove CO_2, thus potentially damping its atmospheric rise. It is uncertain which effect will be stronger. These and other biological interactions are discussed in Chapter 19.

GEOGRAPHICAL DISTRIBUTION OF THE GREENHOUSE EFFECT

Changes in the global average surface temperature do not adequately describe the changes that will occur in specific geographical regions. One attempt to do that more challenging task is that of Syukuro Manabe and colleagues at the Geophysical Fluid Dynamics Laboratory in Princeton, New Jersey. In their 3D global climate model, the researchers have interposed different concentrations of carbon dioxide and then calculated changes in temperature across their model grid. A sample of the temperature change results is shown in Figure 17.6, for calculations in which a quadrupling of the atmospheric concentration was assumed. Depending on the actual trend of emissions of CO_2 and other radiatively active gases, such a condition may develop over the latter part of the twenty-first century.

Figure 17.6 The geographical distribution of mean surface air warming (in °C) for a computation that assumes an atmospheric CO_2 concentration four times the present value. (a) Annual; (b) Dec–Feb; (c) Jun–Aug. (S. Manabe and R. J. Stouffer, Sensitivity of a global climate model to an increase of CO_2 concentration in the atmosphere, *Journal of Geophysical Research, 85,* 5529–5554, 1980.)

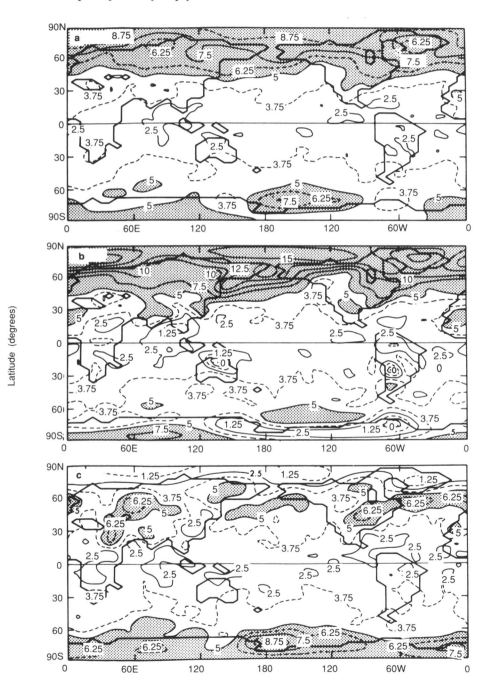

Longitude (degrees)

The figure shows substantial geographical variability in the predicted annual average temperature increases. Near the equator, the anticipated warming trends are two or three degrees Celsius. At the poles, however, the changes are much more severe: up to almost 9 °C at the North Pole. Because most of Earth's water is stored at the poles in ice caps, such changes could result in a significant rise in sea level.

Large geographical variations are also obtained when one examines the model predictions for such parameters as soil moisture, precipitation, continental runoff rates, and other climate-related factors. An example is from the GCM of David Rind, James Hansen, and colleagues at the Goddard Institute for Space Studies, a NASA facility. In their studies, they use the climate change projections for a trace gas scenario in which radiative greenhouse forcing continues to grow at an exponential rate. Figure 17.7 in the color insert reproduces results on the occurrence of drought for four model years. In 1969 there is an equal occurrence of wet and dry regions, with extreme occurrences randomly distributed and infrequent. By 1999, very dry conditions occur over some tropical and subtropical land masses. In 2029 the dry regions have expanded, pushing into a number of midlatitude regions and increasing in drought intensity. By 2059, extreme drought covers most midlatitude locations whereas extreme flood conditions are found at the highest latitudes. The authors point out that if the trace gas emission increases occur at a lower rate than they have assumed, or if the climate sensitivity is less, the effects will be modified accordingly. Nonetheless, if anything approaching this result should occur, the impacts will be very significant indeed, so the warning of the model calculation should be taken very seriously.

The results of the Goddard Institute group have been combined with similar research efforts in a summary by the Intergovernmental Panel on Climate Change on possible changes in five geographical regions. For each region, estimates of changes in temperature, precipitation, and soil moisture between 1990 and 2030 are given (see box). The confidence in these estimates is not very high, because the global models are not complex enough and do not have adequate spatial resolution to accurately predict changes in precipitation and soil moisture. However, they represent the best estimates that can be made at the present time.

Estimates of Regional Changes by 2030 (Intergovernmental Panel on Climate Change Report, 1990)

The numbers given below are based on high-resolution models, scaled to be consistent with our best estimate of global mean warming of 1.8 °C by 2030. For values consistent with other estimates of global temperature rise, the numbers below should be reduced by 30% for the low estimate or increased by 50% for the high estimate. Precipitation estimates are also scaled in a similar way.

Central North America (35°–50° N, 85°–105° W)

The warming varies from 2 to 4 °C in winter and 2 to 3 °C in the summer. Precipitation increases from 0 to 15% in winter but decreases by 5 to 10% in summer. Soil moisture increases by 5 to 10%.

Southern Asia (5°–30° N, 70°–105° E)

The warming varies from 1 to 2 °C throughout the year. Precipitation changes little in winter and generally increases throughout the region by 5 to 15% in summer. Summer soil moisture increases by 5 to 10%.

Sahel Desert, Africa (10°–20° N, 20° W–40° E)

The warming ranges from 1 to 3 °C. Area mean precipitation increases and area mean soil moisture decreases marginally in summer. However, throughout the region, there are areas of both increase and decrease in both parameters.

Southern Europe (35°–50° N, 10° W–45° E)

The warming is about 2 °C in winter and varies from 2 to 3 °C in summer. There is some indication of increased precipitation in winter, but summer precipitation decreases by 5 to 15%, and summer soil moisture by 15 to 25%.

Australia (12°–45° S, 110°–155° E)

The warming ranges from 1 to 2 °C in summer and is about 2 °C in winter. Summer precipitation increases by around 10%, but the models do not produce consistent estimates of the changes in soil moisture. The area averages hide large variations at the subcontinental level.

Another anticipated consequence of regional or global temperature change is on the geographical distribution of various types of biota. Figure 17.8 shows the anticipated range of the eastern hemlock tree, if temperature conditions predicted by GCMs should come to pass. The rates of temperature change anticipated for the coming centuries are so rapid that successful relocation of the trees on short time scales may not be possible. As a consequence, major loss of hemlock forest seems likely, with much of the potential range remaining unoccupied for long periods into the future. In other words, even if the overall effect of future climate changes may be beneficial for some ecosystems and some nations several centuries from now, the rapid transitions anticipated in temperature and climatic conditions are a major problem for the next several centuries, because environmental adaptation will be too slow to transfer healthy ecosystems intact from one geographical location to another.

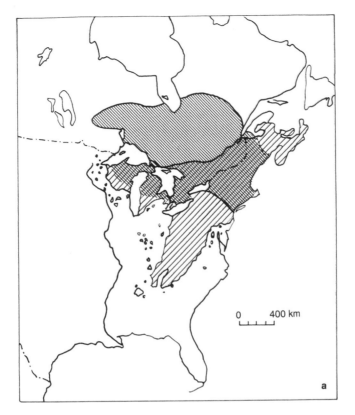

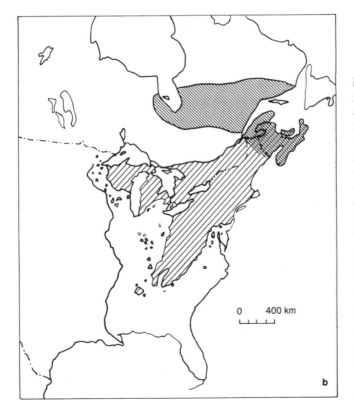

Figure 17.8 The present and future geographical range of the eastern hemlock in northeastern North America. The light shading indicates the present range, the heavier shading the range in about the year 2080, and the crosshatched area where the two ranges overlap. (a) This diagram is constructed by using the temperature results of the computer model at the Goddard Institute for Space Studies, New York, NY. (b) This diagram relies on the temperature results of the computer model formulated at the Geophysical Fluid Dynamics Laboratory, Princeton, NJ. (Committee on Global Change, *Toward an Understanding of Global Change*, National Academy Press, Washington, D.C., pp. 72–73, 1988.)

THE NEGATIVE RADIATION FORCING OF ANTHROPOGENIC SO₂ EMISSIONS

We pointed out in Chapter 14 that the anthropogenic emissions of sulfur now quantitatively surpass those coming from nature. The result is much reduced visibility in polluted regions and major deleterious effects in the biosphere due to the resulting formation of acid precipitation. For those reasons, efforts are under way in many developed nations to reduce sulfur emissions by using fuels with low levels of sulfur and by extensive control of emissions from industrial and combustion processes. It appears, however, that sulfur emissions may have a positive effect, namely, to cool the Earth–atmosphere system. The proposed mechanism is enhanced backscattering to space of solar radiation by sulfate particles, thus reducing the amount of solar energy that can be absorbed by Earth's surface. Recently, the potential significance of this effect has been investigated theoretically in a collaborative effort by scientists from the United States, Sweden, and Germany. The potential influence of anthropogenic SO_2 emissions was explored by calculating the distribution of sulfate aerosol for both natural and anthropogenic sulfur emissions, and then computing the influence of that aerosol on the atmospheric radiation budget. Some of the results are shown in Figure 17.9.

A comparison of the results for anthropogenic sulfur emissions only and for anthropogenic plus natural sulfur emissions clearly shows the anthropogenic effects, especially over and downwind of those parts of the world where the bulk of the SO_2 emissions occur. The calculations also show, however, that the influence of the anthropogenic emissions is restricted almost entirely to the northern hemisphere. This result is not surprising, in view of the atmospheric lifetime of about a week for sulfate aerosol due to precipitation scavenging.

The magnitude of the scattering of radiation as well as its geographical location is of interest. Values due to anthropogenic activity are as much as 4 W m⁻² over southeastern Europe, as much as 2 W m⁻² over the southern United States, and about 1 W m⁻² for the entire northern hemisphere. An increase of 1 W m⁻² is just about the magnitude of greenhouse warming caused by the increase of CO_2 in the atmosphere since preindustrial times. Further, the sulfate aerosol influence is in the opposite sense from that of CO_2. The industrial and combustion-related SO_2 emissions may thus have counteracted an appreciable part of the warming from anthropogenic greenhouse gases. Not withstanding this interesting idea, now being tested by analyses of temperature records and by field experiments of various types, the authors point out that the SO_2/CO_2 interactions with radiation are not spatially uniform and that the anticipated effects differ with hemisphere and with location within each hemisphere.

A final point to make is that the climate effects of sulfate aerosols may involve changes in cloud albedo as well as backscattering from

Figure 17.9 Calculated increase due to sulfate aerosol particles in the reflected flux of radiation to space (in watts per square meter). (a) Both anthropogenic and natural sources of sulfur are included in the calculations. (b) Anthropogenic sources only included. (Reproduced with permission from R. J. Charlson, J. Langner, H. Rodhe, C. B. Leovy, and S. G. Warren, Perturbation of the northern hemisphere radiative balance by backscattering from anthropogenic sulfate aerosols, *Tellus*, 43AB , 152–163. Copyright 1991 by Munksgaard International Publishers Ltd.)

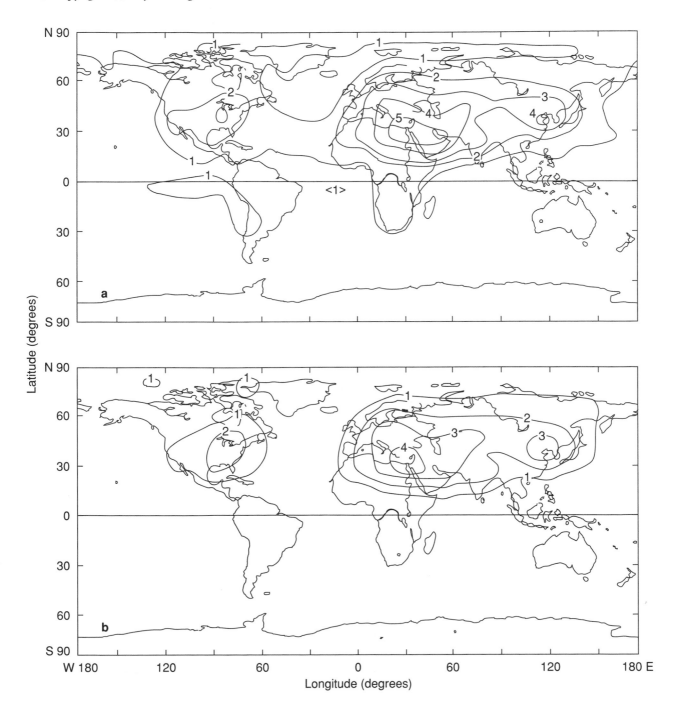

cloud-free regions. The reason for this is that sulfate aerosol particles are extremely efficient cloud condensation nuclei. Thus, an increase in SO_2 emissions followed by the formation of sulfate aerosol could lead to higher concentrations of cloud droplets at the same atmospheric liquid water concentration and thus greater droplet surface areas. The result will be whiter clouds that are more efficient scatterers of solar radiation.

NUCLEAR WINTER, NUCLEAR AUTUMN

Nuclear war is regarded by the population of the world with horror because of the devastation and suffering that would be caused. A related and not unimportant problem is the impact on the atmosphere of the smoke and soot particles that would be generated by the fires caused by a major nuclear exchange. In the event of a nuclear war in which industrial centers and cities are attacked, fires would burn over much of the developed world, fed by the large quantities of combustible materials (wood, oil, gasoline, asphalt, plastics, and so forth) that are maintained in urban and military locations. The black, sooty smoke from these fires would be carried high into the atmosphere, where it would intercept solar radiation. The residence times of smoke particles in the atmosphere are expected to be at least several months, so the process of photosynthesis on which the survival of life depends could be substantially diminished. A lowering of surface temperatures would also occur, not only as a result of less penetration of solar radiation to the surface, but also through the establishment of an "anti-greenhouse effect": because at high soot levels most of the energy of the incoming solar radiation would be deposited at higher altitudes in the atmosphere, there would be much less absorption of upward-going infrared radiation by the greenhouse gases, especially water vapor in the atmosphere above, and thus also less downward radiation from the atmosphere to the surface. This anti-greenhouse effect would add strongly to the direct cooling effect. An additional impact is the effect of the lower surface temperature on the hydrologic cycle: much less water vapor transfer to the atmosphere would be expected and, therefore, much less precipitation would fall. Soot particle lifetimes would be enhanced both by the decrease in precipitation scavenging near the ground and by the transport of heated soot-laden air to the relatively dry upper troposphere and stratosphere.

A number of research groups around the world have attempted to estimate the severity of the atmosphere's perturbation by these processes. The difficulties in doing so are substantial, because the magnitude of the war must be assumed, the locations of the nuclear bursts specified, and the season of the year defined. All of these important factors would be under the control of the nuclear combatants. Other factors to be considered are scientific rather than tactical: how much smoke would be produced from the fires? What would be the optical absorption properties and particle size range in the smoke? How much rain would be induced? How effective would the rain be in washing out the smoke particles? Would coupling between the atmosphere and oceans modify any of these effects?

Some of the original predictions of 1D models attempting to simulate the atmospheric chemistry and physics resulting from a nuclear war suggested temperature reductions of as much as 35 °C for continental interiors during the summer. This difference is more than the summer to winter temperature difference in midlatitudes, and the term *nuclear winter* was therefore coined. Subsequently, advanced 3D model simulations led to somewhat less drastic but still catastrophic predictions. The results from models now suggest that decreases in temperature are likely to average about 10–15 °C, which would still involve the risk of frost conditions even during summer. Altogether, the severe temperature drops and the loss of solar radiation for photosynthesis would cause obvious difficulties for agricultural productivity.

An example of current model results are those of Stephen Schneider and Starley Thompson of the National Center for Atmospheric Research. They use a global climate model with several different smoke assumptions to derive temperature trends. The calculations demonstrate that the temperature change is quite sensitive to the amount of smoke injected into the atmosphere. Even rather small changes in the global average temperature are predicted to produce noticeable consequences; it is of interest in this regard that the temperature differences between glacial and interglacial periods on the Earth are only 4–5 °C.

Like the temperature changes induced by greenhouse warming, temperature changes resulting from nuclear war will show large geographical differences. Schneider and Thompson's predictions for the northern hemisphere are that the lowest temperatures will occur over the high northern latitudes and over the most populous areas of the United States and Russia (the assumed combatants). However, most of the northern hemisphere would find its climate perturbed to at least some extent by nuclear war. In fact, because rice plants are quite sensitive to temperatures below 15 °C for even a few days, there is a risk that noncombatant nations in the developing world would lose most of the food stock and mass starvation could result. According to an international study conducted by the Scientific Committee on Problems of the Environment (SCOPE) and summarized by Schneider and Thompson, "The sum of climate disturbances, radioactive fallout, ozone depletions, and the interruption of basic social services could threaten more people globally than would the direct effects of nuclear explosions. The latter is a sufficiently horrendous prospect to make a nuclear war an unprecedented human catastrophe. The addition of environmental effects deepens the horror for the combatants and significantly extends the catastrophic consequences to noncombatants."

SUMMARY

By this point, three things should be clear about computer models of the atmosphere:

1. They are very detailed and complicated, and it is not always possible to tell whether factors or formulations that must be assumed in the development of any atmospheric model are perfectly correct.

2. The results of well-regarded computer models can differ significantly.

3. Nonetheless, the results from the models are useful tools for predicting the atmospheric properties and climates of the future. At the least, their predictions should be treated as important warning signals.

For the models discussed above and for others, we can summarize the results of this chapter and the preceding one by two tables. The first, Table 17.2, shows the state of knowledge of the impacts of various gases on different environmental perturbations. (This is an expanded but completely qualitative version of Figure 16.9.) It is of interest that many gases have multiple impacts and many impacts multiple causes, as has been seen previously, but also that in a number of cases either negative or positive impacts can occur. For example, the effects of carbon dioxide, the NO_x gases, and nitrous oxide on stratospheric ozone depletion depend on altitude. Methane ameliorates ozone depletion, except in the Antarctic ozone hole, and its tendency to affect the self-cleaning of the atmosphere (by reducing the abundance of hydroxyl) is different in the north and the south: it favors diminished hydroxyl concentrations in the southern hemisphere and increased HO• levels in the northern hemisphere.

Table 17.3 shows current data on gas sources, emission rates, atmospheric lifetimes, and past, present, and potential future concentrations. For gases with lifetimes of years, estimated global averages are listed. The concentrations of the NO_x gases and sulfur dioxide over highly industrial sites may not increase very much in the next 40 yr, but the number of affected locations can be expected to grow, especially in the developing nations. Chlorofluorocarbon concentrations are given in terms of chlorine atoms because the molecules generally contain more than one ozone-destroying chlorine atom. Even though substantial ranges are given for the future concentrations, the trends are clear, and one can use this information as a guide to policy matters related to rates of emission.

It is clear that most of the predictions concerning the future state of the global atmosphere are not encouraging. The very long lifetimes of many of the chemicals being emitted into the atmosphere render long-term change, once begun, difficult to halt. Although we continue to have concern about the absolute accuracy of the global model predictions, it is difficult to avoid the conclusion that the atmosphere will change very noticeably over the next century, no matter what steps are now taken, and that the effects on Earth will be highly significant.

It is also evident that the complexity of atmospheric chemistry results in possible scenarios where one action will have precisely the opposite effect of another on one of the atmosphere's constituents; a good example of this is HO•, which is vital as an atmospheric detergent. Its concentrations clearly can be decreased by added methane and carbon monoxide but can be increased by added nitric oxide. A decision intended to optimize HO• concentrations can clearly not be taken in a vacuum, because changes in CH_4, CO, SO_2, and NO influence not only HO• but also global temperatures, tropospheric and stratospheric ozone, the acidity of precipitation, and a host of other atmospheric properties. Developing a solidly grounded approach to such an overall assessment and receiving support for such an approach by policy-makers worldwide is one of the challenges for the future.

TABLE 17.2. Trace gases and the environmental perturbations with which they are associated. Plus signs indicate a contribution to the effect; minus signs indicate amelioration.

GAS	GREEN-HOUSE EFFECT	STRATOSPHERIC OZONE DEPLETION	ACID DEPOSITION	SMOG	CORROSION	DECREASED VISIBILITY	DECREASED SELF-CLEANS-ING OF ATMOSPHERE
Carbon Monoxide (CO)							+
Carbon Dioxide (CO_2)	+	−					
Methane (CH_4)	+	− except in ozone hole					+ SH − NH
NO_x: Nitric Oxide (NO) and Nitrogen Dioxide (NO_2)		+ above 20 km − below 20 km	+	+		+	−
Nitrous Oxide (N_2O)	+	+ above 20 km − below 20 km					
Sulfur Dioxide (SO_2)	−		+		+	+	
Chlorofluoro-carbons	+	+					−
Ozone (O_3)	+			+			−

TABLE 17.3. Current atmospheric properties, historical concentrations, and ranges of possible futures for several important trace gases. The concentrations are given in parts per billion by volume.

GAS	MAJOR ANTHRO-POGENIC SOURCES	ANTHROPO-GENIC/TOTAL EMISSIONS PER YEAR (MILLIONS OF TONS)	AVERAGE RESIDENCE TIME IN ATMOSPHERE	AVERAGE CONCEN-TRATION 100 YEARS AGO (ppbv)	APPROXIMATE CURRENT CONCENTRA-TION (ppbv)	PROJECTED CONCENTRA-TION IN YEAR 2030 (ppbv)
Carbon Monoxide (CO)	Fossil Fuel Combustion, Biomass Burning	700/2000	Months	20 to 40?, N. Hem. 30 to 100?, S. Hem. (Clean Atmospheres)	100 to 200, N. Hem. 40 to 80, S. Hem (Clean Atmospheres)	Probably Increasing
Carbon Dioxide (CO_2)	Fossil Fuel Combustion, Deforestration	7500/~7500	120 Years	290,000	350,000	400,000 to 550,000
Methane (CH_4)	Rice Fields, Cattle, Landfills, Fossil Fuel Production	300 to 400/ 500 to 600	10 Years	900	1,700	2,200 to 2,500
NO_x Gases	Fossil Fuel Combustion, Biomass Burning	20 to 30/ 30 to 60	Days	0.001 to ? (Clean to Industrial)	0.001 to 50 (Clean to Industrial)	0.001 to 50 (Clean to Industrial)
Nitrous Oxide (N_2O)	Nitrogenous Fertilizers, Deforestration, Biomass Burning	6/20	170 Years	285	310	350 to 370
Sulfur Dioxide (SO_2)	Fossil Fuel Combustion, Ore Smelting	140 to 180/ 200 to 280	Days	0.01 to ? (Clean to Industrial)	0.01 to 50 (Clean to Industrial)	0.01 to 50 (Clean to Industrial)
Chlorofluoro-carbons	Aerosol Sprays, Refrigerants, Foams	~1/1	60 to 120 Years	0	About 3 (Chlorine Atoms)	2 to 4 (Chlorine Atoms)

EXERCISES

17.1 The lifetime of methane in the atmosphere is determined by the HO• abundance. Using the HO• concentrations for the high NO_x case of Figure 17.2, compute local methane lifetimes in years for (a) 1970, 0.1 km, (b) 1970, 2 km; (c) 2035, 0.1 km; (d) 2035, 2 km. See Table 7.1 for applicable data and assume that temperature decreases linearly from 25 °C at the surface to 20 °C at 2 km altitude. Discuss whether these lifetimes have any global significance.

17.2 It was recently reported (C. A. M. Brenninkmeijer, M. R. Manning, D. C. Lowe, G. Wallace, R. J. Sparks, and A. Volz-Thomas, Interhemispheric asymmetry in OH abundance inferred from measurements of atmospheric ^{14}CO, *Nature, 356,* 50–52, 1992) that HO• concentrations in the southern hemisphere may be much higher than those in the northern hemisphere. Referring to that article and to previous chapters in this book, discuss the possible implications should this report prove to be accurate.

17.3 The potential atmospheric impact of a major nuclear exchange has been a subject of vigorous debate since the theory was proposed a decade ago. Prepare a report on the current understanding of this topic, based on the following references and other sources of your choosing: J. W. Birks and P. J. Crutzen, Atmospheric effects of a nuclear war, *Chemistry in Britain*, 927–930, Nov., 1983; C. Sagen, Nuclear war and climatic catastrophe: Some policy implications, *Foreign Affairs*, No. 62202, 257–292, Winter, 1983/84; R. P. Turco, O. B. Toon, T. P. Ackerman, J. B. Pollack, and C. Sagen, The climatic effects of nuclear war, *Scientific American, 251* (2), 33–43, 1984; S. H. Schneider and S. L. Thompson, Simulating the climatic effects of nuclear war, *Nature, 333,* 221-227, 1988; R. P. Turco, O. B. Toon, T. P. Ackerman, J. B. Pollack, and C. Sagen, Climate and smoke: An appraisal of nuclear winter, *Science, 247,* 166–176, 1990.

FURTHER READING

J. T. Houghton, G. J. Jenkins, and J. J. Ephraums, eds., *Climate Change: The IPCC Scientific Assessment*, Cambridge University Press, Cambridge, UK, 1990.

J. F. B. Mitchell, The "greenhouse effect" and climate change, *Reviews of Geophysics, 27,* 115–139, 1989.

V. Ramanathan, L. Callis, R. Cess, J. Hansen, I. Isaksen, W. Kuhn, A. Lacis, F. Luther, J. Mahlman, R. Reck, and M. Schlesinger, Climate-chemical interactions and effects of changing atmospheric trace gases, *Reviews of Geophysics, 25,* 1441–1482, 1987.

Scientific Committee on Problems of the Environment, *Environmental Consequences of Nuclear War,* Vol. 1 (1986), Vol. 2 (1985), John Wiley & Sons. Oxford–New York.

18

The Climate of the Far Future

By this point, it should be clear that rapid population increases and accompanying emissions to air, water, and soil will result in substantial changes in the environment over the next century. On that time scale, we naturally concern ourselves with the possible impacts of those changes on ourselves, our children and grandchildren, and their grandchildren. What can be said if we try to take a longer look than that, focusing on thousands or millions of years in the future? On that time scale, the genus *Homo* will have evolved into new forms, and our concern for our own genus is outweighed by concern for life and the planet in a more general way. On these long time scales, natural forcing functions may assume greater prominance than anthropogenic ones, and a history of the past becomes, at least in part, a useful guide to the future.

The evidence for climates of the distant past was discussed in Chapter 10. Much of the discussion was summarized by Figure 10.15, on which periods of variability in global atmospheric surface temperature were plotted. In Chapter 11, we presented evidence on the chemistry of those epochs. There is every reason to believe that these historic patterns will repeat themselves in the future, perhaps in somewhat altered form. In this chapter, we explore the possible climates of the future, paying especial attention to the climate forcing functions called out in Figure 10.16.

In Chapters 16 and 17 we saw that several of the gases being added to the atmosphere by anthropogenic activity are efficient absorbers of infrared radiation and are expected to produce a gradual warming of the planet. Computer models suggest the possibility that such a warming may now be approaching the lower limit of detectability if measurements are carefully collected, calibrated, and assessed. A limitation to near-term climate predictions is that we have no way at present of anticipating such unpredictable factors as volcanic activity. Ice core records show that volcanic activity has been much greater at certain times in the past than it has been during most of the present century and that the effect of major volcanic eruptions is to cool Earth as the dust injected into the atmosphere scatters a fraction of the incoming sunlight into space. The effects would last for only a short time on geological time scales, but the times would be long enough (a few months to a few years) to be of concern to the human population.

What of the further time horizons? Just as going back in time becomes increasingly uncertain as the times become greater, so going forward becomes increasingly uncertain as we try to look further and further ahead. Nonetheless, the attempt is instructive. We divide the discussion into time periods of the future chosen to mimic those of the most interesting and informative of the epochs of the past.

NEO-HOLOCENE AND NEO-PLEISTOCENE CLIMATE

The Neo-Holocene ("new Holocene") is the term by which we designate the period extending as far into the future as the Holocene extends into the past: about 10 kyr or so. Similarly, the Neo-Pleistocene extends from 0.01 to 2 Myr AP (after Present). One of the driving factors for natural climate during these epochs is expected to be the variations in rotation and revolution of the Earth in its orbit. The anticipated changes in the orbital parameters are shown in Figure 18.1.

It is interesting to compare this figure with Figure 10.8, which showed the changes in the same parameters for the past 250 kyr. Note first that the rotational axis orientation to the Sun–Earth line (the tilt) is roughly at its midpoint at present. For the next 8–10 kyr, the tilt will decrease, thereby resulting in less solar radiation reaching each pole during its summer and increasing the chances for ice and

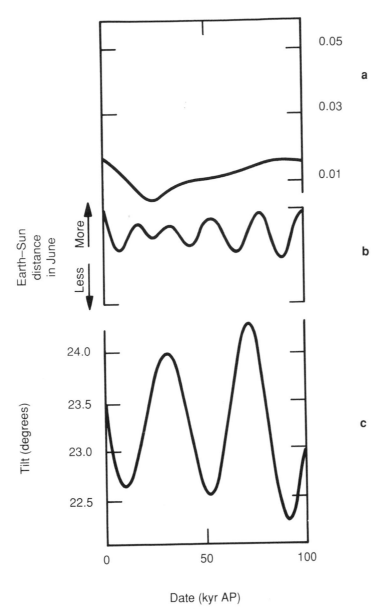

Figure 18.1 Long-term variations in three of Earth's orbital parameters: eccentricity, precession, and tilt, from the present time to 100 kyr AP. (After A. Berger, Milankovich theory and climate, *Reviews of Geophysics, 26,* 624–657, 1988).

snowpack buildup there. The eccentricity of the orbit of revolution is decreasing as well, a development that has a slight tendency to equalize seasonal variations. The axial precession (the measure of the Sun–Earth distance at a particular season) has the shortest period of the three parameters, and the magnitude of the distance in June is decreasing. Because of the time lag between absorption of solar radiation and Earth surface warming, decreasing Earth–Sun distances in June and increasing Sun–Earth distances in December encourage global cooling.

The result of a consideration of all three orbital parameter effects is shown in Figure 18.2, where the total radiation reaching the top of

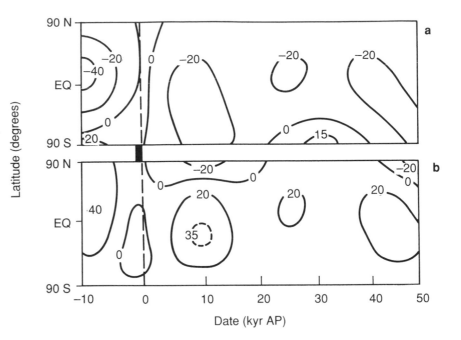

Figure 18.2 Changes from 10 kyr BP to 50 kyr AP in the amount of solar radiation reaching the top of Earth's atmosphere as a function of latitude, from pole to pole. (top) January; (bottom) July. The units are cal cm² day⁻¹, expressed as departures from present values. (The latter are of the order of 700 cal cm² day⁻¹ and depend on latitude and season.) (Adapted with permission from H. H. Lamb, *Climate: Present, Past, and Future*, Vol. 2. Copyright 1977 by Methuen & Co.)

the atmosphere as a function of time, latitude, and season is shown. Because changes are most important in the northern high- and midlatitudes, where the bulk of the land is presently found, and in the winter, when the reductions in radiation interact most directly with the maintenance of the snowline in the interior portions of the continents, the general decrease in irradiation shown in the top portion of the top diagram for the next few tens of kiloyears is most relevant to our considerations.

The ordering of orbital elements by importance is demonstrated nicely by a study of monsoon-related climatic records, in which the average northern hemisphere summer radiation flux during the past few tens of kiloyears is seen to vary largely as a function of ellipticity and precession (Figure 18.3; compare Figure 10.8). In rather close concert with those variations, global average temperature (as measured by the oxygen isotope content of fossils), the occurrence of monsoon-related sediments, and the occurrence of monsoon-transported pollen all show four nearly coincident maxima.

Considerations of the orbital parameters thus predict a global glaciation gathering force over the next few millenia. Historical records suggest that such a temperature decrease would not be gradual, but would occur in an irregular series of transitions of the order of a degree each. (The temperature difference between the Little Ice Age of about 1450–1890 and the present is thought to have been about 1 °C.) Berger of the Universite de Louvain has studied several predictions for climate over the next 100 kyr; an example of those predictions is shown in Figure 18.4a. In the absence of anthropogenic influences,

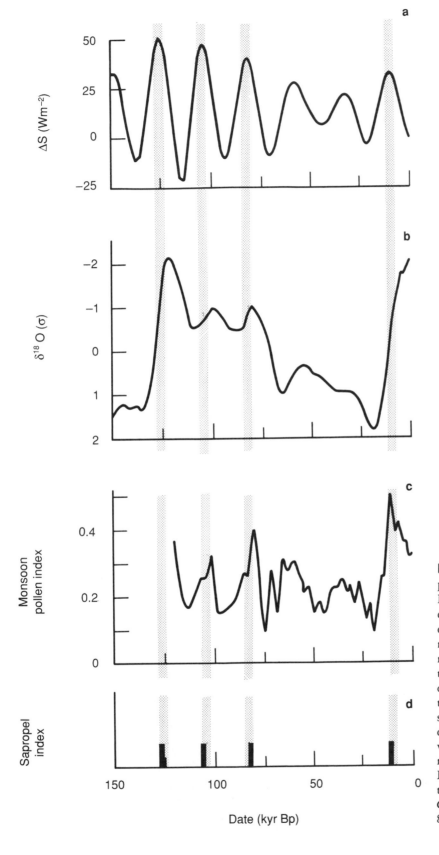

Figure 18.3 Selected
paleoclimatic records from 150 kyr
BP to the present as an illustration
of the impact of orbital variations
on indicators of climate. (a) Average
northern hemisphere summer solar
radiation; (b) composite temperature record (as reflected by $\delta^{18}O$
content of fossils); (c) monsoon-
transported pollen in deep-sea
sediments; (d) occurrence of
organic-rich sediments associated
with tropical African monsoon
runoff. (After W. L. Prell and J. E.
Kutzbach, Monsoon variability over
the past 150,000 years, *Journal of
Geophysical Research, 92,* 8411–
8425, 1987.)

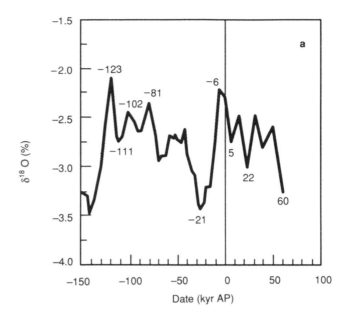

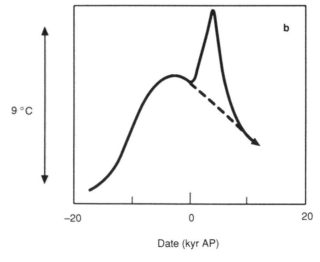

Figure 18.4 (a) Forecast
variations in average long term
future climate of Earth, expressed
as oxygen isotope ratios of deposited sediments. (b) A schematic
picture of the temperature behavior
of Earth during the Holocene and
its possible pattern during the first
part of the Neo-Holocene. The
dotted line indicates the expected
pattern in the absence of anthropogenic greenhouse gas forcing.
(c) Variations in continental ice
volume for the next 80 kyr in the
absence of ice sheet disturbance
(solid line) and if the Greenland ice
sheet is assumed to melt over the
next few centuries (dashed line).
(This figure constucted from
information in A. Berger, The
Earth's future climate at the
astronomical time scale, in *Future
Climate Change and Radioactive
Waste Disposal*, C. Goodess and S.
Pakstikof, eds., Norwich, UK,
1989; and W. Broecker, *How to
Build a Habitable Planet*, Eldigio
Press, Palisades, NY, 1987.)

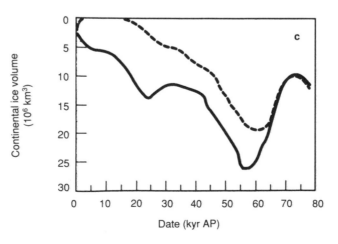

the long-term cooling trend that began some 6 kyr ago is expected to continue for the next 5 kyr. The 5 kyr AP minimum in temperature is expected to be followed by a cold interval centered at about 22 kyr AP and finally by a major glaciation beginning at about 50 kyr AP.

The key phrase in the description of Figure 18.4a is *in the absence of anthropogenic influences*. A factor in the Neo-Holocene significantly different from that in the Holocene will be the concentrations of greenhouse gases. These projections are, of course, very uncertain, depending heavily on long-term patterns of fossil fuel use, the deforestation of the planet, and other factors, but concentrations four times the present are possible for CO_2 and other greenhouse gases. This condition will occur over a period of probably a thousand years or less, by which time the fossil fuel supplies of Earth may well be exhausted. Wallace Broecker of the Lamont-Doherty Geophysical Observatory in New Jersey proposes that this injection of greenhouse gases will result in a brief (on geological time scales) "anthropogenic superinterglacial," with a temperature pattern along the lines shown in Figure 18.4b. The effect of the temperature spike over the next 1000 yr may be enhanced by biospheric feedback, as manifested in increased CO_2 and CH_4 emissions and illustrated in Figure 1.4.

Looking again to longer time scales, we show in Figure 18.4c Berger's prediction of continental ice volumes for the next 80 kyr. The pattern is seen to follow that of temperature rather well, the principal difference being that an asymmetry results from the tendency for ice sheets to shrink faster than they grow and hence for major glaciations to terminate faster than they begin. The dotted curve, however, is for a calculation assuming that the Greenland ice sheet disappears in the next few centuries as a result of greenhouse warming, a possible occurrence. In that case, the northern hemisphere ice sheets would not reappear before 15 kyr AP and the climate would remain different from its natural evolution up to about 65 kyr AP.

Finally, we mention an important and uncertain factor: the degree to which the oceans will serve as a sink for CO_2, thus buffering any warming trend. A warmer ocean will dissolve less CO_2, but a crucial aspect may be the productivity of the marine biota, which serve as a biological CO_2 pump by transporting organic carbon from surface waters to the deeper ocean layers as a rain of detritus. The marine biota are thought to constitute 30% or so of the planet's biomass, and the biological pump reduces surface CO_2 concentrations substantially. If the marine biota were to be substantially decreased in number as a consequence of increased ultraviolet radiation·or some other cause, atmospheric CO_2 may rise and global temperatures may be pushed upward. Other ocean feedback mechanisms, such as changes in ocean circulation patterns, are known to have occurred in the past, but are poorly understood. They are thought to have the potential, however, to modify global climate dramatically.

NEO-TERTIARY CLIMATE

The Neo-Tertiary is the period from the end of the Neo-Holocene, about 10 kyr AP, until 65 Myr AP. It covers a time scale that may encompass changes in glaciation, aridity, and other temperature-related factors, but one that is too short to register tectonic movement of the continents, major changes in solar radiation intensity, and similar very long time scale forcing processes. As seen from Figure 10.16, a significant climatic variational scale within the Neo-Tertiary range is expected to have a frequency of 30–60 Myr.

There appear to be two possibilities for the cause of this periodic climatic variation. One is that 35 Myr is thought to be about the right time scale for tectonic mountain building (as distinct from major continental relocation), and mountain building, being related to glacier formation by the amount of land at high altitudes, may be reflected in climate records. A second possibility is that of bolide (comet or asteroid) impacts.

The anticipated effects of bolide impacts, discussed in Chapter 10, are consistent with what is understood about the appearance and ages of craters on Earth and with time scales of variations in the orbital relationships of Earth, Sun, and nearby planets and stars. If such impacts indeed follow regular patterns dictated by astronomical orbits, Figure 10.6 suggests that the last such encounter occurred about 15 Myr BP and the next may occur about 11–15 Myr AP. However, not all bolide collisions produce major biological impacts.

Recall from Chapter 10 the anticipated consequences of a bolide impact: inhibition of photosynthesis for a year or more as a consequence of solar radiation extinction by NO_2 and perhaps suspended dust; and near-extinction, because of highly acidified rain, of the oceanic microorganisms whose shell-making normally removes CO_2 from the atmosphere–ocean system. In addition, the increased acidity of the oceans will decrease the solubility of CO_2.

For perhaps thousands of years following a major bolide impact, a strong greenhouse warming is likely as a consequence of the much-increased atmospheric concentrations of carbon dioxide and other greenhouse gases, following the short-term initial cooling due to the reflection of solar radiation by the increased amount of dust in the atmosphere.

For the Neo-Tertiary, therefore, we are unable to determine a likely trend in climate. If bolide impacts occur, they will encourage short-term cooling and long-term warming. If not, climate would be expected to follow its standard pattern of glacial-interglacial response to the solar forcing function.

The Neo-Mesozoic and Neo-Paleozoic periods extend over the approximate time range 65–600 Myr AP. The significant forcing function for climate within this time span that applies in less significant ways to shorter periods is the alteration of land mass distribution caused by continental drift. Projections of the manner in which the dispositions of the continents will change are speculative; we follow here a general overview based on the analysis for historical continental drift that resulted in the patterns shown in Figure 2.7.

Continental locations predicted for 100 Myr AP are pictured in Figure 18.5a. This situation, although differing substantially in detail from the present day, retains much of the present character of the land distribution but moves the primary land masses farther to the north. Some cooling can be expected, but climate changes as a consequence of continental location modifications will probably be local and regional rather than global.

Continental positions are expected to change significantly by 150 Myr AP as shown in Figure 18.5b. By this time, projected plate motions will have carried much of the continental mass toward the equator, decreasing the possibility of polar ice caps and making more of the land available to the equatorial sun for the absorption of radiation. Thus, the tendency of continental drift during this period will be to increase the average temperature of the planet by decreasing its albedo. A supermonsoon regime may develop, with a very hot and wet climate over much of the continental area. This situation is even more likely to occur as one looks ahead to 250 Myr AP, when the continents could be nearly joined in a new supercontinent spreading over tropical and temperate latitudes as shown in Figure 18.5c. Little potential will remain under those circumstances for ice and snow formation or retention, and planetary heating will be strongly reinforced.

NEO-ARCHEAN CLIMATE

Our longest look into the future goes as far forward as we are able to look back: some 4 or 5 Gyr. On this time scale, the major additional factor that enters into our assessment is the aging of the Sun and the consequent effect on the intensity of solar radiation. Chapter 10 presented the "Weak Sun Paradox", in which during the early life of the planet the average temperature was suitable for water to exist in the liquid phase despite a presumed lower solar radiation flux than at present. As we discussed earlier, high atmospheric carbon dioxide

a

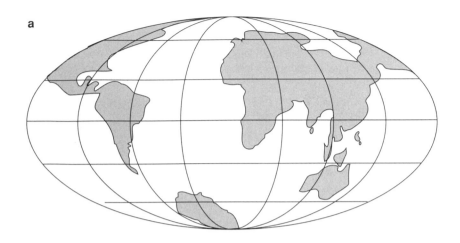

b

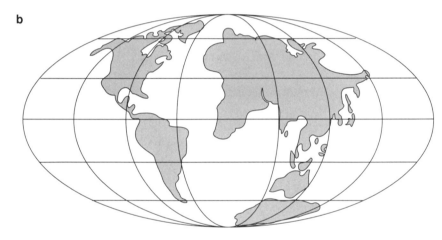

Figure 18.5 Prospective future locations of the continents. Compare with the present condition, as shown in Figure 2.7f, where the American continents are joined, as are Africa and Eurasia, and India has collided with Eurasia to form the Himalayan mountains. (a) 100 Myr AP: Africa and Europe join completely, forming new mountain chains. Australia moves closer to Eurasia. The Atlantic Ocean exceeds the Pacific in size and the land bridge between North America and Asia strengthens. (b) 150 Myr AP: Eastern and western hemispheres separate as the Atlantic Ocean narrows. Land masses move toward the Equator. (c) 250 Myr AP: The closing of the Atlantic forces the continents toward the formation of a new supercontinent, containing a remnant ocean between Africa-Eurasia and South America-Australia-Antarctica. New mountain chains are formed by the continental collisions. (Courtesy of C. R. Scotese, Univ. of Texas.)

c

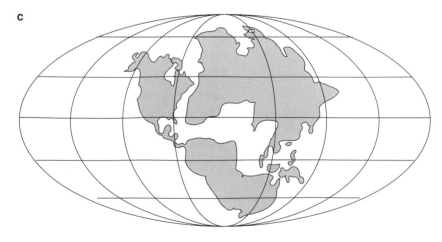

levels and a luminous mass-shedding Sun have both been proposed in efforts to resolve the paradox. What is anticipated for the far future?

Astronomers have established many of the aging characteristics of common stars such as the Sun, and a standard model for luminosity of a Sun-like star as a function of age is shown in Figure 18.6. This model, together with a wealth of other astronomical information, indicates that solar radiation will increase substantially over the next few billion years. Unless this added radiation is reflected back into space or is otherwise offset, it will greatly warm the planet regardless of the atmospheric content of CO_2. New life forms, adapted to the increased temperature, can be expected to dominate the planet, at least that portion of it that is exposed to the surface climate.

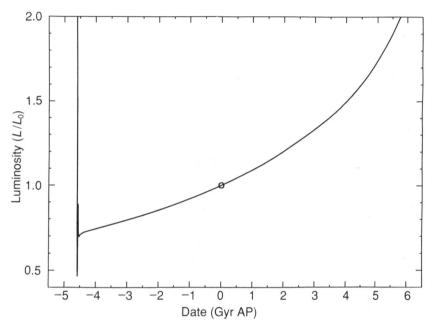

Figure 18.6 Predicted total solar luminosity (erg s^{-1}) relative to that of the present, from the time of the organization of the solar nebula and the assembly of the Sun and planets at 4.7 Gyr BP until more than 5 Gyr AP. (Courtesy of I.-J. Sackmann, California Institute of Technology.)

A SUMMARY OF POSSIBLE CLIMATE FUTURES

A warmer climate in the shorter term as a consequence of increased concentrations of the greenhouse gases, in conjunction with an expected decrease in Earth's albedo as ice and snow areas are decreased, will most likely be sufficient to overcome for a thousand years or so the cooling of the Earth in its natural oscillatory cycle. If this overcompensation does occur, the Earth's temperature will exceed within the Neo-Holocene period its highest level in more than 100 kyr. (See Figure 10.15 and 18.4b). Following that warm spike, climate will resume its cooling trend for the next several thousand years, with periodic glacials and interglacials to follow.

The Neo-Tertiary adds the possibility of bolide impacts to the forces of greenhouse gases and orbital cycles. On a time scale of a few years, bolide impacts would cool the Earth and cause substantial extinctions of many of its life forms. Over a longer term, the additional carbon dioxide and other greenhouse gases added to the atmosphere as a consequence of the impacts would provide increased impetus to global heating. The probability of a bolide impact of the magnitude of that which apparently occurred at the Cretaceous–Tertiary boundary is rather small on the time scale of our genus (perhaps 10 Myr). However, the probability of smaller sized but still huge impacts (corresponding to energies many times larger than that from a full nuclear war) is much larger.

In the Neo-Cenozoic and Neo-Paleozoic the motions of the continents toward the equator are expected to bring a further and increasingly important heating process into play, because most of the land masses will be in positions conducive to absorbing radiation and not conducive to retaining highly reflective ice and snow.

Over the longest time scale of interest, the Neo-Archean, the aging of the Sun will result in substantial increases in solar radiation flux, with intensities at Earth increasing by 50% or more. The tendency toward global heating will be marked as the Sun proceeds along the path to its eventual demise.

EXERCISES

18.1 Glaciers and polar ice have combined volumes of 2.9×10^7 km³. If global warming causes 1% of that ice to melt into the ocean, what will the rise be? What if 10% melts? Finally, suppose the mean sea surface temperature increases from 23 °C to 26 °C as a consequence of global warming. If this increase occurs throughout an ocean depth of 1300 m, what will be the sea level rise attributable to thermal expansion of the oceans? Should any of these processes occur today, which of the following cities would be strongly affected:

Sydney, Australia (altitude, 8.5 m); Rio de Janeiro, Brazil (altitude, 10 m); New York City, USA (altitude, 18.5 m); Rome, Italy (altitude, 30 m); Athens, Greece (altitude, 97 m)?

18.2 What do you think might happen to Earth's climate under the situations depicted in Figure 18.5? Explain your reasoning.

18.3 Discuss the possible influences of increasing solar radiation flux (Figure 18.6) on a regional climate of your choosing.

FURTHER READING

A. Berger, The Earth's future climate at the astronomical time scale, in *Future Climate Change and Radioactive Waste Disposal*, C. Goodess and S. Pakstikof, eds., Norwich, UK, 1989.

R. S. Deitz and J. C. Holden, The breakup of Pangaea, *Scientific American*, 223 (4), 30–41, 1970.

A. S. Endal and K. H. Schatten, The faint young sun-climate paradox: continental influences, *J. Geophys. Res.*, 87, 7295–7302, 1982.

H. H. Lamb, *Climate: Present, Past, and Future*, Metheun and Co., London, 1977.

19

On Change and Sustainability

STIMULI FOR ATMOSPHERIC CHANGE

Throughout this book we have focused on the theme that the environments of Earth have always been subject to change but that at present the changes are occurring at a rate far faster than any in history, transient events such as volcanic eruptions and meteorite impacts excepted. To provide a bit of additional perspective, we plot in Figure 19.1 the time scales for some of those changes. The ordinate of the plot is logarithmic, so distances along that axis are substantial. At the top is the stimulus with the longest time scale, the alterations to the location and morphology of continents produced by plate tectonic motions. Alterations to climate and chemistry as a consequence are indisputable, but the time scale is such that humanity will not be influenced by, nor can it influence, those changes.

Figure 19.1 shows next the possible frequency of bolide impacts, as suggested by both the geological and fossil records. With typical intervals of perhaps 30 Myr they are natural events about which we can probably do nothing. If not intercepted, bolides will produce effects that are extremely prompt and devastating. Less catastrophic but still drastic enough to completely alter the physical and biological character of the planet are the ice ages, whose time scales of 10–20 kyr are not much longer than is the human record left by writings and artifacts.

The effects of emissions attributable to mankind are plotted in three places in Figure 19.1: for those that result in global, regional, and local forcing. In technical terms, we are introducing a surge in the spectrum of Earth climate variability (Figure 10.16) at about 10^2 yr. Only impulsive natural events, such as very large volcanic eruptions and bolide impacts, have a more immediate effect on the atmosphere.

Figure 19.1 Time scales involved in stimuli for atmospheric change. Typical lifetimes of several features and inhabitants of the planet are indicated at the right of the display.

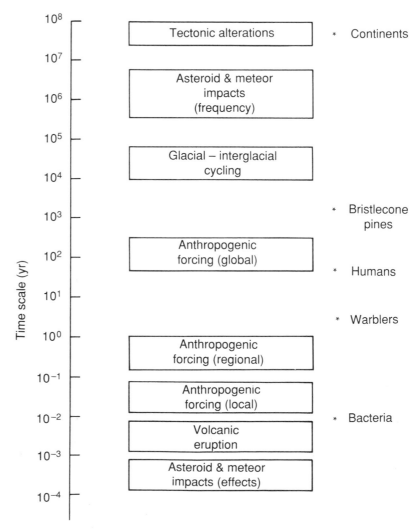

At the right of Figure 19.1 are a few indicators of typical lifetimes for living and nonliving inhabitants or features of the planet. It is striking, though quite coincidental, that humanity's emissions to the atmosphere are causing global change over a time span consistent with typical life spans. As a consequence, both the fruits and the dire consequences of our actions will be visible to us and to our children, grandchildren, and their grandchildren, not delayed until some time too distant to cause concern.

KNOWN UNCERTAINTIES

Scientists do not yet have a good understanding of the Earth system. If this shortfall were a matter of details, the deficiency might not be very important. It may be, however, that many of the present uncertainties do not deal with details but with the foundations of the planet and its climate system. Various studies have predicted catastrophic results as a consequence of planetary forcing not much beyond what is now occurring. Others suggest that the systems are much more stable than such a result would suggest. Clearly, no one can be certain of the outcome, in part because in many ways the planet is being pushed into situations it has never before encountered and because scientists have no satisfactory previous histories to study in an attempt to make accurate predictions.

A specific problem about which scientists are uncertain is the role of particles and ice crystals in the atmosphere in the chemistry that occurs there. As the Antarctic ozone hole demonstrated, such chemical processes can be dramatic and deleterious. The processes that are involved in the ozone hole are now reasonably well understood, but particles appear in many parts of the upper troposphere, not just over Antarctica, and in the troposphere at large their effects are little studied. Similar uncertainties apply to many aspects of atmospheric and surface aqueous phases.

Another important area of uncertainty is the degree of nonlinearity of many Earth system processes, and whether thresholds for sudden change are present. It is worthwhile in this regard to recall the stability diagram of Chapter 1, as well as the strong suspicion that the planet oscillates between several quite different metastable states. It is clear that our present position on the stability diagram, and the locations of the points of inflection, are unknown. In complex systems one is often unsure whether system response will be proportional to the degree of forcing, yet our intuitive expectation of such a response will leave us unprepared if the responses happen to be dramatic.

IMPORTANT AND NOT-SO-IMPORTANT DRIVING FORCES

One of the ultimate goals of predictions about Earth's future is to provide policy-makers with usable techniques for evaluating the environmental effects of local, regional, and global development. For this purpose, we and others have generated new data bases relevant to the transformed planet, made projections concerning future development, evaluated the effects of past, present, and future activities of man, and devised new visual techniques for displaying the results in a clear and compact format.

The results described in earlier chapters have been directed at a restricted set of perceived problems, such as greenhouse warming, ozone depletion, and precipitation acidity. Other potential problems have been considered in less detail, such as the possible degradation of atmospheric visibility, of aquatic populations, or of soil, yet these and many other environmental quality parameters are known to be under developmental pressure. The overriding issue is that policy decisions concerning a specific impact should be made not in isolation but with the best current knowledge of the consequences of each decision across the entire spectrum of possible impacts and the various time and space scales.

The environmental predictions of the last several chapters have identified only a small number of specific processes as playing a major role in the degradation of environmental properties: industrial emissions, fossil fuel production and use, biomass burning, and effects of agricultural activity. Coal production and use is the source having the most substantial impacts over the broad spectrum of environmental quality parameters. (This circumstance poses a major concern for global development planning, because coal and shale oil are by far Earth's most abundant energy resources.) The combustion of petroleum is also of significant concern, as are industrial emissions. The effects of biomass combustion and many facets of agricultural activity also have both global and regional impacts. In planning for a more sustainable world, these dominant and intensifying sources of problems should receive the most attention.

CORRECTIVE COUNTERMEASURES FOR GLOBAL ENVIRONMENTAL PROBLEMS?

In recent years several proposals have been made for the major manipulation of the atmosphere in such a way that some of the negative consequences of previous human activity could be counteracted. Among the problems intended to be addressed by such tactics are global warming, stratospheric ozone depletion, and photochemical smog. Are these ideas worth considering, just as we consider building sea walls or dredging entrances to harbors, or are they ill advised?

Stratospheric Ozone Depletion Countermeasures

In Chapter 8 we noted that the severe ozone depletion observed each spring over the Antarctic continent is a consequence of reactions taking place on polar stratospheric cloud particles. These reactions convert the chlorine atoms sequestered in HCl and ClONO$_2$ into the chlorine-destroying radicals Cl• and ClO•. To counter this process, Ralph Cicerone and Scott Elliott of the University of California, Irvine, and Richard Turco of the University of California, Los Angeles, have proposed the injection of ethane (C$_2$H$_6$) and propane (C$_3$H$_8$) into the Antarctic stratosphere. The concept is that the reactive chlorine radicals will rapidly react with the hydrocarbons to produce unreactive HCl by

$$Cl• + C_2H_6 \text{ or } C_3H_8 \rightarrow HCl + C_2H_5 \text{ or } C_3H_7 . \qquad (19.1)$$

To evaluate this proposal, the researchers carried out computer model calculations to simulate the ozone hole development in the absence and presence of added ethane or propane. As shown in Figure 19.2, the ozone hole, which generally forms around Julian day 240 (the first day of the year is Julian day 1), was suppressed substantially by greater hydrocarbon additions.

So, what are we waiting for? For one thing, a fleet of airplanes. To deliver the 50,000 tons of hydrocarbons that would be needed annually, a fleet of several hundred large airplanes would be needed, and their emissions would need to be considered as would the logistical requirements of generating and transporting the gases to the continent

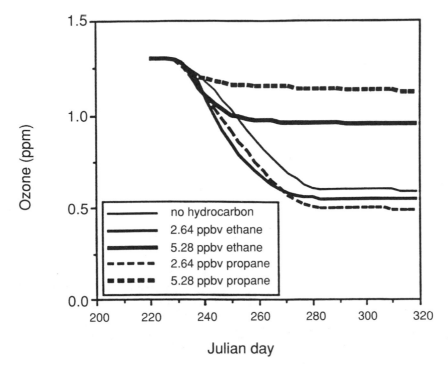

Figure 19.2 The time development of ozone concentrations at 15 km altitude for several models of Antarctic stratospheric chemistry in which inorganic chlorine was set to 1987 levels and different amounts of ethane or propane were injected. (Reproduced with permission from R. J. Cicerone, S. Elliott, and R. P. Turco, Reduced Antarctic ozone depletions in a model with hydrocarbon injections, *Science, 254*, 1191–1193. Copyright 1991 by the AAAS.)

most distant from the developed world. Finally, the authors of this proposal point out that it would be very difficult to assure that the gases could be well mixed in the atmospheric region where they would be needed, because variations in the location and integrity of the Antarctic vortex are substantial.

Greenhouse Warming and Ice Age Countermeasures

Greenhouse warming arises from the emission into the atmosphere of gases that absorb the infrared radiation emitted by Earth, thus preventing its escape into space. To counter this warming, it has been proposed that the amount of radiation reaching Earth from the Sun be reduced. The same idea could, of course, be conceived with respect to the long-term increase in solar luminosity as the Sun approaches old age. If accomplished, such action would result in a "Strong Sun Paradox", in which the planet might be maintained at temperatures suitable for life even as solar radiation increased markedly.

How might the amount of solar radiation be diminished by design? One possible technique might be to place mirrors on space satellites sent to the Lagrangian L1 Point (the point along the Earth–Sun line where the gravitational forces acting on a satellite are in approximate balance). Assessments of this idea are preliminary but suggest that spacecraft stabilization may be difficult, that great expense would be involved, and that political difficulties would be substantial.

An alternative to the reflecting satellite idea was proposed some years ago by Russian scientist Mikhail Budyko, who envisioned the annual injection of some 35 million tons of SO_2 into the stratosphere. He calculated that such an amount, once converted to sulfate aerosol, should be sufficient to significantly enhance the backscattering of solar radiation to space and cause a cooling of Earth's surface. A variation of this plan is to inject soot particles, not SO_2. The soot particles would not reflect solar radiation, but would absorb it, and are so efficient at doing so that only about 1% as much soot would intercept as much radiation as Budyko's sulfate particles. In addition, the heat absorbed at high latitudes in the lower stratosphere may be sufficient to prevent the development of the polar stratospheric clouds and the ozone hole.

What are the prospects for any of these ideas working as intended? We discussed in Chapter 17 the effect of the emissions of anthropogenic sulfur gases, the aerosols from which appear inadvertently to have counteracted some of the heating from elevated CO_2 concentrations. Thus it seems possible that if properly carried out, they might achieve their goals. However, we need to consider the quantitative nature of carrying out these intentional countermeasure proposals. When we do, it turns out that delivering thousands of tons of aerosol particles is a major logistical task. One possibility suggested is to put particles in ballistic shells and have the world's large naval vessels fire several thousand rounds per day each into the stratosphere, day after day, year after year. Another is to fly fleets of hundreds to thousands of planes to deliver the particles to the proper altitude. Either method of delivery could probably be accomplished in principle, at a cost of tens of billions of dollars (U.S.) annually.

A proposal in the opposite direction considers the global climate several thousand years from now, when undisturbed natural cycles would be expected to be moving in the direction of the next ice age and greenhouse gas warming of the atmosphere will have diminished as fuel for greenhouse gas emissions will be exhausted. To warm the planet instead of cool it, one would want to inject greenhouse gases with highly efficient absorption properties into the atmosphere. A promising candidate is carbon tetrafluoride (CF_4), which is virtually unreactive and insoluble in water and thus would be expected to have a very long atmospheric lifetime. Its strongest absorption is centered in the infrared window at a wavelength of about 8 mm. Thus CF_4 has the necessary properties to be a powerful greenhouse gas. It also has the major logistical difficulties of the other suggestions discussed above, so its delivery to the stratosphere would be a major global undertaking.

Bolide Impact Countermeasures

A suggestion quite unlike those above is directed toward the possibility, apparently realized numerous times over the eons, of a collision with Earth by an asteroid or comet. Such a collision, if large and relatively direct, would render habitation of the planet difficult or impossible for many of its biological species. At present, we have surveillance too inadequate to detect such objects routinely and no means of preventing a collision were the orbital paths so aligned. However, the observation and intercept technology is not qualitatively different from that used to detect and intercept intercontinental ballistic missiles, and major progress in doing so has been made over the past several decades. To achieve bolide destruction capability, it would be necessary to continue and enhance antiballistic missile research and countermeasure activities, while ensuring that the employment of the resulting hardware were solely for peaceful purposes. Although unlikely, it is not inconceivable that the world's military hardware could be turned from defense against attack of one nation by another to defense against attack of Earth by random motion of celestial objects.

Discussion

It is important to realize that several of the environmental countermeasures discussed above have three daunting characteristics: they depend on near-perfect knowledge of the effects that would be produced were they to be carried out, they require major logistical efforts, and they are expensive. Furthermore, once started they would have to be continued indefinitely to avert environmental catastrophe, regardless of the world's political situation, economic conditions, and degree of mutual cooperation. These are conditions that only the very bold would be willing to bet on.

Bolide interception requires different approaches. Continuing research and development of military hardware will be needed, and surveillance maintained over long periods of time. Because bolide impacts are expected to occur no more frequently than every few million years, maintaining an appropriate state of readiness would be extraordinarily difficult.

Even given that finances, cooperation, and logistics were in place to implement one of these suggestions, one would want to be very sure that the consequences of the action were well understood. It is at this point that all these ideas founder. To paraphrase a comment by John Firor of the U.S. National Center for Atmospheric Research, if we do not understand how a leaf works, how can we think we can make the atmosphere dance to our tune? In our opinion, environmental counter-measures such as those discussed here should be considered only in the face of very grave and apparently unremitting disasters of our own making, and then with the utmost caution and flexibility.

SURPRISES

Most discussions of future possibilities begin at or near the present time, anticipate moderate changes to important environmental properties, and predict in some analytic or subjective way the consequences of these perturbations. Such techniques appear to be quite satisfactory under most circumstances and on short time scales. For the longer time scales of interest to us here, however, they may be less apposite. All kinds of history, including particularly the history of climate, show that over time scales of several decades or more the important changes are generally not the result of gradual modifications of factors of concern such as the price of oil or the global average temperature, but instead are manifested by rapid and dramatic surprises. C. S. Holling has defined such surprises as those "when causes turn out to be sharply different than was conceived, when behaviors are profoundly unexpected, and when action produces a result opposite to that intended—in short, when perceived reality departs *qualitatively* from expectation."

Surprises can result from thresholds or nonlinearities that have not been taken into account. A classic example, cited by Harvey Brooks of Harvard University, is that of automobile use in urban areas at peak traffic hours, in which participants acting for their own benefit precipitate trouble for all, because the disbenefits of such action set in quite abruptly at a critical traffic density. An example of nonlinear atmospheric response can also be presented as a consequence of heavy automobile traffic: it is the generation of ozone from automotive emissions. Because the ozone is an indirect consequence of reactions among several emittants and because the product flux from the reactions is nonlinear with concentration (see Figure 16.1), high levels of urban ozone appear permanent at some critical level of urban development unless emissions from each automobile are reduced to extremely low levels.

As scientists learn more about the earth system, they are discovering evidence for a number of these surprise scenarios in historical data of one kind or another. Among the examples one might cite are the mass extinctions of the dinosaurs at about 65 Myr BP or the "punctuated" pace of the evolution of species during several other geological periods. We have already mentioned the unexpected decreases in stratospheric ozone over the Antarctic in the austral spring.

Another intriguing example of the episodic nature of many of Earth's systems is provided by the work on ocean circulation of Wallace Broecker of the Lamont-Doherty Geophysical Observatory. Recall from Figure 6.5 that, in rough overview, the major ocean currents at depths well below the surface are initiated by the overturning of cold surface water of high salinity in the North Atlantic, followed by deep water flow into the Pacific. This flow is driven by the temperature differences between the two oceans, the higher temperature of the North Atlantic resulting in evaporation of water (thus promoting high salinity and high density) and the cooler temperatures of the North Pacific resulting in condensation (thus promoting low salinity and low density). The way in which the atmosphere and ocean are coupled is obviously important, but not at all well understood. Broecker emphasizes that Earth's climate does not respond to forcing in a smooth and gradual way, but in sharp jumps. Examples are provided by Figure 19.3, which shows the average surface temperature of the North Atlantic, determined by the deuterium content of marine fossil shells, over the past 30 kyr. The record, except for the most recent 10 kyr, is one of rapid fluctuations, some temperature changes that are

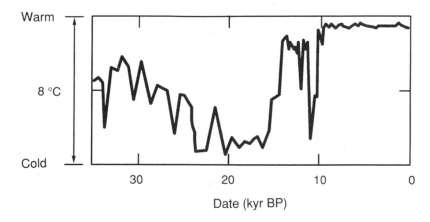

Figure 19.3 A temperature history of the surface waters of the North Atlantic ocean. (Adapted with permission from W. Broecker, Unpleasant surprises in the greenhouse?, *Nature, 328*, 123. Copyright 1987 by Macmillan Magazines Limited.)

several degrees in magnitude taking place in as little as a century. The concern is that increases to the concentrations of CO_2 and the other greenhouse gases will jolt the ocean–atmosphere system out of its current mode and into one more appropriate to Earth's altered temperature state and less appropriate to the well-being of Earth's present population or portions thereof. As Broecker points out, scientists have little basis for answering this question, and surprises, probably disagreeable ones, appear more likely than not.

BIOSPHERE–ATMOSPHERE COUPLING

One of the most controversial ideas of recent environmental science has been the "Gaia hypothesis" developed by James Lovelock of Cornwall, England. This hypothesis suggests that microorganisms, plants, and animals act in such a way that Earth's environment is adjusted to states optimum for the maintenance of living things. As formulated, it requires not that the adjustments by the biosphere be done knowingly but that the adjustments arise from natural selection. The concept is perhaps best illustrated by Lovelock's simple paradigm, paraphrased here. Lovelock pictures Daisyworld, a cloudless planet in which the environment is indicated by a single variable, temperature, and the biota by a single species, daisies. Initially, let us assume that all the daisies are white. Because they are lighter in color than the ground in which they grow, they tend to increase the albedo of their locality, and, as a consequence, are cooler than a comparable area of bare ground. Because the daisies cover a significant fraction of the planetary surface, they influence the mean surface temperature of the planet to some extent. This influence might change if there were a change in some external variable that influenced the planetary temperature. An example of such a variable is the output of radiant energy from Daisyworld's sun.

Like most plant life, the daisies grow best over a restricted range of temperatures. The growth rate peaks near 23 °C and falls to zero below 5 °C and above 40 °C. The effect of daisy populations on planetary temperatures is monotonic, however, and the steady state of the whole system must be specified by the point of intersection of the two curves. If the system is initialized at some arbitrary point, it will normally settle down at a stable solution.

What happens to this stable solution when some change of the external environment alters the planetary temperature? Suppose, for example, that the sun warms up as our Sun is said to be doing. If the daisy population is artificially held constant, the planetary temperature will simply follow the change of heat output of the sun; there will be a much larger temperature change than if we allow the daisy population to grow to its new natural steady state. In this new steady state, the daisy population has changed to oppose the effect of a change in solar output.

Very few assumptions are made in this model. It is not necessary to invoke foresight or purpose on the part of the daisies. It is merely assumed (in this picture) that the growth of daisies can affect the mean planetary temperature and vice versa. Note that the mechanism works equally well whatever direction the effect is. Black daisies would have done as well; as long as the Daisyworld albedo is different from that of the bare ground, some thermostasis will result. The assumption that growth is restricted to a narrow range of temperatures is crucial to the working of the mechanism, but all mainstream life is observed to be limited within this same narrow range; indeed, the peaked growth curve is common to other variables besides temperature, for example, pH and the abundance of nutrients.

The Gaia concept has inspired vigorous scientific discussion, partly because of the difficulty of defining what is meant by *optimum*. The concept can clearly not be applied to a single species in a group of species, because each is constantly involved in winning and losing relationships. Even for Earth's entire biosphere, the idea appears suspect. The most dramatic counterexample to Gaia is that some 600 Myr ago, i.e., relatively recently in the planet's history, the increase in the atmospheric oxygen concentration was lethal to most of Earth's (at that time entirely) anaerobic life forms. The result was the evolution of an almost entirely new biosphere that was adapted to the high oxygen supply and utilized solar energy much more efficiently than before. Such a transformation is "anti-Gaian" in the sense that it was detrimental to the biosphere from which it sprang. In more recent geological history, during the ice ages the atmospheric concentrations of the greenhouse gases CO_2, CH_4, and N_2O were substantially lower than during interglacial periods and thus enhanced the climate forcing implied by the variations in the orbital elements of the planet's orbit.

In response to such arguments, Lovelock has cited possible current examples in which the mediating actions of the biosphere appear to arise naturally from actions taken in local self-interest. In the most recent rejoinder, Robert Charlson of the University of Washington, Lovelock, and coworkers propose that the major source of cloud condensation nuclei (CCN) over the oceans is sulfate particles formed by the oxidation of the dimethyl sulfide (DMS) produced by planktonic algae in seawater. Because the reflectance of clouds (and thus Earth's radiation budget) tends to increase with the number concentration of droplets (for the same liquid water content), a parameter that is in turn sensitive to CCN density, biological regulation of the climate may be possible through the effects of temperature and sunlight on phytoplankton population and dimethyl sulfide production. A schematic diagram of the possible feedback cycle is shown in Figure 19.4. If this mechanism is indeed operative, it would tend to counteract any atmospheric warming arising from increased concentrations of CO_2 because the increased CO_2 leads to higher oceanic productivity and larger populations of DMS-emitting plankton, thus

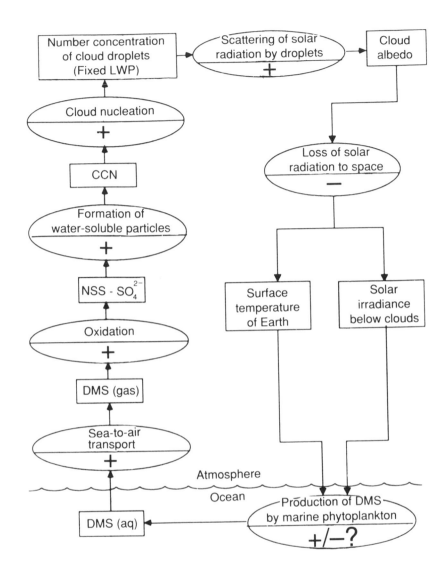

Figure 19.4 Conceptual diagram of a possible climate feedback loop. The rectangles are measurable quantities, and the ovals are processes linking the rectangles. The sign (+ or −) in the oval indicates the effect of a positive change of the quantity in the preceding rectangle on that in the succeeding rectangle. The most uncertain link in the loop is the effect of cloud albedo on DMS emission; its sign would have to be positive in order to regulate the climate. CCN, cloud condensation nuclei; DMS, dimethyl sulfide; LWP, liquid water path; NSS–SO_4^{2-}, non-seasalt sulfate. (Reproduced with permission from (R. J. Charlson, J. E. Lovelock, M. O. Andreae, and S. G. Warren, Oceanic phytoplankton, atmospheric sulphur, cloud albedo and climate, *Nature, 326,* 655–661. Copyright 1987 by Macmillan Magazines Limited.)

higher concentrations of CCN, thus more clouds and higher albedo, thus a cooling of water and climate, and thus increased absorption of CO_2 into the oceans. Indeed, it appears from the higher measured deposition of methane sulfonic acid in Antarctic ice cores (MSA is a DMS derivative) that oceanic DMS emissions were higher during glacial periods than during interglacial periods, as Gaia would propose.

Earlier in this chapter, we discussed the possibility of a Strong Sun Paradox, in which the planet might be maintained at temperatures suitable for life even as solar radiation increased markedly. Could the biosphere respond to create such a paradox by either decreasing the infrared absorbers in the atmosphere or increasing the planet's albedo? Decreasing the concentrations of infrared absorbers would presumably require that the biomass absorb large amounts of CO_2, and it is intriguing that recent studies of the CO_2 budget suggest a strong land-based biospheric sink. A limit to such intervention seems to be the evidence that plant respiration requires about 180 ppmv CO_2. The alternative of an albedo increase might be attempted by the DMS cycle of Figure 19.4. It is very uncertain whether sufficient regulation of temperature could be accomplished in this way. Alternatively, one might imagine a transformation of life on the planet to almost totally new forms suitable to the altered conditions, as occurred at the Precambrian–Cambrian boundary.

Despite the fact that specific examples can be raised against the Gaia hypothesis, there is no doubt that the chemical composition of the atmosphere is regulated to a large degree by the activity of Earth's biosphere. Because Earth's climate and the availability of nutrient elements are, in turn, strongly dependent on the chemical composition of the atmosphere, it is clear that the life forms that adapted best to the environment provided a type of Gaia coupling. Exactly how the coupling is manifested and whether the driving forces and the driven responders can be readily distinguished remains to be determined. It is in this context that we consider again the comparative situations of Earth and Venus. The total amount of carbon is similar on both planets, but on Venus the carbon is mostly present as atmospheric CO_2, whereas on Earth it is mostly stored in biospheric reservoirs. In at least this passive manner, the biosphere provides an environment suitable to its own sustainability.

Lovelock's ideas have raised the expectation with some that the planet has unexpected resources capable of resisting at least some of the perturbations caused by human development of the biosphere. However, Gaia is concerned with much longer time scales than those of human industrial development. The crucial question is not whether Gaia is active, but whether humanity's actions can drive the Earth system beyond any hypothetical Gaia repair capability.

THE THREE TIME SCALES

The information we have presented makes it clear that mankind is causing major changes in the composition of Earth's environments on local, regional, and global scales. These changes are caused by a relatively small number of source types, and the sources of most general concern are almost wholly anthropogenic: fossil fuel combustion, land use changes, biomass burning, deforestation, agricultural practices, and industrial processes.

It is perhaps possible to summarize this book by discussing the stability of the atmosphere on three time scales: near term (a few decades), intermediate term (a climate cycle, perhaps 10 kyr), and ultimate term (the lifetime of the stable planet). As we have discussed here and elsewhere, it is difficult to be optimistic about the environmental quality of the planet over the near term. On local and regional scales, there is abundant evidence that the carrying capacity of the natural systems is being overwhelmed, not in all locales, not for all things measured, but more often than not. Only a return to the world of two centuries ago, with little combustion-generated power usage and little heavy industrial activity, would seem a certain way to mitigate this trend, given the present and predicted global population and current development scenarios, and such a transition is clearly unrealistic. Can this pessimistic prospect be changed? Two unprecedented and seemingly unlikely developments would be required. The first is the invention and quick and global implementation of technology designed to minimize the emissions of trace gases and particles to the atmosphere, hydrosphere, and pedosphere. The second is the willingness of the developed world to make the political and financial commitments necessary to bring that technology promptly to the developing world. As with many unexpected rapid changes in natural processes on Earth, achieving the two developments presented above would demand global cooperation so extensive that its achievement would be a major surprise in its own right.

For the intermediate time scale, one can perhaps be somewhat more hopeful regarding stability (although perhaps stability in a somewhat different state than that at present). Earth has shown itself capable of enduring catastrophic natural events and slower challenges such as the ice ages, with major effects on climate and biology, and coming back to regenerate the amazing diversity of ecology we see today. Thus, even if we as humans are shortsighted enough to induce a surprise with consequences as severe as a full glacial or thermal event, the planet may be able to regenerate itself, although this outcome is in no way guaranteed. It is quite possible instead to imagine some runaway effect such as the release of large quantities of methane from the permafrost and methane hydrate sediments in polar regions as a consequence of a long-lasting super greenhouse warming.

On very long time scales, the planet will surely become incapable of sustaining life as a consequence of the increasing solar radiation. Whether that time scale is of any interest, however, may or may not turn out to be under humanity's control. As a people, we are causing stresses to the biospheric system beyond all previous experience, and hoping that a surprise that cannot be managed is not waiting in the wings. It is unlikely that scientists can be particularly useful in attempting to predict the ultimate stability of the planet under its increasing stress, because such a system is totally outside experience and thus probably beyond reasonable prediction.

Optimism and pessimism are more often products of instinct and temperament than of rigorous logic, and some have foreseen great new eras in the life of our planet and of humanity on it. This optimistic outlook was given eloquent expression in 1902 by H. G. Wells:

It is possible to believe that all the past is but the beginning of a beginning, and that all that is and has been is but the twilight of the dawn. It is possible to believe that all the human mind has ever accomplished is but the dream before the awakening. We cannot see, there is no need for us to see, what this world will be like when the day has fully come. We are creatures of the twilight. But it is out of our race and lineage that minds will spring, that will reach back to us in our littleness to know us better than we know ourselves, and that will reach forward fearlessly to comprehend this future that defeats our eyes. All this world is heavy with the promise of greater things, and a day will come, one day in the unending succession of days, when beings, beings who are now latent in our thoughts and hidden in our loins, shall stand upon this earth as one stands upon a footstool, and shall laugh and reach out their hands amidst the stars.

Can we hope to realize Wells's vision? It will only occur if the planet that is now peopled by the men and women whose decendants will bring that vision to pass can be sustained until the great days of supreme achievements arrive. What, then, should be our actions as the present citizens of "Spaceship Earth"? One overriding guideline seems most prudent: that in seeking to develop Earth's resources for the benefit of all, we seek at the same time to minimize in every way possible the stresses on our planet, for it is, indeed, the only one we have.

EXERCISES

19.1 To reduce the atmospheric concentrations of carbon dioxide, it has been suggested that energy-saving measures be vigorously promoted, perhaps by imposing a tax on CO_2 emissions. Do you think this would be effective? Of the numerous and difficult possible ways to deal with CO_2 emissions, do you think it is the most desirable? (There is a rich literature on the economics of environmental problems. The specific question of a CO_2 tax is treated at length in Chapter 2 of *Changing Climate*, Carbon Dioxide Assessment Committee, National Academy Press, Washington, D.C., 1983. Most libraries will have other references.)

19.2 Nuclear power is regarded by many as a possible alternative to energy production by fossil fuel combustion. The major advantage is that nuclear power produces no atmospheric emissions. The disadvantages include fear of accidents, fear of terrorism, and difficulties of storage of spent fuel. Considering the alternatives, do the advantages outweigh the disadvantages? (Among the readily available short introductions to

aspects of this topic are the following: J. M. Harrison, Disposal of radioactive wastes, *Science, 226*, 11–14, 1984; J. Haggin, New era of inherently safe nuclear reactor technology nears, *Chemical and Engineering News*, 18–22, June 30, 1986; J. F. Ahearne, Nuclear power after Chernobyl, *Science, 236*, 673–679, 1987.)

19.3 In the Neo-Tertiary, a major climate forcing may arise as a consequence of a massive bolide impact with Earth. In recent years, the United States has embarked upon a "Strategic Defense Initiative" (SDI) for the interception of nuclear missiles. It has been proposed that SDI technology might lend itself to bolide interception just as well as nuclear missile interception. Using information on bolides, missiles, and SDI technology, evaluate whether today's SDI technology is adequate for the task. If not, does eventual achievement of such a capability seem possible? If it appears to be possible, either now or later, should such a countermeasure program be implemented? Explain your answer.

FURTHER READING

W. Broecker, *How to Build a Habitable Planet*, Eldigio Press, LDGO Box #2, Palisades, NY, 1985.

R. J. Charlson, J. E. Lovelock, M. O. Andreae, and S. G. Warren, Oceanic phytoplankton, atmospheric sulphur, cloud albedo, and climate, *Nature, 326*, 655–661, 1987.

S. H. Schneider and P. J. Boston, Eds., *Scientists on Gaia*, MIT Press, Cambridge, MA, 1992.

Single topic dedicated issue: Managing Planet Earth, *Scientific American, 261* (3), 1989. (Also available in paperback from W.H. Freeman and Co., New York, 1990.)

World Commission on Environmental Development, *Our Common Future*, Oxford University Press, Oxford–New York, 1987. Copyright 1987 by Macmillan Magazines Limited.)

Glossary

Absorption cross section A quantitative measure of the absorption efficiency of light by a given atom or molecule at a specified wavelength, generally expressed in units of square centimeters per particle (cm^2 particle^{-1}).

Abstraction The removal of an atom from its parent molecule. In atmospheric chemistry, abstraction most often refers to the loss of a hydrogen atom from a molecule undergoing reaction with the hydroxyl radical.

Accumulation mode particles Particles formed in the *atmosphere* by accretion or nucleation of gas-phase molecules.

Acid A species that reacts in liquid water to generate hydrogen ions (actually, hydrated protons: H_3O^+); anions that were associated with the hydrogen ions in the acid are also released. See also *Strong acid*.

Acid deposition The deposition of acidic constituents to a surface. This occurs not only by precipitation but also by the deposition of atmospheric particulate matter and the incorporation of soluble gases.

Acid precipitation Rain, snowfall, or atmospheric moisture with a pH reading below about 5. Natural rain is slightly acidic, so 4.8–5.5 pH is considered natural for most parts of the world.

Activation energy The energy barrier that must be overcome during a collision of two potential reactants in order for a reaction to occur.

Activity The activity a and the activity coefficient f of a substance are defined by the equation $a = cf$ where c is the concentration (mol L^{-1}). The activity coefficient f is usually an empirically determined quantity and depends not only on the concentration of the substance but also on its particular properties and on the concentration and kind of other substances present. As solutions become ideal ($c \rightarrow 0$), $f \rightarrow 1$.

Adiabatic lapse rate See *Lapse rate* .

Advection The transport of a quantity (of material, heat, or other property) due solely to the mean motion of an air mass. Advection relates largely to horizontal transport as opposed to *convection*, which involves local vertical motion, as in a cumulus cloud.

Aerosol In atmospheric chemistry, a gaseous suspension of fine solid particles that may or may not have water comprising a small to moderate fraction of their mass.

Albedo The ratio of the electromagnetic radiation flux reflected from a body (or surface) to that incident upon it. Albedo connotes a broad wavelength band, whereas *reflectivity* more often is used for monochromatic radiation.

Anion A negatively charged ion (e.g., Cl^- or OH^-) or radical (e.g., $O_2^-\bullet$) that migrates toward the anode of an electrochemical cell.

Anthropogenic Derived from human activities.

Anticyclonic flow The air flow produced about a high pressure center by the combination of two forces: the pressure gradient accelerating the air away from the center and the Coriolis force acting inward. Anticyclonic flows are clockwise in the northern hemisphere and counterclockwise in the southern hemisphere.

Aqueous Pertaining to water; e.g., an aqueous solution is a solution of a constituent in water.

Aquifer Any water-bearing rock formation or group of formations, especially one that supplies ground water, wells, or springs.

Asthenosphere The worldwide layer that lies below the *lithosphere* and at the top of the upper *mantle* and is marked by low seismic wave velocities and high *seismic attenuation*. The asthenosphere is a soft layer, probably partially molten. It may be a region of convection .

Atmosphere The envelope of air surrounding Earth and held to it by gravity. It is one of the five main regimes of the environment: atmosphere, biosphere, cryosphere, hydrosphere, and pedosphere. See also *Biosphere; Cryosphere; Hydrosphere; Pedosphere.*

Base A species that, when dissolved in water, generates hydroxide (OH^-) ions, or that is capable of reacting with an *acid* to form a *salt*.

Bimodal distribution   A plot of the frequency of occurrence of a variable versus the variable is a bimodal distribution if there are two maxima of the frequency of occurrence separated by a minimum.

Biosphere   That spherical shell encompassing all forms of life on Earth. It is one of the five main regimes of the environment: atmosphere, biosphere, cryosphere, hydrosphere, and pedosphere. The biosphere extends from the ocean depths to a few thousand meters of altitude in the atmosphere, and includes the surface of land masses. Alternatively, the life forms within that shell. See also *Atmosphere; Cryosphere; Hydrosphere; Pedosphere.*

Budget   A balance sheet of all of the sources and sinks for a particular species or group of species in a single reservoir or in two or more connected reservoirs.

Cation   A positively charged ion (e.g., H^+ or Zn^{2+}) or radical (e.g., $NH_4^+•$) that migrates toward the cathode of an electrochemical cell.

Chain termination   A chemical reaction between two free radicals (the chain carriers) to produce stable products.

Climate   The temperature, humidity, precipitation, winds, radiation, and other meteorological conditions characteristic of a locality or region over an extended period of time.

Cloud   In general, a recognizable, visible aerosol parcel. Usually, *cloud* refers to a water droplet aerosol with relative humidity slightly above 100%.

Coagulation   The process by which small particles in a colloidal system collide with and adhere to one another to form bigger particles. *Brownian motion* is the usual mechanism that causes collision, although others exist, including sound waves and electrical forces.

Coarse particles   Particles with diameters equal to or greater than 2.5 μm.

Complex   A compound formed by the union of a metal ion, often a transition metal, with a nonmetallic ion or molecule (termed a *ligand* or complexing agent). The bonding is neither covalent nor electrostatic, but intermediate between the two. Also called *coordination compound.*

Condensation nuclei   Aerosol particles that are capable of initiating liquid drop formation at low water vapor supersaturations. The most effective particles are those consisting of water-soluble, hygroscopic substances.

Continent   One of the seven great bodies of land on Earth: Africa, Antarctica, Asia, Australia, Europe, North America, and South America.

Continental drift   The horizontal displacement or rotation of continents relative to one another.

Continental shelf   The gently sloping submerged edge of a *continent*, extending commonly to a depth of about 200 m or the edge of the continental slope.

Convection   A mechanism of rapid heat transfer through a liquid in which hot bubbles of material from the bottom rise because of their lesser density, while cool surface material sinks.

Convergence zone   A band along which moving tectonic *plates* collide and area is lost either by shortening and crustal thickening or subduction and destruction of *crust*. The site of volcanism, earthquakes, trenches, and mountain-building. See also *Subduction zone*.

Coordination compound   See *Complex*.

Cordillera   (If capitalized) The continuous mountain system extending from Alaska to extreme South America and ranging up to 1500 km in width. (If not capitalized) Any similar chain of parallel mountain ranges.

Coriolis force   The deflecting effect of Earth's rotation whereby freely moving air masses are deflected to the right in the northern hemisphere and to the left in the southern hemisphere, relative to an observer on Earth's surface.

Corrosion   The chemical or electrochemical reaction between a material, usually a metal, and its environment that produces a deterioration of the material and its properties.

Crust   The outermost layer of the *lithosphere*, consisting largely of relatively light materials with low melting points.

Cryosphere   The frozen portion of Earth's surface. It is one of the five main regimes of the environment: atmosphere, biosphere, cryosphere, hydrosphere, and pedosphere. See also *Atmosphere; Biosphere; Hydrosphere; Pedosphere*.

Cycle   A system consisting of two or more connected *reservoirs*, where a large part of the material of interest is transferred through the system in a cyclic manner.

Cyclonic flow   The air flow produced about a low pressure center by the combination of two forces: the pressure gradient accelerating the air toward the center and the Coriolis force acting outward. Cyclonic flows are clockwise in the southern hemisphere and counterclockwise in the northern hemisphere.

Daughter element   Also daughter product. An element that occurs in a rock as a product of the radioactive decay of another element.

Deliquescence   The process that occurs when the vapor pressure of the saturated aqueous solution of a substance is less than the vapor pressure of water in the ambient air. Water vapor is collected until the substance is dissolved and in equilibrium with its environment.

Diffusion A process by which substances, heat, or other properties of a medium are transferred by individual or small-scale motions from regions of higher concentration to regions of lower concentration. Molecular diffusion is caused by the brownian motion of molecules. *Eddy diffusion* is due to transport by fluid eddies.

Disproportionation A chemical reaction in which a single compound serves as both oxidizing and reducing agent.

Divergence zone A belt along which *plates* move apart and new *crust* and *lithosphere* are created: the site of *midocean ridges*, earthquakes, and volcanism.

Dry deposition The transfer of trace species (gases or particles) from the *atmosphere* to surfaces as a consequence of molecular diffusion, brownian diffusion, or gravitational settling, in the absence of active precipitation. The term refers to the transfer process and not to the surfaces themselves, which may be moist.

Eddy In turbulent fluid motion, a blob of the fluid that has some definitive character and moves in some way differently from the main flow.

Eddy diffusion The process by which substances are mixed in the *atmosphere* or in any fluid system due to the eddy motion. See also *Diffusion*.

Efflorescence The reverse process of *deliquescence*: the drying of a *salt* solution when the vapor pressure of the saturated solution of a substance is greater than that of the ambient air.

Electrolyte An ionic conductor, usually an aqueous solution of various common *salts*.

Elementary reaction A reaction that cannot be subdivided into two or more simpler reactions.

Eon The largest division of geological time, embracing several *eras* (for example, the Phanerozoic, 600 Myr ago to Present); also any span of 1 billion years.

Epoch One subdivision of a geological period, often chosen to correspond to a stratigraphic series. Also used for a division of time corresponding to shorter intervals of interest.

Equilibrium constant A number that relates the final concentrations of starting materials and products of a reversible chemical reaction to one another.

Fine particles Particles with diameters less than or equal to 2.5 μm.

Flux The rate of emission, absorption, or deposition of a substance from one *reservoir* to another. Often expressed as the rate per unit area of surface.

Fog Water droplet aerosol in contact with or close to Earth's surface. By international definition, the aerosol is called a fog if it reduces visibility below 1 km. Fog differs from cloud in its location and mode of formation, and in generally smaller drop sizes.

Fossil An impression, cast, outline, track, or body part of any animal or plant that is preserved in rock after the original organic material is transformed or removed.

Fossil fuel A general term for combustible geological deposits of carbon in reduced (organic) form and of biological origin, including coal, oil, natural gas, oil shales, and tar sands.

Free energy An exact thermodynamic quantity used to predict the maximum work obtainable from the spontaneous transformation of a given physical or chemical system.

Free radical An atom or group of atoms that does not normally exist in stable forms. Radicals possess unpaired electrons, which are denoted by a centered dot following the formula or superimposed on a chemical structure at the site of unsatisfied bonding. They are produced in the atmosphere or surface waters when a solar photon is absorbed by a molecule and the energy is used to break a chemical bond.

Gondwanaland A hypothetical supercontinent comprising approximately the present *continents* of the southern hemisphere. See also *Pangaea*.

Greenhouse effect The trapping by atmospheric gases of outgoing infrared energy emitted by Earth. Part of the radiation absorbed by the atmosphere is returned to Earth's surface, causing it to warm.

Greenhouse gas A gas with absorption bands in the infrared portion of the spectrum. The principal greenhouse gases in Earth's atmosphere are H_2O, CO_2, O_3, CH_4, N_2O, CF_2Cl_2, and $CFCl_3$.

Half-life The time required for half of a sample of a given radioactive isotope to decay. The half-life of an isotope is inversely related to its decay constant.

Haze An atmospheric aerosol of sufficient concentration to be visible. The particles are so small that they cannot be seen individually but are still effective in visual range restriction. See also *Visibility*.

Henry's law The equilibrium relationship between the concentration of a species in the gas phase, [C(g)], and in the associated liquid phase, [C(aq)]: [C(aq)] = H_c[C(g)]. The Henry's law coefficient H_c is independent of concentration but dependent on temperature.

Hydrometeor Any condensed water particle in the atmosphere of size much larger than the individual water molecule. Fog, cloud, some hazes, rain, and snow are all hydrometeors.

Hydrophilic   A substance having a strong tendency to bind or absorb water.

Hydrophobic   A substance having a strong tendency to reject association with water.

Hydrosphere   The water portion of Earth (oceans, ice caps, lakes, rivers, etc.). It is one of the five main regimes of the environment: atmosphere, biosphere, cryosphere, hydrosphere, and pedosphere. See also *Atmosphere*; *Biosphere*; *Cryosphere*; *Pedosphere*.

Hypsometric diagram   A graph that shows in any way the relative amounts of Earth's surface at different elevations with regard to sea level.

Inversion   An atmospheric condition in which temperature increases with height. Inversions produce very stable atmospheric conditions.

Ion   An electrically charged atom, or molecular constituent, e.g., Na^+, Al^{3+}, Cl^-, S^{2-}, or a group of atoms, e.g., NH_4^+, SO_4^{2-}, PO_4^{3-}.

Ionic strength   A measure of the ionic concentrations and charge of a solution, defined by $I = {}^1\!/_2 \sum_i X_i Z_i^2$, where X_i is the molar concentration of ion i and Z_i is its charge.

Ion pair   A compound formed by the transient association of a cation and an anion. The bonding is due to electrostatic forces.

Kinetics   The study of the rates at which reactions occur and the influence of physical and chemical conditions on these rates.

Lapse rate   The rate of decrease of temperature with increasing height in the atmosphere. If heat is neither gained nor lost from the air parcel under consideration, then the lapse rate is said to be adiabatic.

Latent heat   The heat released into the atmosphere upon condensation of water vapor into water droplets.

Laurasia   A hypothetical supercontinent comprising approximately the present *continents* of the northern hemisphere. See also *Pangaea*.

Lifetime   In environmental chemistry, the time τ required for the decay of a species from a concentration $[C_0]$ to a concentration $[C_0]/e$, where $e = 2.7183$. The term is often applied either to species that decay by a dominant first-order process such as deposition to surfaces (in such cases, $\tau = 1/k$, where k is the first-order rate coefficient for the decay process) or to processes that can be represented as first order.

Ligand   A molecule, ion, or atom that is attached to the central atom of a coordination compound.

Lithosphere   The outer, rigid shell of the Earth, situated above the *asthenosphere* and containing the *crust, continents*, and *plates*.

Magma   Molten rock material that forms igneous rocks upon cooling. Magma that reaches the surface is referred to as lava.

Mantle   The main bulk of the Earth, between the *crust* and the core, ranging from depths of about 40 to 2900 km. It is composed of dense silicates and divided into concentric layers by phase changes that are caused by the increase in pressure with depth.

Midocean ridge   A major, elevated, linear feature of the seafloor, consisting of many small, slightly offset segments, with a total length of 200 to 20,000 km. This type of plate boundary occurs in a *divergence zone*, which is a site where two *plates* are being pulled apart and new oceanic *lithosphere* is being created.

Mixing ratio   The dimensionless ratio of the mass or volume of a substance (such as water vapor) in an air parcel to the mass or volume of the remaining substances in the air.

NO_x   The sum of the common pollutant gases NO and NO_2.

Nuclear winter   The phenomenon postulated as a possible consequence of nuclear warfare: fires involving petroleum, biomass, and other materials create heavy smoke cover, shielding solar radiation from Earth's surface and causing rapid cooling regardless of season of the year.

Ophiolite suite   An assemblage of igneous rocks with deep-sea sediments supposedly associated with *divergence zones* and the seafloor environment.

Outgassing   The release of juvenile gases to the atmosphere and oceans by volcanism.

Oxidant   A species having a greater oxidation potential than does molecular oxygen. Also, a species that readily accepts electrons.

Oxidized, oxidation   In general, the state of an atom that has given up electrons in forming a molecule. Oxide formation is a specific case, as in the case of CO or CO_2, which are oxidized forms of carbon. See also *Reduced*.

Pangaea   A great protocontinent, thought to have existed about 200 Myr ago, from which all present *continents* have broken off by the mechanism of *seafloor spreading* and *continental drift*.

Parent element   An element that is transformed by *radioactive decay* to a different (*daughter*) *element*.

pE   A measure of the tendency of an aqueous solution to transfer electrons, given by the negative logarithm of the (hypothetical) electron activity. Solutions with pE values higher than about 5 are strongly oxidizing, those with negative pE values are strongly reducing.

Pedosphere   The soil-bearing or solid portion of Earth's surface. It is one of the five main regimes of the environment: atmosphere, biosphere, cryosphere, hydrosphere, and pedosphere. See also *Atmosphere; Biosphere; Cryosphere; Hydrosphere.*

pH   A measure of the acidity or alkalinity (basicity) of a solution, or of the tendency of the solution to transfer protons. A value of 7 is neutral; numbers less than 7 are acid, numbers higher than 7 are alkaline. Strictly speaking, pH is the negative base-10 logarithm of the hydrogen ion concentration in moles per liter.

Phase   A physically distinct, homogeneous portion of a heterogeneous mixture. There are three main phases: solid, liquid, and gas.

Photochemical smog   See *Smog.*

Photodissociation   The fragmentation of a molecule as a consequence of the absorption of solar radiation and subsequent breaking of one or more chemical bonds.

Photolysis   Chemical change resulting from the irradiation of a molecule. For atmospheric chemistry, the most important photolysis process is *photodissociation.*

Plate   One of seven or eight major and several minor segments of the *lithosphere* that are internally rigid and move independently over the interior, meeting in *convergence zones* and separating at *divergence zones.*

Plate tectonics   The processes of plate formation, movement, interaction, and destruction; the attempt to explain seismicity, volcanism, mountain-building, and paleomagnetic evidence in terms of plate motions.

Precipitation   (a) In meteorology, rain or snowfall. (b) In chemistry, the sedimentation of a solid material from a liquid solution in which the material is present in amounts greater than permitted by its solubility.

Preexponential factor   The constant of proportionality in the Arrhenius expression for the temperature dependence of a chemical reaction rate constant.

Primary particles   Particles emitted or injected directly into the atmosphere.

Quantum yield   For a given molecule at a specified wavelength, the number of photons absorbed that result in a specific chemical or physical process divided by the total number of photons absorbed. The sum of the quantum yields for all processes involving photon absorption by a molecule is unity.

Radical   See *Free radical.*

Radioactive decay   The spontaneous breakdown of certain kinds of atomic nuclei into one or more nuclei of different elements, involving the release of energy and subatomic particles.

Radiometric dating   The method of obtaining ages of geological materials by measuring the relative abundances of radioactive parent and daughter isotopes in them.

Rate coefficient   The coefficient of proportionality in the expression relating the rate of a reaction to the concentrations of reactants and/or products. Also called *rate constant*.

Reaction rate   The change with respect to time of the concentration of a reactant or product as a consequence of the occurrence of a specific reaction.

Reductant   A species that readily donates electrons.

Reduced, reduction   In general, the state of an atom or molecule that has taken up electrons. Reduced carbon compounds are things like coal, oil, and other hydrocarbons. Reduced sulfur compounds are molecules like methyl mercaptan (CH_3SH), hydrogen sulfide (H_2S), or elemental sulfur (S_8).

Reflectivity   See *Albedo*.

Relative humidity   The ratio of the partial pressure of water to the saturation vapor pressure, $P_{H_2O}/P_{s\,H_2O}$; also called the saturation ratio. Relative humidity is often expressed as a percentage.

Reservoir   A receptacle defined by characteristic physical, chemical, or biological properties that are relatively uniformly distributed.

Residence time   The average time spent in a reservoir by an individual atom or molecule. Also, the age of a molecule when it leaves the reservoir.

Respiration   The chemical process by which animals and plants convert their food into energy, through oxidation of carbohydrates; the process consumes oxygen and produces carbon dioxide.

Response time   A time scale that characterizes the adjustment to $1/e$ of the initial perturbation value in a reservoir after a sudden change in the system.

Salinity   A measure of the quantity of dissolved salts in seawater. In practice, salinity is computed from chlorinity, electrical conductivity, or some other property whose relationship to salinity is well established.

Salt   The compound formed when a hydrogen atom of an acid is replaced by a metal atom or its equivalent.

Scavenging   The removal of materials from the gas or *aerosol* in the *atmosphere* into *hydrometeors* by cloud, rainout, or washout processes. Often called precipitation scavenging.

Seafloor Spreading The mechanism by which new seafloor *crust* is created at ridges in *divergence zones* and adjacent *plates* are moved apart to make room. This process may continue at the rate of 0.5 to 10 cm yr^{-1} throughout many geological periods.

Secondary particles See *Accumulation mode particles.*

Sedimentation The removal of particulate matter from the *atmosphere* as a result of the effect of gravity. Also called fallout, dry fallout, or dustfall.

Sink In environmental chemistry, the sink is the process causing loss, or the receptor for material when it disappears or is removed from the reservoir of interest.

Smog Classically, a mixture of smoke plus fog. Today the term smog has the more general meaning of any anthropogenic haze. Photochemical smog involves the production, in stagnant, sunlit atmospheres, of oxidants such as by the photolysis of NO_2 and other substances, generally in combination with haze-causing particles.

Solubility The mass of a substance that can be dissolved in a certain amount of solvent, if chemical equilibrium is attained.

Solute The substance dissolved in a *solvent.*

Solution A mixture in which the components are uniformly distributed on an atomic or molecular scale. Although liquid, solid, and gaseous solutions exist, common nomenclature implies the liquid phase unless otherwise specified.

Solvent A medium, usually liquid, in which other substances can be dissolved.

Sorption A class of processes by which one material is taken up by another. Absorption is applied to the penetration of one material into another; adsorption is applied to a surface phenomenon.

Source In environmental chemistry, the process or origin from which a substance is injected into a reservoir.

Stratosphere The atmospheric shell lying just above the *troposphere* and characterized by a stable *lapse rate.* The temperature is approximately constant in the lower part of the stratosphere and increases from about 20 km to the top of the stratosphere at about 50 km.

Strong acid A hydrogen-containing substance that, in water, is completely dissociated into one or more hydrogen ions and a conjugate cation.

Subduction zone A planar zone descending away from a trench and defined by high seismicity, interpreted as the shear zone between a sinking oceanic plate and an overriding plate. See also *Convergence zone.*

Tectonics   See *Plate tectonics* .

Thermodynamics   The science of heat and temperature and of the laws governing the conversion of heat into mechanical, electrical, or chemical energy.

Tropopause   The boundary between the *troposphere* and the *stratosphere*, identified as the point at which temperature stops decreasing with altitude.

Troposphere   The lowest layer of the atmosphere, ranging from the ground to the base of the stratosphere at 10–15 km altitude, depending on latitude and weather conditions. About 85% of the mass of the atmosphere is in the troposphere, where most weather features occur. Because its temperature decreases with altitude, the troposphere is dynamically unstable.

Turnover time   The ratio between the content of a specific substance in a *reservoir* and the total flux of the substance in or out of the reservoir.

Vadose zone   The top layer of soil in which the pores contain significant amounts of both air and water.

Vapor   A gaseous substance below its critical temperature that may be liquefied by pressure alone. It is important to note that vapors are gases and that they are generally not visible in air.

Visibility   The degree to which the atmosphere is transparent to light in the visible spectrum, or the degree to which the form, color, and texture of objects can be perceived. In the sense of visual range, visibility is the distance at which a large black object just disappears from view as a recognizable entity.

Volatiles   Gaseous materials that are readily lost from a system if not confined; also, substances such as water and carbon dioxide, which are loosely bound into a mineral structure and can escape from a rock if the mineral breaks down during metamorphism.

Volcano   Any opening through the *crust* that has allowed *magma* to reach the surface.

Weak acid   A hydrogen-containing substance, which, in water, is only partially dissociated into one or more hydrogen ions and a conjugate cation.

Wet deposition   The transfer of trace species (gases or particles) from the *atmosphere* to surfaces as a consequence of their absorption into or onto *hydrometeors*.

Units of Measurement in Environmental Chemistry

The most common way of expressing the abundance of a gas phase species is as a fraction of the number of molecules in a sample of air. The units in common use are *parts per million* (ppm), *parts per billion* (thousand million; ppb), and *parts per trillion* (million million; ppt), all expressed as volume fractions and therefore abbreviated ppmv, ppbv, and pptv to make it clear that one is not speaking of fractions in mass. Any of these units may be called the *volume mixing ratio* or *mole fraction*. Mass mixing ratios can be used as well (hence, ppmm, ppbm, pptm), a common example being that meteorologists use mass mixing ratios for water vapor. Since the pressure of the atmosphere changes with altitude and the partial pressures of all the gaseous constituents in a moving air parcel change in the same proportions, mixing ratios are preserved as long as mixing between air parcels can be neglected.

Particles can be mixtures of solid and liquid, so a measure based on mass replaces that based on volume, the usual units for atmospheric particles being micrograms per cubic meter (μg m^{-3}) or nanograms per cubic meter (nm m^{-3}). For particles in liquids, micrograms per cubic centimeter (μg cm^3) is common. It is sometimes convenient to compare quantities of an element or compound present in more than one phase, say as both a gas and as a particle constituent. In that case, the gas concentration in volume units is converted to mass units prior to making the comparison.

In the case of very reactive gaseous species, often present at extremely low concentrations, the number of molecules per unit volume of air (molecules per cubic centimeter) is generally used. This practice reflects the fact that for these species the concentrations are much less affected by dynamical processes than by chemical ones.

For constituents present in aqueous solution, as in seawater, the convention is to express concentration in units of moles per liter (designated M) or some derivative thereof [one mole (abbreviated mol) is 6.02×10^{23} molecules]. Common concentration expressions in environmental chemistry are millimoles per liter (mM), micromoles per liter (μM), and nanomoles per liter (nM), Sometimes one is concemed with the "combining concentration" of a species rather than the absolute concentration. A combining concentration, termed an equivalent, is that concentration which will react with 8 grams of oxygen or its equivalent. For example, one mole of hydrogen ions is one equivalent of H^+, but one mole of calcium ions is two equivalents of Ca^{2+}. Combining concentrations have typical units of equivalents, milliequivalents, or microequivalents per liter, abbreviated eq/l, meq/l, and μeq/l.

Acidity in solution is expressed in pH units, pH being defined as the negative of the logarithm of the hydrogen ion concentration in moles per liter. In aqueous solutions, pH = 7 is neutral at 25 °C; lower pH values are characteristic of acidic solutions, higher values are characteristic of basic solutions.

Earth Data and Physical Constants

Mass of the Earth	5.98×10^{27} g
Mass of the Hydrosphere	1.66×10^{24} g
Mass of the Atmosphere	5.14×10^{21} g
Mass of the Biosphere	1.15×10^{19} g
Area of Earth's Surface	5.10×10^{8} km^2
Area of Earth's Continents	1.49×10^{8} km^2
Human Population (1987)	5.0×10^{9}
Non dairy Cattle Population (1984)	1.1×10^{9}
Gas Constant	$R = 8.314$ J °K^{-1}mole^{-1}
Boltzmann Constant	$k = 1.3807 \times 10^{23}$ J °K^{-1}
Avogadro Constant	6.02×10^{23} molecules mole^{-1}
Molecular Density at STP	2.45×10^{19} cm^{-3}

Answers to Selected Exercises

For many of those exercises where numerical answers are requested, we have provided either the entire answer or enough to indicate whether the correct approach is being taken. Precise agreement with these numbers is not necessarily expected, since the results may be a function of how many significant figures can be extracted from a graph, or some similar complication.

1.2 1.0×10^7 molecules NH_3 cm^{-3}.

1.3 0.63 μg CO_2 cm^{-3}.

1.4 Gas phase tetradecane has a concentration of 2.0×10^3 ng m^{-3} and that phase is the more abundant.

2.1 (a) ^{222}Rn; (b) ^{214}Bi.

2.3 3.5×10^{-13} J cm^{-3} s^{-1}.

2.5 (a) 1.5×10^{-10} yr^{-1}; (b) 550 atoms.

3.2 (a) 33.7 μm; (b) 76.3 μm.

3.3 It would have little effect, since the atmosphere is already almost totally absorbant at that wavelength.

3.4 1.2×10^{-7} absorbed photons cm^{-3} s^{-1}.

3.5 5.4×10^{-3} J/s.

5.2 Φ (fine particles) = 3.1×10^{-11} g cm^{-2} s^{-1}; Φ (coarse particles) = 1.1×10^{-10} g cm^{-2} s^{-1}.

5.3 N = 6.9×10^3 particles cm^{-3}; V = 80 μm cm^{-3}; TPM = 140 μg cm^{-3}.

5.4 $[SO_4^{2-}]$ (fine) = 28 ug cm^{-3}; $[Cl^-]$ (coarse) = 2 μg cm^{-3}.

5.5 8.9 km.

6.1 1.6×10^9 J.

6.2 Cloud water density: 0.86 g H_2O m^{-3}. Rainfall: 1.2×10^3 g m^{-2}; large drops are more important than small ones.

7.1 1×10^{-4} M H$^+$.

7.2 (a) Yes; (b) Yes; (c) Disproportionation, chain termination.

7.3 $[H^+] = [HCO_3^-] = 2.9 \times 10^{-6}$ M; $[CO_3^{2-}] = 1.1 \times 10^{-8}$ M.

7.4 $R_{O_3} = 9.5 \times 10^4$ molec cm^{-3} s^{-1}; $R_{HO\bullet} = 9.8 \times 10^5$ molec cm^{-3} s^{-1}; the latter reaction results in the higher propene loss.

7.5 $[C_2H_4]_{aq} = 4.3 \times 10^{-11}$ M; $[H_2O_2]_{aq} = 4.7 \times 10^{-4}$ M.

7.6 3.8×10^6 molec cm^{-3} s^{-1}.

7.7 SO_2 gas: 6.7×10^{-6} μg S cm^{-2} s^{-1}; particles: 2.5×10^{-8} μg S cm^{-2} s^{-1}.

7.8 $R_{HO\bullet} = 3.4 \times 10^{-6}$ s^{-1}; $R_{diss} = 9.6 \times 10^{-6}$ s^{-1}; $R_{depn} = 5.0 \times 10^{-6}$ s^{-1}. The photodissociation loss is limiting regardless of the H_2O_2 concentration.

8.1 $\tau_{O_3} = 7.4$ s; $\tau_{CO} = 1.2$ s; $\tau_{CH_4} = 0.3$ s; $\tau_{SO_2} = 2.7$ s. The HO• lifetime is limited by methane.

8.2 pH = 6.6.

8.3 pH = 4.0.

8.4 $[COS]_{aq} : [H_2S]_{aq} : [SO_2]_{aq} = 1 : 12 : 56$.

8.5 $I_{seawater} = 0.70$; $k_{25} = 2.2 \times 10^5$ M^{-1} s^{-1}; $k_{10} = 6.6 \times 10^4$ M^{-1} s^{-1}.

9.1 I = 0.0035.

9.2 pH = 5.10; $[Zn^{2+}] = 6.6 \times 10^{-6}$ M.

9.3 pH = 6.7; $[Cl^-] = 3.4 \times 10^{-4}$ M.

9.4 Visibility = 8.9 km.

10.1 Magma: 0.1 mm.

10.2 -26.8 $^o/_{oo}$.

12.2 $[CO_2]_{1850} = 2.2 \times 10^{18}$ g. $[CH_4]_{1750} = 2.0 \times 10^{15}$ g.

12.3 71 kg km^{-2}; 2.4×10^{10} kg.

13.1 $q_{75} = 37.8$; $q_{50} = 27.8$.

13.2 Calcutta: 3.0 mg day^{-1}; Kuwait: 92 g/50 yr.

13.3 Pb (median, 1975) = 3.3 mg yr^{-1}.

13.4 $[SO_4^{2-}]_{1900} = 0.7$ μM cm^{-2} yr^{-1}.

14.1 $\Delta CO = 2.4 \times 10^{12}$ g yr^{-1}.

15.2 $\dfrac{d[NO_2]}{dt} = -9.9 \times 10^{-3} [NO_2]$
$- 3.2 \times 10^{-17} [NO_2] [O_3]$
$+ 1.0 \times 10^{-14} [NO] [O_3]$

16.1 (a) 125 ppbv O_3.

16.2 Φ_{NO_x} (8 AM) = 4400 g hr^{-1}.

16.3 [CO] (9 AM) = 19 ppmv.

16.4 Copenhagen, 1980: 170 yr.

16.5 Coal needed = 1.4×10^7 g/yr; Φ_{CH_4} (coal) = 7.0×10^4 g/hr; Φ_{CO} (petroleum) = 5.7×10^5 g/hr; Φ_{CO_2} (natural gas) = 1.9×10^7 g/hr.

17.1 τ (1970, 0.1 km) = 3.4 yr.

18.1 Ice caps (1%): 8 mm; temp. change: 1.0 m.

Index